Studies in Computational Intelligence

Volume 1176

The series "Studies in Computational Intelligence" (SCI) publishes new developments and advances in the various areas of computational intelligence—quickly and with a high quality. The intent is to cover the theory, applications, and design methods of computational intelligence, as embedded in the fields of engineering, computer science, physics and life sciences, as well as the methodologies behind them. The series contains monographs, lecture notes and edited volumes in computational intelligence spanning the areas of neural networks, connectionist systems, genetic algorithms, evolutionary computation, artificial intelligence, cellular automata, self-organizing systems, soft computing, fuzzy systems, and hybrid intelligent systems. Of particular value to both the contributors and the readership are the short publication timeframe and the world-wide distribution, which enable both wide and rapid dissemination of research output.

Indexed by SCOPUS, DBLP, WTI AG (Switzerland), zbMATH, SCImago.

All books published in the series are submitted for consideration in Web of Science.

Bikram Pratim Bhuyan · Amar Ramdane-Cherif ·
Thipendra P. Singh · Ravi Tomar

Neuro-Symbolic Artificial Intelligence

Bridging Logic and Learning

Springer

Bikram Pratim Bhuyan
LISV Laboratory
University of Versailles—Paris-Saclay
Université Paris-Saclay
Vélizy, France

Amar Ramdane-Cherif
LISV Laboratory
University of Versailles—Paris-Saclay
Université Paris-Saclay
Vélizy, France

Thipendra P. Singh
School of Computer Science Engineering and Technology
Bennett University
Greater Noida, Uttar Pradesh, India

Ravi Tomar
Persistent Systems
Pune, Maharashtra, India

ISSN 1860-949X ISSN 1860-9503 (electronic)
Studies in Computational Intelligence
ISBN 978-981-97-8170-6 ISBN 978-981-97-8171-3 (eBook)
https://doi.org/10.1007/978-981-97-8171-3

This Springer imprint is published by the registered company Springer Nature Singapore Pte Ltd.
The registered company address is: 152 Beach Road, #21-01/04 Gateway East, Singapore 189721, Singapore

Contents

Part VIII Conclusion

Part I
Introduction

Chapter 1
The Emergence of Neuro-Symbolic Artificial Intelligence

"I' m not a robot. I'm more like a human with a processor."

–Ava, *Ex Machina*

1.1 Introduction

Consider the scenario of playing a game of chess. You know the rules: how each piece moves, the mechanics of capturing pieces, and the strategies to win. These rules are clear and explicit, much like the rules used in symbolic artificial intelligence. However, merely knowing the rules is not enough to become a chess master. Mastery requires extensive gameplay, learning from mistakes, and recognizing patterns—paralleling how neural networks learn from data.

Imagine combining these two approaches: having both the game's rules and the ability to learn from experience. This combination allows for the use of rules as a framework to guide learning, avoid illegal moves, and develop strategies. Simultaneously, experiential learning enhances the application of these rules, helping to understand when to follow them strictly and when to adapt. This integration of rule-based reasoning and experiential learning is the essence of Neuro-Symbolic AI [1].

Neuro-Symbolic AI can be likened to a skilled chess player who understands the rules and continually improves through practice. A sequence diagram is shown in Fig. 1.1. By integrating the rule-based knowledge representation of symbolic AI with the learning capabilities of neural networks, Neuro-Symbolic AI aims to create systems that can understand and learn from their environment while explaining their decision-making processes in a manner comprehensible to humans.

B. P. Bhuyan et al., *Neuro-Symbolic Artificial Intelligence*, Studies in Computational Intelligence 1176, https://doi.org/10.1007/978-981-97-8171-3_1

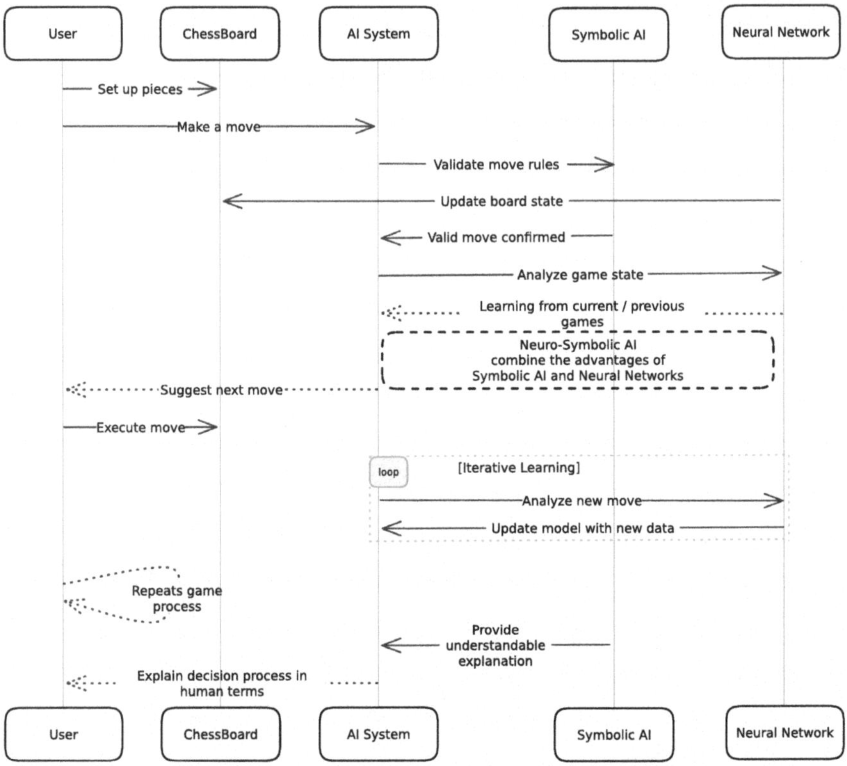

Fig. 1.1 A sequence diagram of the chess board example where the user interacts with an AI system capable of rule-based reasoning by the symbolic AI component and learning through extensive gameplay by the neural network component

Just as learning chess can seem daunting at first, studying Neuro-Symbolic AI might initially appear complex. However, as we progress through each chapter, we will gain a deeper understanding of this fascinating field. Let us now embark on this journey into the realm of Neuro-Symbolic Artificial Intelligence.

1.2 The Dawn of Symbolic AI

Symbolic AI, often referred to as "good old-fashioned AI," emerged as the foundation of artificial intelligence research in the mid-twentieth century [2]. This approach focused on encoding human knowledge into explicit, rule-based systems that could perform logical reasoning and problem-solving tasks.

The field of AI research was formally established at the Dartmouth Conference in 1956 [3], where John McCarthy, Marvin Minsky, Nathaniel Rochester, and Claude

Shannon proposed the idea of creating machines that could simulate every aspect of human intelligence. This conference marked the birth of AI as a distinct scientific discipline and symbolic AI quickly became its primary focus.

Symbolic AI is based on the manipulation of symbols and the use of rules to represent knowledge. Early AI researchers like Allen Newell and Herbert A. Simon [4] developed systems such as the Logic Theorist [5] and the General Problem Solver (GPS) [6], which were designed to mimic human problem-solving processes through the application of logical rules. GPS was designed to solve a wide range of problems by breaking them down into smaller subproblems and applying rules to solve these subproblems sequentially. This approach demonstrated the potential of symbolic reasoning for AI, laying the groundwork for future developments.

The 1960s and 1970s saw the rise of expert systems, which were designed to emulate the decision-making abilities of human experts. One of the earliest and most influential expert systems was Dendral [7], developed in the mid-1960s by Edward Feigenbaum, Bruce Buchanan, Joshua Lederberg, and Carl Djerassi. Dendral was used to analyze chemical compounds and identify molecular structures, showcasing the power of rule-based systems in specialized domains.

Another significant development during this period was the creation of ELIZA [8], a natural language processing program developed by Joseph Weizenbaum in 1966. ELIZA could engage in simple conversations with users by applying pattern-matching rules to their input, demonstrating the potential for AI to interact with humans using natural language.

Despite early successes, symbolic AI faced significant challenges. One major issue was scalability: symbolic systems required extensive manual encoding of knowledge and rules, making them difficult to scale to more complex domains. Additionally, these systems struggled with tasks requiring common sense reasoning or dealing with uncertainty and incomplete information.

As an example, symbolic AI systems encountered difficulties as the level of complexity in the chess game escalated, similar to that experienced by a beginner chess player. Although they demonstrated proficiency in problem domains characterized by explicit rules and clear boundaries, they encountered difficulties when confronted with situations that entailed ambiguity, uncertainty, or incomplete information. As an illustration, a symbolic AI system may encounter challenges when faced with a game such as poker, where elements of bluffing and uncertainty hold substantial influence, thereby rendering the determination of the optimal move beyond the purview of the game's rules alone.

In addition, symbolic AI systems were constrained by their dependence on explicit programming. The extent of their knowledge was limited to the information they were programmed to possess, and their capacity for acquiring new knowledge was contingent upon being programmed with additional instructions. The absence of adaptability and learning capacity posed a substantial constraint for symbolic AI systems.

In 1973, the "Lighthill Report" [9], authored by James Lighthill, critically assessed the progress of AI research and highlighted its limitations. The report led to a reduction in funding for AI research in the UK, contributing to a period known as the "AI Winter," characterized by decreased interest and investment in AI.

Despite these challenges, symbolic AI laid the foundational principles for the field of artificial intelligence, emphasizing the importance of logic, rules, and explicit knowledge representation. These principles continue to influence modern AI research, even as the field has shifted towards data-driven approaches like machine learning and neural networks.

The symbolic AI approach also paved the way for the development of hybrid systems, which combine symbolic reasoning with other AI techniques to create more robust and adaptable systems. As AI continues to evolve, the legacy of symbolic AI remains a critical component of the field's history and ongoing development.

1.3 The Rise of Neural Networks

While symbolic AI was advancing through the use of rule-based systems, an alternative approach inspired by the human brain began to gain traction [10]. This approach, known as neural networks, aimed to replicate the structure and function of biological neural networks to process information and learn from experience.

In order to comprehend the emergence of neural networks, it is instructive to examine an illustrative example (Fig. 1.2). Consider a scenario in which a young child is in the process of acquiring the ability to identify and distinguish various species of animals. Rather than commencing with a predetermined set of rules, they acquire knowledge through experiential learning. The individual visually perceives a canine entity, subsequently receiving verbal confirmation that it is indeed a canine. Over a period of time, through repeated exposure to numerous canines exhibiting diverse physical attributes such as shapes, sizes, and colors, the individual gradually develops a comprehension of the defining characteristics that classify an entity as

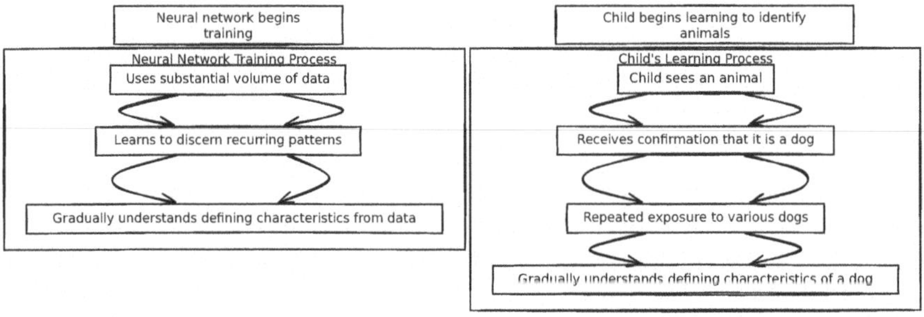

Fig. 1.2 Initial thoughts about one-to-one mapping between the child's learning process and the neural network learning process

a canine. This bears a resemblance to the learning process of neural networks. The process commences by utilizing a substantial volume of data and acquiring the ability to discern recurring patterns within said data.

The conceptual foundation of neural networks can be traced back to the 1940s and 1950s [11]. In 1943, Warren McCulloch and Walter Pitts introduced a computational model for neural networks [12], demonstrating that artificial neurons could perform logical operations. Their work laid the groundwork for future developments in neural network theory and its applications.

In 1958, Frank Rosenblatt developed the perception [13], an early type of artificial neural network designed for image recognition. The perceptron algorithm represented a significant step forward, showing that machines could learn from data rather than rely solely on pre-programmed rules.

Despite early promise, neural networks faced significant challenges in the following decades. In 1969, Marvin Minsky and Seymour Papert published "Perceptrons" [14], highlighting limitations in the perceptron model, particularly its inability to solve non-linear problems like the XOR function. This led to a decline in interest and funding, contributing to the first AI winter.

However, the development of backpropagation in the 1970s revived interest in neural networks. Backpropagation, a method for efficiently training multilayer neural networks, allowed for the adjustment of weights in the network to minimize errors. This breakthrough, popularized by researchers like Paul Werbos [15], enabled neural networks to learn more complex patterns and functions.

The 1980s and 1990s saw significant advancements in neural network research. John Hopfield introduced Hopfield networks [16], which used recurrent connections to model associative memory. At the same time, Geoffrey Hinton and others developed Boltzmann machines [17], which used stochastic processes to model probabilistic relationships.

In 1989, Yann LeCun and his colleagues applied backpropagation to a neural network for recognizing handwritten digits, leading to the development of convolutional neural networks (CNNs) [18]. CNNs, designed to recognize visual patterns directly from pixel images, revolutionized image recognition and became the foundation for many modern computer vision systems.

The 2000s and 2010s marked the era of deep learning, characterized by neural networks with many layers (deep neural networks). The success of AlexNet [19], a deep convolutional neural network, in the 2012 ImageNet competition showcased the power of deep learning for image classification tasks and sparked widespread interest in the approach.

Key developments during this period included the introduction of new architectures, such as recurrent neural networks (RNNs) for sequence data and generative adversarial networks (GANs) for generating realistic synthetic data. The invention of the transformer model in 2017 further advanced the field by improving the handling of long-range dependencies in natural language processing tasks [20]. The transformer model leverages the attention mechanism to improve performance on a variety of natural language processing tasks. Transformers consist of encoder and decoder layers that use self-attention to process input data, making them highly effective for tasks that require understanding the context of the entire input sequence.

Building on the success of transformers, large language models (LLMs) like OpenAI's GPT-3 and GPT-4 have demonstrated unprecedented capabilities in generating coherent and contextually relevant text. These models are trained on vast amounts of data and can perform a wide range of tasks, from language translation to creative writing, showcasing the transformative potential of neural networks in AI.

Today, neural networks are used in various applications, from image and speech recognition to natural language processing and autonomous systems. The development of specialized hardware, such as graphics processing units (GPUs), has significantly accelerated the training and deployment of deep neural networks, enabling real-time processing and analysis of large datasets.

Neural networks have transformed many industries, demonstrating the potential of machine learning to solve complex problems and perform tasks previously thought to require human intelligence. As research continues, the integration of neural networks with symbolic AI and other approaches promises to further enhance the capabilities and applications of artificial intelligence.

1.4 The Birth of Neuro-Symbolic AI

The emergence of Neuro-Symbolic AI represents a pivotal moment in the evolution of artificial intelligence. This approach seeks to combine the strengths of symbolic AI, with its rule-based reasoning and logical manipulation of symbols, and neural networks, which excel in learning from data and recognizing patterns as shown in Fig. 1.3. The goal is to create AI systems that are both highly capable of learning from vast amounts of data and adept at performing complex reasoning tasks.

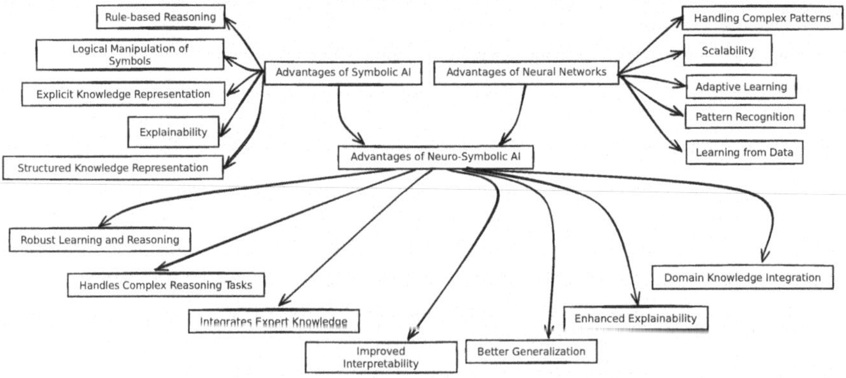

Fig. 1.3 Advantages of symbolic AI, neural-based AI, and neuro-symbolic AI

The proposed system possesses the capability to dynamically adjust its strategies in response to the actions of its adversary, acquire knowledge from previous instances, and elucidate its decision-making processes within the framework of regulations. A prevalent methodology involved employing a symbolic AI framework to produce a collection of potential solutions or actions, subsequently employing a neural network to assess and appraise these alternatives. In the context of a game-playing artificial intelligence, the symbolic system has the capability to generate a comprehensive compilation of permissible moves. Subsequently, the neural network can assign a ranking to these moves by utilizing its acquired knowledge through the evaluation function.

The concept of integrating symbolic and neural approaches can be traced back to the early days of AI research, but significant strides were made in the 1980s and 1990s [21]. During this period, researchers like Paul Smolensky proposed models such as Tensor Product Representations (TPRs) [22], which aimed to encode symbolic data in a format suitable for manipulation by neural networks. These early efforts laid the groundwork for more sophisticated Neuro-Symbolic systems.

In the late 1980s and early 1990s, significant progress was made in understanding how to combine symbolic reasoning with neural learning. Researchers explored various methods to bridge the "semantic gap" between high-level symbolic representations and the low-level numeric data processed by neural networks. This era saw the development of hybrid systems that began to show the potential of Neuro-Symbolic AI.

One notable advancement was the creation of Logic Tensor Networks (LTNs) [23], which integrate logical formulas directly into neural network architectures. This approach allows the systems to learn and reason simultaneously, leveraging the strengths of both paradigms.

In recent years, there has been a resurgence of interest in neuro-symbolic AI, driven by the limitations of purely neural approaches in terms of explainability and reasoning. Modern implementations often use sophisticated techniques to integrate symbolic knowledge into neural networks. For instance, systems like DeepProbLog [24] combine neural networks with probabilistic logic programming, allowing for more robust and interpretable AI models.

Applications of Neuro-Symbolic AI span various domains, including natural language processing, robotics, and cognitive modeling. For example, Neuro-Symbolic Concept Learners (NSCL) [25] use neural networks to parse images and extract features, while symbolic reasoning modules interpret these features to answer complex questions about the images.

Neuro-Symbolic AI offers several advantages over purely neural or symbolic systems. By combining the two, these systems can perform complex reasoning tasks while also learning from data, leading to more flexible and powerful AI. They can leverage the structured knowledge of symbolic AI to improve the interpretability and robustness of neural models, addressing some of the key limitations of deep learning.

Furthermore, Neuro-Symbolic systems possess the capability to execute cognitive processes that encompass the utilization of symbolic rules in conjunction with acquired knowledge. One potential approach is to employ symbolic rules as a means of directing the search process during a reasoning task. Additionally, acquired knowledge can be utilized to assess the credibility of various reasoning pathways.

However, integrating these approaches is not without challenges. One significant issue is ensuring seamless communication between the symbolic and neural components, as they operate on fundamentally different principles. Additionally, developing efficient training algorithms that can leverage both symbolic rules and neural data remains an ongoing area of research.

Recent research has focused on enhancing the scalability, efficiency, and interpretability of Neuro-Symbolic systems. For example, IBM's SOFAI [26] architecture combines fast and slow AI processes inspired by Daniel Kahneman's cognitive framework, "Thinking, Fast and Slow." This system uses multiple planning approaches categorized as fast or slow and incorporates a metacognitive process for governance, balancing solving speed and plan optimality.

Another significant advancement is the integration of attention mechanisms into Graph Neural Networks (GNNs), enabling these models to incorporate symbolic rules selectively. This approach enhances the explainability and robustness of GNNs, making them suitable for complex reasoning tasks.

1.5 The Evolution of Neuro-Symbolic AI

The progression of Neuro-Symbolic AI is characterized by notable advancements and practical implementations during the contemporary epoch. The integration of rule-based reasoning from symbolic AI and the learning capabilities of neural networks has resulted in the emergence of advanced systems capable of effectively addressing intricate tasks, acquiring knowledge from extensive datasets, and generating outputs that can be readily understood and interpreted.

In recent years, there has been notable progress in the development of contemporary Neuro-Symbolic systems across multiple domains [27]. These systems effectively combine the advantageous features of symbolic AI and neural networks, enabling them to effectively tackle intricate tasks, acquire knowledge from extensive datasets, and generate outputs that are easily interpretable.

An important development in this field involves the utilization of Neuro-Symbolic methodologies for the purpose of fault diagnosis in intricate systems such as intelligent power grids [28]. These systems employ Artificial Neural Networks (ANNs) to detect patterns in data, such as voltage waveforms, and subsequently utilize symbolic reasoning to deliver a conclusive evaluation of the system's state. The integration of high-fidelity pattern recognition and rule-based decision-making facilitates the enhancement of power grid quality.

Neuro-Symbolic classifiers have been devised within the domain of network security to oversee security alerts in network traffic [29]. The aforementioned classifiers integrate the learning capabilities of deep neural networks with the interpretability offered by symbolic AI. The fusion of different technologies not only improves the accuracy of identifying security breaches but also offers a transparent rationale for the results, thereby minimizing the requirement for human involvement in the process of cyber-threat hunting.

Neuro-Symbolic AI has demonstrated its efficacy in the healthcare domain, specifically within the realm of mental healthcare [30]. By integrating a wide range of clinical knowledge, these systems have the ability to effectively educate Neural-AI systems, thereby enhancing interventions in the field of mental healthcare. The incorporation of reasoning capabilities into the model's outcomes, along with the application of deductive reasoning on data, has given rise to the notion of interactive accuracy, thereby augmenting the efficacy of these systems.

An additional noteworthy advancement pertains to the rise of Neuro-Symbolic computing through the utilization of spiking neural networks [31]. These networks possess the capability to encode both symbolic and multi-relational information, thereby enabling the process of reasoning over symbolic structures such as knowledge graphs. This technological progress presents novel opportunities for Neuro-Symbolic AI, encompassing capabilities such as the handling of intricate tasks, acquisition of knowledge from extensive datasets, and generation of outputs that can be comprehended and interpreted.

In hist, contemporary Neuro-Symbolic systems signify a noteworthy advancement within the realm of artificial intelligence. The researchers are expanding the limits of technological feasibility by integrating the deductive reasoning methods of symbolic artificial intelligence with the adaptive learning capacities of neural networks, resulting in the development of advanced, scalable, and comprehensible artificial intelligence systems.

1.6 The Future of Neuro-Symbolic AI

As the field of Neuro-Symbolic AI progresses, a promising landscape emerges characterized by ample opportunities for innovation and progress. The amalgamation of symbolic reasoning and neural learning has demonstrated its efficacy as a formidable methodology, with ample potential for further investigation and revelation.

One of the primary domains of prospective advancement pertains to the enhancement and augmentation of various categories of Neuro-Symbolic Artificial Intelligence. Categorizing Neuro-Symbolic systems presents a challenge due to their complex nature. However, a general classification reveals six distinct types [32].

Symbolic Systems Guided by Neural Networks: The utilization of neural networks to direct the search or decision-making procedures of symbolic systems.

Neural Networks Guided by Symbolic Systems: The utilization of symbolic systems in guiding neural networks involves employing a symbolic framework to organize and direct the learning process of the neural network. This may entail specifying the network's architecture or determining the structure of the objective function.

Symbolic and Neural Systems Working in Parallel: The parallel systems under consideration encompass a symbolic system and a neural network, operating in tandem, wherein both systems engage in the processing of identical input data and jointly contribute to the ultimate output.

Hybrid Symbolic-Neural Systems: Hybrid Symbolic-Neural Systems refer to integrated systems that combine symbolic and neural components, wherein a neural network is utilized to process symbolic data directly.

Symbolic Systems Augmented with Neural Components: These systems predominantly rely on symbolic processing, yet incorporate neural components to address specific tasks, such as perception or pattern recognition.

Neural Systems Augmented with Symbolic Components: These systems are predominantly neural in nature, yet incorporate symbolic elements to facilitate specialized functions, such as logical reasoning or decision-making processes.

Each of these categories possesses distinct advantages and disadvantages, and the selection among them is contingent upon the specific demands of the given task (to be covered in depth in forthcoming chapters). It is anticipated that forthcoming advancements will give rise to novel Neuro-Symbolic systems that amalgamate the respective merits of these distinct types while alleviating their inherent limitations.

The utilization of graph neural networks (GNNs) represents a prospective avenue for the advancement of Neuro-Symbolic AI. GNNs are a specific category of neural networks that have been specifically developed to effectively process and analyze data that is organized in a graph structure. Tasks that require relational reasoning, such as social network analysis, recommendation systems, and knowledge graph completion, are particularly suitable for them. Graph Neural Networks possess the capability to encode symbolic knowledge into a format that is amenable to processing by a neural network. An example of encoding a knowledge graph involves utilizing a graph neural network, where entities are represented as nodes, relations are represented as edges, and attributes are represented as node and edge features. GNNs have the capability to acquire the ability to engage in logical reasoning using symbolic knowledge. This is achieved by learning the process of propagating and modifying features across the graph.

The utilization of Graph Neural Networks within the domain of Neuro-Symbolic AI represents a burgeoning field of study that exhibits considerable potential. The integration of symbolic systems' relational reasoning capabilities with the learning capabilities of neural networks has the potential to facilitate the development of advanced AI systems. These systems would possess the ability to acquire knowledge from data, engage in logical reasoning using rules, and provide explanations for their decision-making processes in a manner comprehensible to humans.

In the foreseeable future, the field of Neuro-Symbolic systems is anticipated to witness significant progress due to continuous advancements in machine learning techniques, computational hardware, and our comprehension of both symbolic and neural systems. This progress is expected to result in the development of Neuro-Symbolic systems that are more robust, adaptable, and capable of being interpreted.

1.7 Conclusion

The advent of Neuro-Symbolic AI represents a noteworthy milestone in the progression of artificial intelligence. The concept embodies the integration of two separate paradigms: the rule-based reasoning characteristic of symbolic AI and the learning abilities exhibited by neural networks. The integration of various components has facilitated the emergence of novel opportunities in the development of AI systems. These systems possess the capability to acquire knowledge from data, engage in logical reasoning based on predefined rules, and provide comprehensible explanations for their decision-making processes to human users.

The inception of Neuro-Symbolic AI can be traced back to the independent advancements of symbolic AI and neural networks as distinct disciplines. Symbolic artificial intelligence demonstrated exceptional performance in tasks characterized by explicit rules and well-defined logic, whereas neural networks exhibited remarkable capabilities in tasks necessitating data-driven learning and pattern recognition. Nevertheless, both methodologies exhibited certain constraints, prompting the realization that a comprehensive and all-encompassing approach was imperative.

The emergence of Neuro-Symbolic AI arose as a reaction to this exigency. The initial Neuro-Symbolic systems were characterized by the integration of symbolic systems and neural networks, albeit in a rudimentary manner. Nonetheless, these early systems served as the building blocks for the development of more advanced and intricate systems. In recent times, the field of Neuro-Symbolic systems has witnessed notable progress due to the development of advanced machine learning techniques and increased computational capabilities. These advancements have enabled the systems to effectively tackle intricate tasks, acquire knowledge from extensive datasets, and generate outputs that can be readily understood and interpreted.

When considering the future, it becomes evident that the potential of Neuro-Symbolic AI is extensive. The continuous progress in research and development is anticipated to yield novel Neuro-Symbolic systems that will incorporate advanced methodologies such as graph neural networks. These systems have the potential to expand the limits of what is achievable in the field of artificial intelligence, resulting in the development of AI systems that are more robust, adaptable, and comprehensible.

1.8 Questions for Readers

1. How can the integration of symbolic AI and neural networks transform the future of artificial intelligence?
2. What are the key limitations of symbolic and connectionist AI that have led researchers to explore alternative approaches?
3. How can advancements in hardware, like GPUs, influence the development and application of Neuro-Symbolic AI?

4. What are the main challenges in making neural networks more interpretable and transparent?
5. How can Neuro-Symbolic AI systems ensure ethical alignment and compliance with regulatory standards?
6. What are the most promising advancements in Neuro-Symbolic AI, and how might they shape the next decade of AI development?
7. How can interdisciplinary collaboration accelerate the development of Neuro-Symbolic AI?
8. What might be the long-term societal impacts of the widespread adoption of Neuro-Symbolic AI?

References

1. Bhuyan, B.P., Ramdane-Cherif, A., Tomar, R., Singh, T.: Neuro-symbolic artificial intelligence: a survey. Neural Comput. Appl. 1–36 (2024)
2. Smolensky, P.: Connectionist AI, symbolic AI, and the brain. Artif. Intell. Rev. **1**(2), 95–109 (1987)
3. McCarthy, J., Rochester, N., Shannon, C.: Dartmouth workshop (1956)
4. Newell, A., Simon, H.A.: The simulation of human thought. Rand Corporation Santa Monica, CA, USA (1959)
5. Newell, A., Simon, H.: The logic theory machine-a complex information processing system. IRE Trans. inf. Theory **2**(3), 61–79 (1956)
6. Newell, A., Shaw, J.C., Simon, H.A.: Report on a general problem solving program. In: IFIP Congress, vol. 256, p. 64. Pittsburgh, PA (1959)
7. Buchanan, B.G., Feigenbaum, E.A.: Dendral and meta-dendral: their applications dimension. In: Readings in Artificial Intelligence, pp. 313–322. Elsevier (1981)
8. Weizenbaum, J.: Eliza-a computer program for the study of natural language communication between man and machine. Commun. ACM **9**(1), 36–45 (1966)
9. Agar, J.: What is science for? the lighthill report on artificial intelligence reinterpreted. The Brit. J. History Sci. **53**(3), 289–310 (2020)
10. Abiodun, O.I., Jantan, A., Omolara, A.E., Dada, K.V., Mohamed, N.A., Arshad, H.: State-of-the-art in artificial neural network applications: a survey. Heliyon **4**(11) (2018)
11. Hecht-Nielsen, R.: Theory of the backpropagation neural network. In: Neural Networks for Perception, pp. 65–93. Elsevier (1992)
12. McCulloch, W.S., Pitts, W.: A logical calculus of the ideas immanent in nervous activity. The Bull. Math. Biophys. **5**, 115–133 (1943)
13. Rosenblatt, F.: The perceptron: a probabilistic model for information storage and organization in the brain. Psychol. Rev. **65**(6), 386 (1958)
14. Minsky, M., Papert, S.A.: Perceptrons, Reissue of the 1988 Expanded Edition With a New Foreword by Léon Bottou: An Introduction to Computational Geometry. MIT Press (2017)
15. Werbos, P.J.: Backpropagation through time: what it does and how to do it. Proc. IEEE **78**(10), 1550–1560 (1990)
16. Hopfield, J.J.: Hopfield network. Scholarpedia **2**(5), 1977 (2007)
17. Hinton, G.E.: Boltzmann machine. Scholarpedia **2**(5), 1668 (2007)
18. LeCun, Y., Bengio, Y., Hinton, G.: Deep learning. Nature **521**(7553), 436–444 (2015)
19. Alom, M.Z., Taha, T.M., Yakopcic, C., Westberg, S., Sidike, P., Nasrin, M.S., Van Esesn, B.C., Awwal, A.A.S., Asari, V.K.: The history began from alexnet: a comprehensive survey on deep learning approaches. arXiv preprint arXiv:1803.01164 (2018)

20. Han, K., Xiao, A., Wu, E., Guo, J., Xu, C., Wang, Y.: Transformer in transformer. Adv. Neural Inf. Process. Syst. **34**, 15908–15919 (2021)
21. Shavlik, J.W.: Combining symbolic and neural learning. Machine Learn. **14**, 321–331 (1994)
22. Smolensky, P.: Tensor product variable binding and the representation of symbolic structures in connectionist systems. Artif. intell. **46**(1–2), 159–216 (1990)
23. Badreddine, S., Garcez, A.d., Serafini, L., Spranger, M.: Logic tensor networks. Artif. Intell. **303**, 103649 (2022)
24. Manhaeve, R., Dumančić, S., Kimmig, A., Demeester, T., De Raedt, L.: Neural probabilistic logic programming in deepproblog. Artif. Intell. **298**, 103504 (2021)
25. Mao, J., Gan, C., Kohli, P., Tenenbaum, J.B., Wu, J.: The neuro-symbolic concept learner: Interpreting scenes, words, and sentences from natural supervision. arXiv preprint arXiv:1904.12584 (2019)
26. Fabiano, F., Pallagani, V., Ganapini, M.B., Horesh, L., Loreggia, A., Murugesan, K., Rossi, F., Srivastava, B.: Plan-sofai: a neuro-symbolic planning architecture. In: Neuro-Symbolic Learning and Reasoning in the Era of Large Language Models (2023)
27. Garcez, A.d., Lamb, L.C.: Neurosymbolic AI: the 3rd wave. Artif. Intell. Rev. **56**(11), 12387–12406 (2023)
28. Bohne, T., Windler, A.K.P., Atzmueller, M.: A neuro-symbolic approach for anomaly detection and complex fault diagnosis exemplified in the automotive domain. In: Proceedings of the 12th Knowledge Capture Conference 2023. pp. 35–43 (2023)
29. Onchis, D., Istin, C., Hogea, E.: A neuro-symbolic classifier with optimized satisfiability for monitoring security alerts in network traffic. Appl. Sci. **12**(22), 11502 (2022)
30. Roy, K., Lokala, U., Gaur, M., Sheth, A.P.: Tutorial: neuro-symbolic AI for mental healthcare. In: Proceedings of the Second International Conference on AI-ML Systems. pp. 1–3 (2022)
31. Dold, D., Soler Garrido, J., Caceres Chian, V., Hildebrandt, M., Runkler, T.: Neuro-symbolic computing with spiking neural networks. In: Proceedings of the International Conference on Neuromorphic Systems 2022, pp. 1–4 (2022)
32. Kautz, H.: The third AI summer: aaai robert s. engelmore memorial lecture. AI Mag. **43**(1), 105–125 (2022)

Chapter 2
Neuro-Symbolic AI: The Fusion of Symbolic Reasoning and Machine Learning

"It is not our abilities that show what we truly are... it is our choices.."

–Albus Dumbledore, *Harry Potter and the Chamber of Secrets by J. K. Rowling*

2.1 Introduction

Artificial Intelligence (AI) has achieved remarkable advancements over the past few decades, largely driven by the development of deep learning and neural networks. These technologies excel in tasks such as pattern recognition, image classification, and natural language processing, enabling applications from autonomous driving to voice-activated assistants. However, despite their successes, neural networks often struggle with explainability and fail to handle tasks requiring logical reasoning and abstract thinking [1].

Neuro-Symbolic AI aims to bridge this gap by integrating the learning capabilities of neural networks with the reasoning power of symbolic AI. This hybrid approach seeks to develop AI systems that not only learn from vast amounts of data but also reason with the precision and structure of symbolic logic. The goal is to create AI that is both adaptable and interpretable, capable of handling complex, real-world problems with human-like cognitive abilities [2].

To illustrate the concept, consider the process of teaching a child to identify and categorize animals. By showing pictures and providing descriptions, the child learns to recognize patterns—such as fur and four legs for dogs, feathers and beaks for birds—akin to how neural networks learn from data. This learning from examples is a fundamental principle of machine learning.

B. P. Bhuyan et al., *Neuro-Symbolic Artificial Intelligence*, Studies in Computational Intelligence 1176, https://doi.org/10.1007/978-981-97-8171-3_2

Next, imagine explaining the food chain to the same child. Using examples and logical relationships (e.g., herbivores eat plants, carnivores eat herbivores), the child understands the interconnectedness of ecosystems. This is symbolic reasoning, where explicit rules and knowledge are used to infer new information and make decisions.

In AI, neural networks (resembling quick, intuitive System 1 thinking) are excellent for tasks requiring pattern recognition and rapid decision-making. In contrast, symbolic AI (analogous to deliberate, logical System 2 thinking) excels in tasks that require detailed reasoning and understanding of complex rules.

Neuro-symbolic AI integrates these two modes of thinking, enabling AI systems to benefit from both approaches. Neural networks provide the ability to learn from data and adapt to new situations, while symbolic AI offers robustness in logical reasoning and knowledge representation.

This integration allows for the development of AI systems that can:

1. Learn from extensive datasets and identify patterns (neural networks).
2. Apply logical rules and reason about relationships (symbolic AI).
3. Address tasks that require both intuitive and deliberate thinking, enhancing their overall capability and flexibility.

By combining these cognitive processes, Neuro-Symbolic AI systems can operate efficiently in familiar contexts while also engaging in deep, logical reasoning when faced with novel or complex scenarios. This dual capability is crucial for creating AI systems that can handle the intricacies of real-world applications.

This chapter sets the stage for a deeper exploration of Neuro-Symbolic AI, examining various forms of reasoning and their roles in developing sophisticated AI systems. By understanding the foundational principles of Neuro-Symbolic AI, we prepare to delve into more advanced concepts and applications in the subsequent chapters, moving closer to the goal of developing AI that can think and reason like humans.

2.2 Forms of Reasoning: Laying the Groundwork for Combinatorial and Common Sense Reasoning

The process of reasoning is a fundamental component of human cognitive abilities. Cognitive processing is the cognitive mechanism through which individuals derive meaning from their environment, analyze information to form logical inferences, and engage in decision-making processes. Within the field of artificial intelligence, the process of reasoning holds significant importance as it serves as a fundamental element that empowers systems to effectively analyze data, formulate predictions, and engage in intelligent interactions with their surroundings.

There exist multiple forms of reasoning, each possessing distinct characteristics and applications, as shown in Fig. 2.1. A comprehensive grasp of these forms is essential in order to fully comprehend the integration of symbolic reasoning and machine learning, as well as to establish the fundamental basis for more advanced

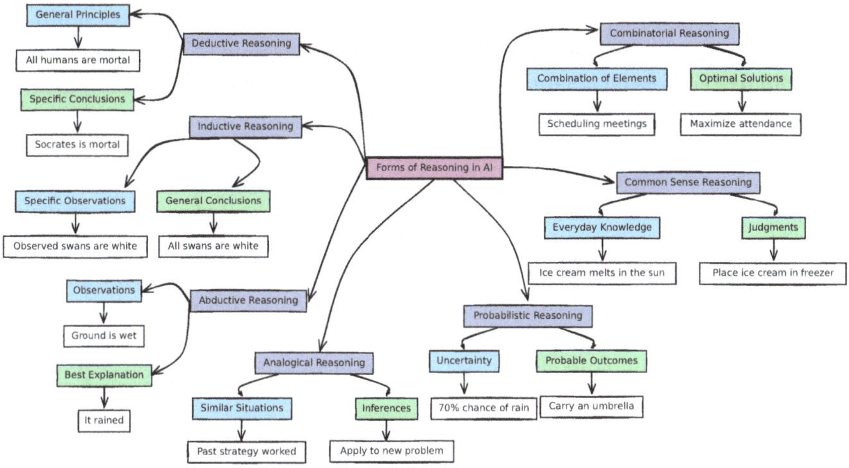

Fig. 2.1 Various forms of reasoning

concepts such as combinatorial reasoning and common sense reasoning. Furthermore, it establishes a foundation for comprehending the dichotomy between System 1 and System 2 cognitive processes, as expounded upon in Daniel Kahneman's seminal work "Thinking, Fast and Slow" [3].

Deductive reasoning is a logical process that involves drawing conclusions based on premises or evidence. Deductive reasoning, alternatively referred to as top-down reasoning, commences by establishing a broad principle or premise and subsequently employs it to anticipate specific outcomes. Deduction is a rational procedure in which inferences are made by relying on assumed true premises. As an illustration, given the general principle that "All birds possess wings" and the specific instance that "A sparrow is classified as a bird," it follows logically that "A sparrow possesses wings." Deductive reasoning is characterized by its deterministic nature, wherein the truthfulness of the premises and the validity of the logical structure ensure the guaranteed truthfulness of the conclusion. Symbolic artificial intelligence frequently employs deductive reasoning as a means to derive logical conclusions from a given set of rules or statements.

Inductive reasoning refers to a logical process of deriving general conclusions or principles based on specific observations or instances. Inductive reasoning, also known as bottom-up reasoning, refers to the cognitive process of formulating generalized conclusions or predictions by analyzing a set of specific instances or empirical data. As an illustration, in the event that a sample of 100 swans is observed and found to be exclusively white, one might infer that all swans possess this characteristic coloration. In contrast to deductive reasoning, inductive reasoning is characterized by its probabilistic nature. The veracity of the conclusions cannot be definitively ascertained, yet they are probable under the condition that the observations are representative and the reasoning is logically valid. Machine learning, specifically supervised learning, is a type of inductive reasoning in which a model acquires the ability to make predictions by analyzing a collection of training examples.

Abductive reasoning is a form of logical inference that involves making educated guesses or hypotheses based on limited information or observations. Abductive reasoning refers to the cognitive process by which the most optimal explanation is generated in response to a given set of observations. The utilization of this approach is frequently observed in situations where there exists a lack of comprehensive or clear-cut information. As an illustration, upon observing dampened streets, one may engage in the process of abduction to infer that precipitation has occurred, notwithstanding the existence of alternative explanations such as the passage of a street cleaning vehicle. Abductive reasoning is frequently employed in diagnostic endeavors, wherein a collection of symptoms is utilized to ascertain the most probable underlying cause.

Analogical reasoning is a cognitive process that involves drawing conclusions or making inferences based on similarities between different objects or situations. Analogical reasoning is a cognitive process that entails making inferences or reaching conclusions by identifying and comparing similarities between two distinct situations or cases. The utilization of analogical reasoning is frequently observed in the context of problem-solving and decision-making processes. This cognitive approach involves identifying a previously resolved problem that bears a resemblance to the current problem at hand and subsequently employing the same solution to address the present issue. For instance, if a particular approach demonstrated efficacy in addressing a comparable issue in prior instances, one may opt to employ the same approach in tackling the present problem.

Probabilistic reasoning is a method of reasoning that incorporates uncertainty and probability in order to make informed decisions or draw conclusions. Probabilistic reasoning entails the process of making inferences and reaching conclusions by taking into account probabilities. This concept is frequently employed in scenarios characterized by indeterminacy or stochasticity. For instance, in the event that the probability of precipitation is 70%, one might opt to bring along an umbrella. Probabilistic reasoning facilitates the formulation of nuanced decisions by incorporating the consideration of various outcome probabilities.

Each of these forms of reasoning plays a pivotal role in the integration of symbolic reasoning and machine learning, thereby contributing to the advancement of sophisticated artificial intelligence systems that possess the ability to acquire knowledge from data and engage in logical reasoning. Logic languages facilitate the application of deductive reasoning, encompassing activities such as consistency verification, materialization, and query resolution. Additionally, these languages support inductive and abductive reasoning processes. Acquiring knowledge about these modes of reasoning also establishes the foundation for delving into more complex notions such as combinatorial reasoning and common sense reasoning, as well as comprehending the dichotomy between System 1 and System 2 cognitive processes.

2.3 Common Sense and Combinatorial Reasoning: In the Context of Neuro-Symbolic AI

2.3.1 Common Sense Reasoning

Common sense reasoning refers to a type of reasoning wherein individuals make assumptions or derive conclusions based on their general knowledge or beliefs about the world. These assumptions and conclusions are often so ingrained in human cognition that individuals may not consciously recognize their utilization. This form of reasoning is essential for comprehending and maneuvering through the world; however, it is widely acknowledged that replicating it in AI systems poses significant challenges.

As an illustration, let us examine the assertion: "John placed the ice cream within the confines of the cupboard." As individuals, we promptly discern that there is an anomaly in this situation. According to conventional wisdom, it is generally advised to store ice cream in a freezer rather than a cupboard, as this practice helps to prevent its melting. Our comprehension of this matter is facilitated by our broad understanding of the characteristics of ice cream as well as the distinction between a cupboard and a freezer. This exemplifies the application of reasoning based on commonly accepted knowledge and understanding.

In an alternative illustration, when perusing a narrative that alludes to a character's act of closing an umbrella prior to entering a dwelling, our innate reasoning leads us to infer that inclement weather, likely rainfall, was prevailing outside despite the absence of explicit mention in the story. This inference is derived from our collective understanding that individuals typically employ umbrellas in instances of precipitation.

The application of common sense reasoning is of utmost importance in various AI tasks, particularly those that pertain to the comprehension of natural language or the interaction between humans and AI systems. In order for an AI system to comprehend textual information or engage in meaningful human interactions, it is imperative that it possesses the capacity to generate common sense inferences akin to those made by humans.

Nevertheless, the ability to employ common sense reasoning poses a considerable obstacle for artificial intelligence. In contrast to alternative modes of reasoning, common sense reasoning frequently entails grappling with information that is incomplete or ambiguous, necessitating the formulation of assumptions and the derivation of inferences from general knowledge that is not explicitly articulated. Moreover, the scope of our collective intuitive understanding is extensive, varied, and frequently challenging to codify in a manner suitable for utilization by an artificial intelligence system.

Notwithstanding these challenges, extensive research is currently being conducted with the objective of endowing AI systems with the capacity for common sense reasoning. This encompasses endeavors to construct extensive repositories of general knowledge, devise algorithms capable of logical inference using said knowledge, and establish educational systems capable of acquiring general knowledge from textual sources or other forms of data. The advancement in this domain will be pivotal for the advancement of artificial intelligence systems that possess the capability to comprehend and engage with the world in a manner akin to human beings.

Symbolic AI, characterized by its utilization of rule-based systems and logical structures, exhibits a strong aptitude for representing and manipulating explicit knowledge, which serves as the foundation for a significant portion of our intuitive reasoning. As an illustration, one could employ a symbolic artificial intelligence framework to encode the proposition "Ice cream undergoes a phase change to a liquid state when exposed to ambient temperature." Subsequently, this encoded rule could be utilized by the system to engage in logical inference, thereby deducing that the act of placing ice cream within a cupboard would probably result in its melting.

Nevertheless, it is important to note that not all knowledge derived from common sense can be readily transformed into explicit rules. A portion of our knowledge is derived from tacit sources, acquired through personal experiences and interactions with the external environment. This is the point at which the cognitive abilities of neural networks are engaged. Neural networks, when exposed to substantial datasets, possess the ability to acquire knowledge of patterns and draw inferences that prove challenging to capture through explicit rule-based systems.

As an illustration, a neural network that has undergone training on an extensive collection of textual data may acquire the knowledge that the expression "place the ice cream in the cupboard" is linked to unfavorable consequences, albeit lacking a comprehensive comprehension akin to that of a human.

The integration of both neurocomputational and symbolic approaches within a Neuro-Symbolic AI system enables the effective addressing of common sense reasoning challenges. The symbolic component is capable of processing the explicit and rule-based aspects of common sense knowledge, whereas the neural component has the ability to acquire knowledge from data and handle the implicit aspects. Through collaborative efforts, individuals can facilitate the AI system in generating the type of logical deductions that come naturally to humans but have historically posed difficulties for artificial intelligence.

2.3.2 *Combinatorial Reasoning*

Combinatorial reasoning refers to a cognitive process wherein individuals engage in the generation and evaluation of various combinations of elements as a means to address a given problem. This form of reasoning is frequently employed in problem-solving endeavors that encompass a substantial array of potential solutions and necessitate the identification of the optimal solution based on specific criteria.

One illustrative instance to contemplate is the issue of arranging a sequence of meetings. The task at hand involves managing a roster of participants, each with their respective availability, with the objective of devising a schedule that maximizes the number of participants able to attend each meeting. The nature of this problem can be classified as combinatorial, as it entails the identification of the optimal arrangement of meeting times from a vast array of potential combinations.

The issue of combinatorial reasoning poses a substantial hurdle for AI systems due to the exponential growth in the number of potential combinations as the problem size increases. This phenomenon, commonly referred to as combinatorial explosion, presents a significant challenge. In the context of complex problems, the generation and evaluation of all conceivable combinations are impractical. Consequently, AI systems must employ heuristics or alternative strategies to efficiently identify satisfactory solutions within a reasonable temporal framework.

Within the domain of Neuro-Symbolic AI, the amalgamation of symbolic reasoning and machine learning holds the potential to enhance combinatorial reasoning. Symbolic AI systems exhibit proficiency in the representation and manipulation of structured data, a capability that proves particularly valuable in addressing combinatorial problems. As an example, a symbolic AI system can depict a scheduling problem using a graph. In this representation, the participants are represented as nodes, while the potential meeting times are represented as edges.

In contrast, machine learning, specifically reinforcement learning, can be employed to acquire proficient tactics for combinatorial reasoning through the analysis of data. An instance of a reinforcement learning algorithm has the capability to acquire a policy for arranging meetings in a manner that optimizes the count of attendees in each meeting, utilizing historical scheduling data as a basis.

Neuro-Symbolic AI has the capability to address combinatorial reasoning problems with enhanced effectiveness through the integration of symbolic reasoning and machine learning techniques. The symbolic component of the system serves to symbolically represent the problem at hand and generate a set of potential solutions. On the other hand, the machine learning component leverages the power of data to acquire knowledge and inform the search for solutions, thereby enhancing the efficiency of the overall process. The amalgamation of these capabilities renders Neuro-Symbolic AI a potent instrument for conducting combinatorial reasoning.

2.4 System 1 and System 2 Thinking: In the Context of Neuro-Symbolic AI

The framework proposed by psychologist Daniel Kahneman in his book "Thinking, Fast and Slow" encompasses the concepts of System 1 and System 2 thinking. This framework offers valuable insights into the distinct cognitive processes that can manifest in the human mind, as well as in AI systems.

2.4.1 System 1 Thinking: The Fast, Intuitive Mind

System 1 thinking, as conceptualized by psychologist Daniel Kahneman, refers to the cognitive process that operates automatically, intuitively, and rapidly. The cognitive processes that operate subconsciously and continuously analyze our surroundings, facilitating the interpretation of sensory information, are referred to as background mental processing.

As an example, upon perceiving a red light at a traffic signal, one does not necessitate deliberate cognitive processing to comprehend its significance; rather, an individual intuitively recognizes the imperative to halt. The prompt comprehension and subsequent response can be attributed to the cognitive process known as System 1 thinking. This cognitive function pertains to the ability to rapidly make decisions, identify patterns, and react to stimuli by drawing upon acquired associations and past experiences.

In an alternative example, let us contemplate the elementary arithmetic quandary "What is the sum of 2 and 2?" The answer "4" can be determined immediately without the need for computational calculations. The instantaneous and involuntary reaction described here can be attributed to the cognitive process known as System 1 thinking.

In the domain of artificial intelligence, System 1 cognition is intricately linked to the functionalities exhibited by machine learning models, specifically neural networks. These models are specifically designed to acquire knowledge from extensive datasets and generate rapid predictions by leveraging identified patterns. As an example, a machine learning model that has been trained on a dataset of images has the capability to promptly categorize an image as either a cat or a dog, akin to the human brain's ability to swiftly identify a familiar face or object.

Similar to System 1 thinking, these models may exhibit errors in the presence of unfamiliar scenarios or insufficient data. Individuals may also encounter difficulties when engaging in activities that necessitate logical reasoning or comprehension of explicit rules, as these tasks typically fall within the purview of System 2 cognitive processes. Notwithstanding these constraints, the swiftness and inherent understanding of System 1 cognition, and consequently machine learning models, render them indispensable for numerous artificial intelligence endeavors.

2.4.2 System 2 Thinking: The Slow, Deliberate Mind

According to Daniel [4], System 2 thinking can be characterized as a cognitive process that is more intentional, demanding mental effort, and characterized by conscious awareness. The cognitive faculty that becomes active when individuals encounter intricate problems, make decisions, or confront scenarios that necessitate meticulous analysis and logical deduction is commonly referred to as critical thinking.

As an example, when presented with the task of determining the product of 17 and 24, one would not possess the ability to promptly generate the solution. To successfully perform multiplication, one must undertake a systematic procedure that involves focused attention and deliberate exertion. The current cognitive process being observed is indicative of System 2 thinking.

Another instance of System 2 cognition can be observed in the process of planning a trip. When making a decision, it is important to take into account a range of factors, including financial considerations, the timing of travel, options for lodging, modes of transportation, and potential activities. It may be necessary to engage in a comparative analysis of various alternatives, weigh the associated pros and cons, and formulate a strategic course of action that aligns most effectively with one's individual requirements and inclinations. Each of these tasks necessitates a deliberate and conscious thought process, which are characteristic features of System 2.

Symbolic AI systems are representative of System 2 thinking within the domain of artificial intelligence. These systems employ explicit rules and logical structures in order to engage in reasoning processes pertaining to the world. Individuals possess the cognitive ability to engage in intricate patterns of logical thinking, comprehend and implement regulations, and render judgments grounded in rational principles.

An illustrative example involves the utilization of a symbolic AI that is endowed with the rules of chess. This system employs these rules to conduct an analysis of the chessboard, assess various potential moves, and ultimately make a selection of the most optimal move. The aforementioned cognitive process, characterized by logical reasoning and decision-making, bears a resemblance to the intentional and cognitively demanding thinking associated with System 2.

Similar to System 2 cognition, symbolic AI systems exhibit the drawback of being characterized by sluggishness and demanding computational resources. Individuals in this context also encounter difficulties when engaging in tasks that necessitate learning from data or managing uncertainty, both of which are areas where neural networks and System 1 thinking demonstrate proficiency. Notwithstanding these obstacles, the cognitive abilities of System 2 thinking, and consequently symbolic AI systems, play a vital role in numerous AI undertakings.

The concept of Neuro-Symbolic AI aims to integrate the cognitive advantages of System 1 and System 2 modes of thinking. Neuro-symbolic AI endeavors to combine neural networks (referred to as System 1) and symbolic AI (referred to as System 2) in order to develop artificial intelligence systems that possess the dual qualities of efficiency and intuitiveness. These systems are designed to acquire knowledge from data and discern patterns while also exhibiting thoughtful and logical behavior by comprehending and employing rules.

The integration of Neuro-symbolic AI systems enables a broader spectrum of tasks to be addressed and greater adaptability to diverse scenarios, surpassing the capabilities of standalone neural networks or symbolic AI systems. Furthermore, this advancement brings us closer to achieving the objective of developing AI architectures that possess the capability to genuinely comprehend and engage with the world in a manner comparable to human beings.

2.5 Conclusion

The integration of symbolic reasoning and machine learning in Neuro-Symbolic AI signifies a notable progression within the domain of artificial intelligence. The objective of Neuro-Symbolic AI is to integrate the advantageous aspects of System 1 and System 2 cognitive processes. This approach aims to develop artificial intelligence systems that possess both rapid and intuitive capabilities, enabling them to learn from data and identify patterns. Additionally, these systems are designed to exhibit deliberate and logical reasoning abilities, enabling them to comprehend and apply rules effectively.

This chapter provides an in-depth examination of the principles underlying symbolic reasoning and machine learning and investigates their respective roles in the integration of Neuro-Symbolic AI. During our discussion, we explored different types of reasoning, such as common sense and combinatorial reasoning, and examined their role in establishing the foundation for advanced artificial intelligence (AI) systems. The examination of the concepts of System 1 and System 2 thinking, as well as their interrelation with various facets of artificial intelligence, was also undertaken.

The integration of symbolic reasoning and machine learning will remain pivotal in the advancement of artificial intelligence systems. The potential of Neuro-Symbolic AI is extensive, encompassing the facilitation of common sense inferences, resolution of intricate combinatorial problems, and seamless integration of intuitive and deliberate thinking.

In the subsequent chapters, a more comprehensive examination of Neuro-Symbolic AI will be undertaken, encompassing its various applications, challenges, and prospective avenues for development. Throughout the course of our exploration, we will progressively acquire a more profound comprehension of this captivating discipline and its capacity to fundamentally transform the realm of artificial intelligence.

2.6 Exercise Questions

1. Consider the following symbolic rules used in a knowledge base for medical diagnosis: a. If a patient has a fever and a cough, they might have the flu. b. If a patient has the flu, they should rest and drink fluids. Using deductive reasoning, explain the steps an AI system would take to recommend treatment for a patient with a fever and a cough.
2. You are given a dataset of customer purchase histories from an online store. The neural network has observed that customers who buy baby products are likely to buy home safety equipment next. Explain how inductive reasoning is applied in this scenario and describe how the neural network generalizes from specific instances to make predictions about future purchases.

3. A manufacturing robot detects a drop in the quality of produced items. Possible causes include: a. Calibration issues with the robot. b. Poor quality of raw materials. c. Errors in the production process. Use abductive reasoning to hypothesize the most likely cause of the quality drop. Discuss how the AI system might gather further evidence to support or refute this hypothesis.
4. An AI system successfully used a heuristic to optimize the layout of a warehouse. The heuristic involved clustering frequently co-accessed items closer together. How can the AI system use analogical reasoning to apply this heuristic to optimizing the layout of a library?
5. An AI system is designed to decide whether to send a reminder for an upcoming appointment. The system knows that sending reminders increases the likelihood of attendance from 50 to 80%, but also that 10% of recipients find reminders annoying. Use probabilistic reasoning to decide whether the AI should send a reminder. Consider both the likelihood of attendance and the annoyance factor.
6. An AI assistant is programmed to manage household chores. The assistant receives the command: "Clean the dishes and put them away." Discuss how common sense reasoning helps the AI understand that "putting away" implies that dishes should be dried first if they are wet.
7. You need to schedule tasks for an AI research team. Each task has a different priority and requires specific team members who have varied availabilities: Task A: High priority, requires Alice and Bob. Task B: Medium priority, requires Carol and Dave. Task C: Low priority, requires Alice and Carol. Use combinatorial reasoning to propose a schedule that maximizes the number of high and medium-priority tasks completed within a week. Explain your approach.

References

1. Garcez, A.D., Gori, M., Lamb, L.C., Serafini, L., Spranger, M., Tran, S.N.: Neural-symbolic computing: an effective methodology for principled integration of machine learning and reasoning. arXiv preprint arXiv:1905.06088 (2019)
2. Bhuyan, B.P., Ramdane-Cherif, A., Tomar, R., Singh, T.: Neuro-symbolic artificial intelligence: a survey. Neural Comput. Appl. 1–36 (2024)
3. Daniel, K.: Thinking, Fast and Slow (2017)
4. Allan, J.: An analysis of Daniel Kahneman's thinking, fast and slow. Macat Library (2018)

Chapter 3
Neuro-Symbolic AI: The Integration of Continuous Learning and Discrete Reasoning

"The measure of intelligence is the ability to change."

Albert Einstein, *The World As I See It*

3.1 Introduction

Imagine teaching a child to recognize different animals. This often involves showing them images and explaining characteristics like fur, feathers, or scales. Over time, the child learns to identify animals based on these continuous sensory inputs, forming patterns and making predictions. This process exemplifies learning in continuous domains, where the child uses ongoing data to make sense of their environment.

On the other hand, consider teaching the child rules about animals: birds have feathers, mammals have fur, and fish live in water. With these discrete pieces of knowledge, the child can make logical deductions, such as recognizing an animal with feathers as a bird. This process is illustrated in Fig. 3.1, where reasoning happens in discrete domains, and explicit rules are applied to draw conclusions.

Traditionally, AI systems have approached these two types of learning separately. Neural networks excel in continuous learning from data, such as recognizing images or sounds. Symbolic AI, however, excels in logical reasoning using predefined rules and structured knowledge. Neuro-symbolic AI aims to merge these approaches, creating systems that can both learn from data and reason logically.

Neuro-symbolic AI seeks to unify the strengths of neural networks and symbolic AI. Neural networks are adept at handling large datasets and learning intricate

B. P. Bhuyan et al., *Neuro-Symbolic Artificial Intelligence*, Studies in Computational Intelligence 1176, https://doi.org/10.1007/978-981-97-8171-3_3

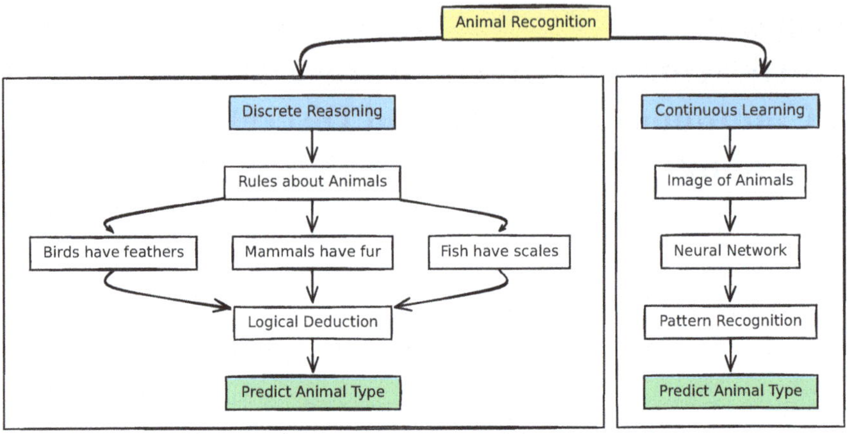

Fig. 3.1 An example to show the difference between continuous learning and discrete reasoning

patterns in continuous spaces. Symbolic AI, on the other hand, is proficient in manipulating symbols and applying logical rules in discrete spaces. By integrating these capabilities, Neuro-Symbolic AI aspires to develop systems that can navigate both continuous and discrete domains seamlessly.

For instance, Graph Neural Networks (GNNs) [1] are one approach to this integration. GNNs can represent knowledge in graph structures, enabling the processing of both continuous and discrete data. They can learn continuous vector representations of nodes and edges in a knowledge graph, which can then be used for tasks like node classification or link prediction, bridging the gap between neural and symbolic methods.

Another promising approach involves Spiking Neural Networks (SNNs) [2], which mimic the discrete spike-based communication of biological neurons. SNNs are well-suited for tasks that require both continuous learning and discrete reasoning, making them ideal for neuro-symbolic integration.

Additionally, methodologies like Neuro-Symbolic Goal and Plan Recognition leverage [3] both neural and symbolic components to understand and predict agents' goals and plans, even in noisy and incomplete environments. This integration enables robust AI systems capable of sophisticated decision-making and reasoning.

The integration of continuous learning and discrete reasoning in Neuro-Symbolic AI represents a significant advancement in the field. It aims to develop AI systems that can learn from their environment, reason logically, and interact with the world in more human-like, nuanced ways. As we explore these methodologies further in subsequent sections, we will uncover the potential applications and future directions of Neuro-Symbolic AI, paving the way for more intelligent and adaptable AI systems.

3.2 Continuous Learning in Neural Networks: The Power of Perception

3.2.1 Understanding Continuous Space: The Realm of Perception

In the field of artificial intelligence and machine learning, the concept of continuous space pertains to a specific realm wherein variables or parameters possess the ability to assume an unbounded multitude of values within a defined interval. The concept being discussed is frequently juxtaposed with that of discrete space, wherein variables are limited to assuming distinct and separate values.

The term "continuous" is derived from the mathematical principle of continuity, which denotes a state of unbroken and uninterrupted progression of values. In a space that is continuous, it is possible to establish a connection between any two points by means of an infinite set of intermediary points, without encountering any discontinuities or abrupt transitions. This phenomenon bears resemblance to our cognitive perception of the tangible environment. When examining a color gradient, it is observed that discrete and distinct colors are not perceived, but rather a seamless progression from one hue to another.

Within the field of machine learning, specifically in the context of neural networks, the term "continuous space" commonly denotes the expansive and multidimensional space wherein the optimization of a model's parameters takes place. Each dimension within this spatial context corresponds to a specific parameter of the model, such as the weights utilized within a neural network. The parameters in question possess the capability to assume any real value, thereby facilitating a seamless and gradual adaptation throughout the process of learning.

Consider a neural network being trained to recognize images. The pixel intensities within an image exhibit a continuous range of values, spanning from 0 to 255, thereby constituting a continuous input space. In a similar vein, the weights and biases within the network, which undergo adjustments during the training process, constitute a continuous parameter space. The training procedure entails traversing a high-dimensional continuous space in order to identify the parameter values that minimize the discrepancy between the predictions made by the network and the true labels assigned to the images.

A comprehensive comprehension of the notion of continuous space is imperative in order to fully grasp the mechanisms by which neural networks acquire knowledge from data and adjust their parameters accordingly. The capacity to function and acquire knowledge within unbroken domains is what enables neural networks to demonstrate exceptional performance in tasks related to perception, such as image and speech recognition. These tasks typically involve inputs that are characterized by continuous and high-dimensional attributes. Although the forthcoming chapters will provide a more comprehensive analysis, let us now examine a rudimentary example of a feedforward neural network comprising a solitary input layer, one hidden layer, and one output layer. The mathematical operations occurring within this network can be delineated as follows.

The input received by the neurons in the hidden layer is a summation of the inputs, each multiplied by their respective weights. The representation of the input to a neuron in the hidden layer can be denoted as follows:

$$z = \sum_{i=1}^{n} (w_i \times x_i) + \beta \tag{3.1}$$

where z represents the dependent variable, w_i represents the weights assigned to each independent input variable x_i, and β represents the bias term. In this context, n represents the number of inputs, also referred to as neurons within the input layer. Correspondingly, "$w_1, w_2, \ldots, w_n$" symbolize the weights that are linked to these inputs.

The input to the neuron is subsequently subjected to an activation function. The activation function is a crucial component in neural networks, and it is important to carefully consider the choice of function. Common options for the activation function include the sigmoid function, the hyperbolic tangent function, and the Rectified Linear Unit (ReLU) function [4]. Let f represent the activation function. The output of the neuron in the hidden layer can be denoted as

$$h = f(z) \tag{3.2}$$

The aforementioned procedure is subsequently iterated from the hidden layer to the output layer. The weights from the hidden layer to the output layer will be denoted as "v," while the bias will be denoted as "c." Consequently, the input to a neuron in the output layer can be mathematically represented as follows:

$$y' = \sum_{j=1}^{m} v_j \times h_j + c \tag{3.3}$$

In this context, m represents the number of neurons present in the hidden layer, while "$v_1, v_2, \ldots, v_m$" denote the corresponding weights assigned to these neurons. The output activation function is applied to the input of the output neuron, resulting in the final output of the network. This activation function may be the same as or different from the one utilized in the hidden layer.

$$y = g(y') \tag{3.4}$$

The equations presented herein depict the forward propagation process of a neural network operating within a continuous space. The process of learning entails the manipulation of weights (w and v) and biases (β and c) in order to minimize the discrepancy between the output (y) generated by the network and the desired target output. The process commonly employed for this purpose is known as backpropagation, wherein the gradient of a loss function is computed in the continuous parameter space. The parameters are then adjusted in a manner that minimizes the loss by moving in the direction opposite to the gradient.

Nevertheless, although neural networks exhibit significant capabilities in learning within continuous domains, they frequently encounter difficulties when confronted with tasks that necessitate reasoning within discrete domains. In contrast, symbolic AI systems excel in such scenarios, showcasing their proficiency in this particular area.

3.2.2 The Role of Continuous Learning in Neural Networks: Learning from the Environment

Neural networks, which are fundamental components of contemporary machine learning, are specifically engineered to acquire knowledge from data within continuous domains. The learning process is frequently characterized as the network's capacity to acquire knowledge from its surroundings, as it entails modifying the network's parameters in response to the data it encounters.

The learning process within a neural network comprises two fundamental stages: the forward pass and the backward pass.

During the forward pass, the neural network receives input data and proceeds to propagate it through the network, resulting in the generation of an output. The process entails executing a sequence of mathematical computations, including weighted sums and activations, as elucidated in the preceding section. The network's output is a prediction that is derived from the present configuration of the network's parameters, which include the weights and biases.

The backward pass, commonly referred to as backpropagation, is the stage in which the actual learning process takes place. During the process of backpropagation, the neural network computes the gradient of a given loss function in relation to its parameters. The loss function quantifies the disparity between the predictions made by the neural network and the true target values. The direction in which the parameters should be adjusted to minimize the loss is indicated by the gradient of the loss function.

After the computation of the gradients, the neural network proceeds to update its parameters by employing a technique known as gradient descent. This process entails modifying the parameters in accordance with the negative gradient, thereby diminishing the loss. The aforementioned procedure is iteratively performed, wherein the neural network consistently assimilates knowledge from the dataset and modifies its parameters to enhance its predictive capabilities.

The capacity of neural networks to acquire knowledge from data in continuous domains enables them to demonstrate exceptional performance in tasks that encompass pattern recognition, prediction, and generalization based on exemplars. As an illustration, a neural network can acquire the ability to identify images depicting felines through the process of training on an extensive collection of cat images. Through the manipulation of its parameters using the pixel values of the images and their corresponding labels, the neural network acquires the ability to discern patterns within the data that are indicative of the existence of a feline entity.

Nevertheless, despite the proficiency of neural networks in assimilating continuous data, they frequently encounter difficulties when confronted with tasks requiring logical reasoning or comprehension of explicit rules, which are commonly expressed in discrete domains. The integration of ongoing learning and distinct reasoning in Neuro-Symbolic AI is a significant aspect that will be explored in subsequent sections.

3.3 Discrete Reasoning in Symbolic AI: The Power of Symbol Manipulation

3.3.1 Understanding Discrete Space: The Realm of Symbols

In contrast to continuous space, discrete space refers to a domain in which variables or parameters are limited to assuming distinct and separate values. Within the realm of artificial intelligence, specifically symbolic AI, the term "discrete space" commonly denotes the domain encompassing potential states or solutions that the system takes into account during the process of reasoning or decision-making.

The term "discrete" originates from the mathematical notion of discreteness, which connotes a clear and independent sequence of values. In the context of a discrete space, variables are constrained to assume only distinct and predetermined values, rather than being able to encompass a continuous range of values. This is comparable to the manner in which symbols are employed to denote particular concepts or entities within the realm of reality.

For example, let us consider the game of chess. The representation of the game state can be expressed as a collection of distinct variables, specifically the spatial arrangement of the game pieces on the board. Every position within the game can be filled by a distinct category of chess piece, such as a pawn, rook, knight, bishop, queen, or king, or alternatively, it may remain unoccupied. The game undergoes discrete changes in its state, wherein each move leads to a distinct and separate state.

Within the domain of symbolic artificial intelligence, the process of reasoning frequently entails the manipulation of symbols and the examination of various states or solutions within a discrete space. As an illustration, a symbolic AI engaged in a game of chess would analyze various potential moves, each resulting in a distinct game state, and subsequently select the move that leads to the most advantageous state based on an evaluation function.

Logic, specifically formal logic, is an efficacious instrument for rational thinking within discrete domains. The system offers a well-organized framework for the representation and manipulation of symbols, enabling the modeling of diverse types of reasoning. In this discourse, we shall examine several fundamental forms of logic and their treatment of discrete spaces.

Propositional logic, alternatively referred to as Boolean logic, is recognized as the most elementary variant of logic. The subject matter pertains to propositions, which are declarative statements that possess the capacity to be evaluated as either true or false. Propositional logic employs logical connectives, such as conjunction ($\wedge$), disjunction ($\vee$), and negation ($\neg$), to construct compound propositions by combining simpler ones. To illustrate, let us consider proposition P as "The occurrence of rainfall" and proposition Q as "My current location being indoors." Consequently, the conjunction "$P \wedge Q$" represents the proposition "The occurrence of rainfall and my current location being indoors."

Predicate logic, also referred to as first-order logic, expands upon propositional logic by incorporating predicates and quantifiers. A predicate is a declarative statement that possesses the ability to be either true or false, contingent upon the values assigned to its variables. An instance of a predicate that exhibits a true value for certain values of x and a false value for others is the statement "x is greater than 3."

Quantifiers, namely the universal quantifier ($\forall$) and the existential quantifier ($\exists$), enable us to formulate assertions pertaining to either all or some elements within a given set. As an illustration, the expression "$\forall$x, x is greater than 3" can be reformulated as the sentence "For all x, x is greater than 3," which is deemed false. Conversely, the expression "$\exists$x, x is greater than 3" can be reformulated as the sentence "There exists an x such that x is greater than 3," which is considered true.

Modal logic expands upon predicate logic by incorporating modal operators, such as the "necessarily" operator ($\Box$) and the "possibly" operator ($\diamond$), enabling the formulation of assertions pertaining to necessity and possibility. An illustration of this can be seen in the statement "$\Box$P," which denotes the proposition "P is necessarily true," and "$\diamond$P," which represents the proposition "P is possibly true."

These various forms of logic offer robust tools for engaging in reasoning within discrete spaces. Graphs enable us to effectively depict intricate connections among entities, facilitate logical deductions, and establish mathematical propositions. Nevertheless, although logic proves to be highly effective in the realm of explicit rules and structured knowledge, it frequently encounters difficulties when confronted with tasks that necessitate learning from data or managing uncertainty. These particular challenges, however, are areas in which neural networks and probabilistic models excel. The integration of logical reasoning and machine learning in Neuro-Symbolic AI is a topic that will be explored in subsequent chapters.

3.3.2 *The Role of Discrete Reasoning in Symbolic AI: Manipulating Symbols for Reasoning*

Symbolic AI, alternatively referred to as classical AI, is primarily focused on the manipulation of symbols for the purpose of executing reasoning tasks. The symbols depicted in this context serve as representations of distinct entities or concepts within the realm of reality, and the connections between them are established based on explicit rules or logical principles.

3.3.2.1 Symbol Manipulation

Symbol manipulation, which is also referred to as symbolic computation [5], constitutes a fundamental component of symbolic AI. Symbolic representation is a fundamental aspect of this process, whereby entities or concepts are denoted by symbols. Additionally, the application of rules or operations to these symbols is employed to facilitate reasoning tasks.

Symbolic AI employs symbols as a means of representing diverse entities or concepts within the realm of cognition. In the context of a chess game, it is possible to assign distinct symbols to each piece (such as the king, queen, or pawn) and every position on the board. In a medical diagnostic system, distinct symbols can be employed to represent various diseases and symptoms.

Symbols are conceptual representations that lack inherent meaning in and of themselves. The semantic significance of a symbol is derived from its contextual utilization within a given system, as well as the interconnections it establishes with other symbols.

Symbol manipulation encompasses the utilization of diverse operations to manipulate these symbols. The operations in question can be derived from logical principles, arithmetic calculations, or any other predetermined set of rules established within the system. In the context of rule-based systems, an operation can be described as the application of a rule of inference to a given set of symbols, resulting in the derivation of a new symbol.

In the context of a chess game, an operation can be defined as the act of relocating a piece, represented by a symbol, from one position, also represented by a symbol, to another position on the board. Within the context of the medical diagnostic framework, an operative procedure may encompass the process of deducing a specific ailment (represented by a symbol) based on a collection of observable indications or symptoms (also represented by symbols).

Symbolic reasoning refers to the systematic application of operations on symbols in order to carry out cognitive tasks involving logical and deductive reasoning. This may encompass the process of drawing logical conclusions, resolving complex issues, making informed choices, or engaging in any cognitive activity that necessitates the utilization of symbolic reasoning.

An example of a symbolic AI system engaged in chess would employ symbolic reasoning to determine the optimal move. The algorithm takes into account the present condition of the game, which is represented by a collection of symbols. It then applies the established regulations of chess, which can be seen as operations, in order to contemplate potential forthcoming conditions. Finally, it selects the move that results in the most advantageous state.

In a similar vein, a medical diagnostic system would employ symbolic reasoning in order to ascertain and identify various diseases. The system would take into account the symptoms reported by a patient, which are represented by a set of symbols. It would then utilize its knowledge of diseases, which is represented by rules that establish connections between symptoms and diseases. Based on this information, the system would make inferences to determine the most probable diseases.

Symbol manipulation and symbolic reasoning are highly effective tools for tasks that necessitate the utilization of explicit rules and structured knowledge. Nevertheless, individuals frequently encounter difficulties when faced with tasks that require the acquisition of knowledge from unprocessed data or the handling of uncertain situations, domains in which neural networks and probabilistic models demonstrate exceptional performance. The integration of symbolic reasoning and machine learning within the field of Neuro-Symbolic AI endeavors to leverage the respective advantages of these two methodologies.

3.3.2.2 Discrete Reasoning

The navigation and exploration of a discrete space of states or solutions is a fundamental aspect of symbolic AI, known as discrete reasoning. Every state within the system represents a distinct arrangement of symbols and their corresponding relationships, and the system employs logical principles or rules to facilitate the transition between states.

Within the domain of symbolic AI, a state can be defined as a distinct configuration of symbols that serves as a representation of a specific circumstance or state of affairs. In the context of a game of chess, a state can be defined as the configuration that encompasses the present arrangement and locations of all the pieces on the chessboard.

The rules of the system dictate the transitions between states. In the context of chess, the regulations of the game establish the permissible actions that can be taken, and each action signifies a shift from one state (the present configuration of chess pieces) to another (the configuration of chess pieces following the action).

The process of discrete reasoning entails the examination and navigation of a given state space in order to attain a specific objective. The process of exploration can be directed through the implementation of different strategies, including depth-first search, breadth-first search, or heuristic search. The choice of strategy is contingent upon the particular task at hand and the characteristics of the state space.

In the context of chess, the objective could be defined as achieving checkmate against the adversary's king. The AI system engages in an exploration of the state space by evaluating different sequences of moves and their corresponding states. It then selects the sequence that it anticipates will have the highest probability of resulting in a checkmate.

Discrete reasoning encompasses the process of drawing logical conclusions and making informed judgments by considering the existing state and the established principles of the system. This frequently entails a procedure of deductive reasoning, wherein the system utilizes established principles to derive additional information or anticipate the potential consequences of various actions based on the present state.

In the context of chess, the AI system may deduce that executing a particular move would result in its own king being exposed to a check, consequently leading to the determination to refrain from making said move. Alternatively, it could anticipate the occurrence of a checkmate by deducing that a specific sequence of moves will result in such an outcome, subsequently opting to execute said sequence.

3.3.2.3 Knowledge Representation and Reasoning

Knowledge representation and reasoning are considered to be fundamental components within the field of symbolic artificial intelligence [6]. These systems encompass the process of transforming human knowledge into a format that can be processed by computers in order to tackle intricate tasks. Subsequently, this encoded knowledge is utilized to draw logical conclusions and make informed judgments. Having discussed the introductory aspects of reasoning in the preceding chapter, we now aim to delve into the fundamental principles of knowledge representation in this section, with further elaboration to be provided in subsequent chapters. Knowledge representation plays a pivotal role in the field of symbolic artificial intelligence, as it offers a systematic approach to encoding information about the world in a manner that enables computers to tackle intricate tasks effectively. The objective of knowledge representation is to effectively capture and organize information pertaining to the world, thereby facilitating logical reasoning and empowering artificial intelligence systems to exhibit intelligent behavior.

There exist multiple methodologies for the representation of knowledge, each possessing distinct advantages and disadvantages. The following are several frequently employed techniques:

Logical representation refers to the utilization of formal logic as a means of representing knowledge. Factual information is typically conveyed through logical statements, while rules are typically expressed as logical implications. As an example, the proposition "Socrates is a man" can be expressed in logical terms as "Man(Socrates)," while the principle "All men are mortal" can be represented as the logical implication "$\forall$ x, Man(x) $\Longrightarrow$ Mortal(x)."

Semantic networks are graphical structures that depict knowledge in the form of interconnected nodes and edges. In this context, nodes serve as representations of entities or concepts, while edges serve as representations of the relationships that exist between these entities or concepts. An instance of a semantic network could include a vertex denoting "Socrates," a vertex denoting "man," and a directed edge originating from the "Socrates" vertex and terminating at the "man" vertex, symbolizing the relationship that Socrates is classified as a man.

Frames are a type of data structure that is used to represent objects or concepts, along with their associated properties. The framework of an object encompasses attributes, also known as slots, which represent the properties of the object, along with corresponding values assigned to those attributes. As an illustration, the frame of an automobile may encompass characteristics such as its color, model, and manufacturer, along with their respective values denoting attributes such as red, Model S, and Tesla.

Rule-based systems are a type of knowledge representation that employs collections of if-then rules. Every rule delineates a specific condition and corresponding action, wherein the action is executed if and only if the condition evaluates to true. An illustrative instance of a rule within a medical diagnostic system could be formulated as follows: "In the event that a patient exhibits symptoms of fever and cough, it is plausible to infer that they could potentially be afflicted with influenza."

An **ontology** can be defined as a structured and systematic representation of knowledge within a specific domain. It comprises a collection of concepts and the interconnections that exist between them. Ontologies are frequently employed in the realm of semantic web technologies, serving as a comprehensive framework for the organization and integration of information.

The selection of a knowledge representation method is contingent upon the particular task at hand and the inherent characteristics of the knowledge that needs to be portrayed. Various methodologies can be employed either independently or in conjunction, contingent upon the specific demands of the task at hand. The crucial factor in problem-solving is the selection of a representation that effectively encompasses the fundamental characteristics of the problem domain and enables the cognitive processes required for task resolution.

3.4 The Integration of Continuous and Discrete Spaces in Neuro-Symbolic AI: The Best of Both Worlds

The field of Neuro-Symbolic AI is an intriguing area of study within the realm of artificial intelligence research. Its primary objective is to integrate the advantageous aspects of neural networks and symbolic AI, with the aim of developing systems that possess the ability to learn from data while also engaging in logical reasoning based on existing knowledge. This entails the integration of the continuous learning capabilities exhibited by neural networks with the discrete reasoning capabilities demonstrated by symbolic AI, thus establishing a connection between these two fundamentally distinct approaches to artificial intelligence.

The incorporation of continuous and discrete spaces within the field of Neuro-Symbolic AI entails the development of systems capable of functioning in both spaces and seamlessly transitioning between them as required. The task at hand necessitates the exploration and creation of novel methodologies and frameworks capable of effectively accommodating the gradient-based learning techniques employed by neural networks, as well as the rule-based reasoning approaches employed by symbolic artificial intelligence.

One potential method for achieving this integration involves the utilization of neural networks to acquire knowledge representations of data that can be effectively employed for symbolic reasoning. As an example, a neural network has the capability to acquire the ability to identify and extract distinctive attributes from unprocessed data that align with symbols or concepts. These extracted features can subsequently serve as inputs to a symbolic reasoning system.

An alternative methodology involves integrating symbolic knowledge into the learning mechanism of neural networks. As an illustration, a neural network could undergo training to make predictions regarding the results obtained from the application of specific rules, or to generate symbolic representations based on its predicted outcomes.

Graph Neural Networks (GNNs) are a highly effective tool within the Neuro-Symbolic AI framework, especially in the context of seamlessly integrating continuous and discrete domains. Symbolic AI often employs data structured as a graph, which is a prevalent format for various forms of knowledge representation. Consequently, these systems are specifically designed to operate effectively with such graph-structured data.

A graph is a set of vertices (also known as nodes) and edges. Within the domain of knowledge representation, nodes frequently serve as representations of entities or concepts, while edges serve as representations of the relationships that exist between these entities or concepts. In the context of a knowledge graph that encompasses medical knowledge, it is plausible to have nodes that correspond to diseases and symptoms, while edges can be employed to denote relationships such as "is a symptom of."

GNN functions by facilitating the exchange of messages between nodes through edges, thereby modifying the states of the nodes in accordance with the received messages. This enables individuals to acquire knowledge of intricate patterns within the graph's structure, which can be applied to various tasks such as node classification (forecasting node attributes), link prediction (anticipating the presence of connections between nodes), and graph classification (predicting attributes of entire graphs).

Within the domain of Neuro-Symbolic Artificial Intelligence, Graph Neural Networks possess the capability to acquire continuous representations of symbolic knowledge. As an illustration, a GNN has the potential to undergo training in order to acquire a continuous vector representation for every individual node within a knowledge graph. This representation would encompass the node's placement within the graph's structure as well as its associations with other nodes. The utilization of these ongoing representations can subsequently serve as input to a neural network, enabling the accomplishment of tasks such as prediction or classification. This process effectively establishes a connection between the discrete and symbolic nature of the knowledge graph and the continuous and perceptual nature of the neural network.

Furthermore, GNNs have the capability to directly execute reasoning tasks on the graph. For instance, knowledge graphs can be utilized to deduce undisclosed connections or to forecast the consequences of implementing specific rules or actions depicted within the graph.

Spiking Neural Networks (SNNs) are a class of artificial neural networks that exhibit a higher degree of resemblance to the functioning of biological neurons. In contrast to conventional artificial neurons that generate continuous output, spiking neurons engage in communication through discrete signals commonly referred to as "spikes." This characteristic renders them highly suitable for the representation and analysis of discrete symbolic data.

In the context of a biological neuron, the conveyance of information occurs through a sequence of electrical impulses, commonly referred to as spikes. The temporal occurrence and rate of these spikes convey information. SNNs endeavor to replicate this phenomenon by employing artificial neurons that aggregate input signals and elicit a spike when the cumulative input surpasses a specific threshold. Following a spike event, the neuron undergoes a process of resetting its state, subsequently initiating the accumulation of input signals once more.

The phenomenon of spiking behavior enables SNNs to engage in information processing that is fundamentally distinct from that of conventional neural networks. Due to their discrete temporal nature and binary state, neural networks possess an inherent capacity to effectively encode and manipulate symbolic data. For instance, various symbols can be denoted by distinct configurations of spikes.

Within the domain of Neuro-Symbolic Artificial Intelligence, Spiking Neural Networks present a prospective methodology for the fusion of continuous learning and discrete reasoning. Spiking neural networks, similar to conventional neural networks, possess the ability to acquire knowledge from continuous data through the adaptation of connection weights in response to observed spike patterns. However, they are also capable of engaging in logical reasoning using discrete symbolic information, which involves the generation and manipulation of patterns of spikes.

Furthermore, due to their discrete time step operation, SNNs exhibit a natural aptitude for the processing of sequential or temporal data. This characteristic renders them well-suited for tasks that necessitate the ability to reason about sequences of events or actions, a prevalent requirement in numerous symbolic reasoning tasks.

The **Neuro-Symbolic Goal and Plan Recognition** approach integrates the capabilities of artificial intelligence learning and planning techniques to address the challenges posed by noise and incomplete observations. This is achieved by leveraging prior data. This methodology proves to be highly advantageous in scenarios where the observations are limited or subject to noise, a common occurrence in practical applications.

The objective of this approach is to deduce the intended objective of an agent based on a series of observations. The achievement of this objective is facilitated through the creation of neuro-symbolic recognition models, which possess the capability to acquire knowledge from available data and generate forecasts pertaining to the objectives of the agent. The models undergo training using a dataset comprising previous observations and their corresponding objectives, enabling them to acquire knowledge about the typical patterns that result in the achievement of each objective.

After inferring the agent's goal, the model can subsequently be utilized to identify the probable plan that the agent will adopt in order to accomplish said goal. This process entails the determination of the sequence of actions or decisions that the agent is anticipated to undertake, relying on the patterns that have been acquired by the model through the training data.

The Neuro-Symbolic Goal and Plan Recognition approach demonstrates a high level of robustness and reliability, as it possesses the ability to deduce goals and generate accurate plans, even when confronted with observations that are noisy or incomplete. This feature renders it a potent instrument for a diverse array of applications, spanning from autonomous systems requiring comprehension and anticipation of other entities' actions, to intelligent tutoring systems necessitating comprehension and responsiveness to the objectives of their users.

A recent study has provided evidence of the efficacy of this methodology, illustrating its ability to consistently deduce objectives and generate accurate strategies across a range of experimental datasets.

In the subsequent chapters, we will conduct a comprehensive examination of the various techniques employed in Neuro-Symbolic AI, thereby facilitating a more profound understanding of this field. This book aims to investigate the utilization of Graph Neural Networks along with various other methodologies in order to establish a connection between continuous learning and discrete reasoning. Additionally, an examination will be conducted to explore the application of these techniques across diverse domains, including healthcare and social sciences. This analysis will shed light on their efficacy in addressing intricate challenges that encompass both continuous and discrete spatio-temporal dimensions.

3.5 Conclusion

Within this chapter, we have commenced an intriguing exploration into the realm of Neuro-Symbolic AI, delving into the amalgamation of ongoing learning and discrete reasoning. The aforementioned approach seeks to integrate the respective advantages of neural networks and symbolic AI, facilitating the development of systems capable of both acquiring knowledge from data and engaging in logical reasoning.

The exploration of continuous and discrete spaces has been undertaken, with an examination of their distinct attributes and their significance in the processes of learning and reasoning. The proficiency of neural networks in acquiring knowledge from data in continuous domains has been observed, contrasting with the strength of symbolic AI systems in conducting logical reasoning within discrete domains.

We have additionally investigated several methodologies employed to establish a connection between these two domains. The techniques demonstrate encouraging prospects for the advancement of Neuro-Symbolic AI, presenting possible resolutions to the complex task of amalgamating continuous learning and discrete reasoning within a unified system.

As we progress, we will further investigate each of these techniques, examining their fundamental principles, evaluating their advantages and disadvantages, and analyzing their utility in the field of Neuro-Symbolic AI. Additionally, an examination will be conducted to explore the application of these techniques across diverse domains, encompassing healthcare to social sciences. This analysis will shed light on their efficacy in addressing intricate challenges that involve both continuous and discrete spatial dimensions.

3.6 Exercise Questions

1. You are developing a neural network to classify images of animals (cats, dogs, and birds). The network uses pixel values as input and predicts the animal type. Describe the process of training the neural network. Explain how the continuous space of pixel values and the gradient descent algorithm are used to adjust the network's parameters.
2. You have a set of rules describing animals: a. Birds have feathers. b. Mammals have fur. c. Fish have scales. d. Penguins are birds. e. Whales are mammals. Using these rules, create a symbolic reasoning system that can deduce the type of animal given specific attributes. For example, deduce that a penguin is a bird, and a whale is a mammal.
3. Consider an AI system designed to identify animals in images and provide logical explanations for its classifications. The system combines a neural network for image recognition with a symbolic AI for reasoning. Outline the steps required to integrate the neural network's continuous learning capabilities with the symbolic AI's discrete reasoning rules. How does the system transition between processing continuous data and applying discrete rules?
4. You are using a Graph Neural Network (GNN) to process a knowledge graph that includes relationships between different animal species. Nodes represent species, and edges represent relationships (e.g., "is a type of," "has a characteristic"). Explain how GNNs can learn continuous vector representations of nodes and edges. How can these representations be used to predict new relationships or classify species within the knowledge graph?

References

1. Scarselli, F., Gori, M., Tsoi, A.C., Hagenbuchner, M., Monfardini, G.: The graph neural network model. IEEE Trans. Neural Netw. **20**(1), 61–80 (2008)
2. Tavanaei, A., Ghodrati, M., Kheradpisheh, S.R., Masquelier, T., Maida, A.: Deep learning in spiking neural networks. Neural Netw. **111**, 47–63 (2019)
3. Amado, L., Pereira, R.F., Meneguzzi, F.: Robust neuro-symbolic goal and plan recognition. In: Proceedings of the AAAI Conference on Artificial Intelligence, vol. 37, pp. 11937–11944 (2023)
4. Nair, V., Hinton, G.E.: Rectified linear units improve restricted Boltzmann machines. In: Proceedings of the 27th International Conference on Machine Learning (ICML-10), pp. 807–814 (2010)
5. Meurer, A., Smith, C.P., Paprocki, M., Čertík, O., Kirpichev, S.B., Rocklin, M., Kumar, A., Ivanov, S., Moore, J.K., Singh, S., et al.: Sympy: symbolic computing in Python. PeerJ Comput. Sci. **3**, e103 (2017)
6. Brachman, R., Levesque, H.: Knowledge Representation and Reasoning. Morgan Kaufmann (2004)

Part II
Symbolic AI and Knowledge Representation

Chapter 4
Knowledge Representation in Artificial Intelligence

"The only true wisdom is in knowing you know nothing."

Socrates

4.1 Introduction

During the initial phase of 2020, when reports of an unfamiliar virus started to circulate, the global community encountered an unparalleled and formidable predicament. The COVID-19 pandemic, subsequently recognized as such, encompassed not only a healthcare emergency but also an information crisis. Hospitals, research institutions, and governmental bodies experienced a significant influx of data encompassing infection rates, patient medical records, genomic sequences, and additional relevant information. In the midst of this overwhelming influx of information, a crucial inquiry emerged: How can this extensive and frequently contradictory collection of data be effectively converted into practical knowledge? How can we effectively identify patterns, forecast outbreaks, and develop efficient interventions? The aforementioned situation, which serves as a representative example of the intricate nature of the information era, acts as an appropriate introduction to our examination of knowledge representation within the field of artificial intelligence.

The fundamental concept of knowledge representation revolves around the organization and structuring of unorganized information, with the aim of converting disordered and chaotic data into a cohesive and understandable form. The objective is to establish a framework akin to the epidemiological models employed during the

B. P. Bhuyan et al., *Neuro-Symbolic Artificial Intelligence*, Studies in Computational Intelligence 1176, https://doi.org/10.1007/978-981-97-8171-3_4

pandemic, which facilitates the processes of reasoning, decision-making, and learning. Within the field of artificial intelligence, this concept pertains to the capacity of machines to comprehend, analyze, and respond to their surroundings in a manner that emulates human cognitive abilities.

Throughout this chapter, we will undertake an exploration that spans from the philosophical foundations of knowledge to the complex techniques utilized in artificial intelligence for its representation. Our objective is to present a comprehensive overview of the domain, encompassing various methodologies such as semantic networks, frames, and ontologies, as well as the hierarchical progression from data to wisdom. Each section of the text explores its respective topic in depth, demonstrating interconnections that mirror the multitude of information pieces that have emerged during the pandemic. These interconnections contribute to the development of a comprehensive and integrated understanding.

4.2 Defining Knowledge

The notion of "knowledge" is complex and has been delineated through diverse perspectives within distinct disciplines and settings. Presented below are several definitions:

The field of philosophy commonly defines knowledge as justified true belief [1]. Epistemologically speaking, the acquisition of knowledge necessitates the presence of three essential conditions: belief in the truth of a proposition, the possession of adequate justification for holding that belief, and the actual truthfulness of the belief itself.

Within the field of cognitive psychology, knowledge is conceptualized as the comprehension and consciousness of factual content, verities, or information acquired through experiential or educational means (a posteriori), or through introspective reflection (a priori).

In the field of information theory, knowledge is conceptualized as the process of diminishing uncertainty [2]. The information possesses a specific significance within a given context.

Within the realm of artificial intelligence, knowledge is commonly characterized as the collection of information pertaining to a specific domain, which can subsequently be employed to address challenges within the said domain. The term "knowledge" refers to the collection of information that an AI system has been specifically programmed to comprehend or has acquired through learning processes. This knowledge encompasses a wide range of elements, such as factual data, conceptual frameworks, procedural guidelines, models, and heuristics, among others.

In the realm of knowledge management, knowledge is frequently delineated from both data and information. Data refers to unprocessed and unorganized facts and figures. Information, on the other hand, is the result of processing data, wherein meaning is attributed to it. Knowledge, however, goes beyond information by being personalized or contextualized, allowing individuals to utilize it in making informed decisions and taking appropriate actions.

Epistemology, a branch of philosophy concerned with the nature of knowledge, commonly characterizes knowledge as a belief that possesses both truth and justification. This definition emphasizes the necessity of not only truth but also the rational basis for holding a belief.

The aforementioned definitions elucidate the multifaceted nature of knowledge and its contextual variability in terms of comprehension. Within the domain of Neuro-Symbolic AI, knowledge encompasses the amalgamation of data-driven insights obtained from neural networks and the rule-based comprehension derived from symbolic AI.

4.2.1 From Data to Wisdom

Within the domain of artificial intelligence, it is advantageous to conceptualize knowledge as a hierarchical construct commencing with unprocessed data and advancing through stages of information, knowledge, and wisdom. This concept is commonly known as the "**DIKW** Pyramid" [3]. In the following analysis, we will deconstruct each tier of this pyramid by providing illustrative instances.

Data refers to the unprocessed and raw facts and statistics that are gathered for the purpose of reference or analysis. The lack of organization and processing renders it incapable of independently yielding a significant conclusion. An instance of data can be exemplified by a collection of random numbers.

The term "**Information**" refers to data that has undergone processing, organization, structuring, or presentation within a specific context in order to enhance its utility. Information refers to data that has undergone interpretation, resulting in the acquisition of meaning for the individual utilizing it. For instance, if the aforementioned figures represent the monetary values assigned to a compilation of merchandise, that constitutes factual data.

The concept of **knowledge** refers to the information and understanding that individuals possess. Strategic utilization of information refers to the capacity to effectively employ information in order to accomplish specific objectives. The process entails the analysis and comprehension of data by utilizing pre-existing knowledge structures, commonly referred to as "schema." Knowledge can be defined as information that has undergone cognitive processing within the human mind. An instance of knowledge can be observed when an individual possesses the information that a particular product is priced lower at one retail establishment compared to another and subsequently exercises their judgment to procure the product from the more affordable store.

Wisdom can be defined as the cognitive capacity to assess and determine appropriate judgments and decisions accurately. Epistemology refers to the understanding or comprehension of objective truth or moral correctness. Wisdom is attained when knowledge is consistently and judiciously applied in practical contexts, with a discerning awareness of its significance. For instance, the exercise of discerning that the most economical choice may not invariably be the most advantageous one necessitates the contemplation of variables such as caliber, sustainability, and equity. Consequently, basing purchasing determinations on these criteria can be regarded as a manifestation of sagacity.

Within the realm of artificial intelligence, our primary focus lies in the progression from the acquisition and processing of raw data to the attainment of meaningful knowledge. AI systems are specifically engineered to effectively analyze structured data and subsequently produce valuable knowledge in the form of insights, predictions, and decisions. The primary objective of numerous researchers in the field of AI is to develop systems that possess the capability to not only generate knowledge but also effectively apply it, akin to human cognition.

4.3 Knowledge Representation in AI

Consider the scenario of entering a library, wherein each book encompasses a vast reservoir of knowledge pertaining to a particular subject matter. The library's cataloging system facilitates efficient book retrieval by organizing books according to various criteria such as subjects and authors. This enables users to promptly locate books pertaining to specific topics, such as Renaissance art or quantum physics. The cataloging system in question can be classified as a method of knowledge representation, as it effectively arranges extensive quantities of information in a structured format, thereby enhancing its accessibility and usability. In the field of artificial intelligence, knowledge representation refers to the process of organizing and classifying information in a manner that facilitates reasoning, learning, and decision-making for AI systems.

4.3.1 The Essence of Knowledge Representation

Knowledge representation is the heart and soul of artificial intelligence. Much like the human brain organizes and categorizes information to make sense of the world, AI systems require a structured format to process, understand, and utilize the vast amounts of data they encounter. But what does it truly mean to represent knowledge, especially in the context of machines?

At its core, knowledge representation is about creating a bridge between the raw, unstructured data and the meaningful insights that can be derived from it. It's about translating the complexities of the real world into a format that a machine can understand. This involves not just storing data, but organizing it, categorizing it, and establishing relationships between different pieces of information.

Consider the example of a simple map. A map is a representation of geographical knowledge. It takes the vast complexities of landscapes, roads, and landmarks and distills them into a format that we can easily understand and use. Similarly, in AI, knowledge representation is about creating such "maps" for machines, but these maps are not just geographical; they can be of any domain, from medical diagnoses to financial forecasting.

There are several key aspects to consider.

Structure: Just as our brain uses neurons and synaptic connections to structure information, AI systems use data structures, graphs, and matrices. The way knowledge is structured can significantly influence how effectively an AI system can retrieve and process it.

Semantics: Beyond just storing data, knowledge representation is about understanding the meaning behind the data. This involves defining what different pieces of information represent and how they relate to each other.

Inferencing: One of the primary goals of representing knowledge is to enable machines to draw conclusions or make predictions based on the information they have. This requires a system that can not only store information but also reason with it.

Learnability: As AI systems encounter new data, they should be able to integrate this into their existing knowledge base, learning and adapting over time.

In essence, knowledge representation is the foundation upon which all AI reasoning and decision-making processes are built. It's the difference between a machine merely storing data and a machine that can think, learn, and make informed decisions. It's the process of turning raw data into actionable intelligence, bridging the gap between simple information and meaningful knowledge.

4.3.2 Why Is Knowledge Representation Crucial?

In the vast realm of artificial intelligence, knowledge representation stands as a pivotal cornerstone. Its significance can be likened to the importance of a solid foundation when constructing a building. Without a sturdy foundation, no matter how impressive the architecture or design, the structure is vulnerable to collapse. Similarly, without effective knowledge representation, even the most advanced AI algorithms would struggle to function efficiently. But why is it so indispensable?

1. Facilitating Reasoning and Decision-Making: At its core, AI is about making informed decisions based on available data. Knowledge representation provides the structured format that allows AI systems to reason about the world, draw conclusions, and make predictions. It's the difference between a machine that can merely store data and one that can think and decide.

2. Bridging the Gap Between Data and Meaning: Raw data, in its unprocessed form, lacks context and meaning. Knowledge representation provides the necessary context, turning abstract data into meaningful information. It's about translating the complexities of the real world into a format that a machine can understand, thereby enabling it to derive actionable insights.

3. Enhancing Learnability and Adaptability: The world is dynamic, with new information emerging continuously. An effective knowledge representation system allows AI models to integrate new data seamlessly, adapting and evolving their understanding over time. This ensures that the AI system remains relevant and up-to-date, capable of handling novel situations.

4. Improving Efficiency and Performance: A well-structured knowledge representation scheme can significantly boost the computational efficiency of AI systems. By organizing information in a logical and accessible manner, AI algorithms can retrieve and process data faster, leading to quicker responses and better performance.

5. Enabling Human-AI Collaboration: For AI systems to collaborate effectively with humans, they need to "think" and "understand" in ways that are relatable to us. Knowledge representation provides a structured way for AI to process information that mirrors human cognitive processes, facilitating more natural and intuitive human-AI interactions.

6. Providing a Foundation for Advanced AI Techniques: Many advanced AI techniques, such as deep learning and reinforcement learning, rely on a structured knowledge base to function effectively. Knowledge representation provides the groundwork, ensuring that these techniques have the necessary data structures and relationships to operate on.

4.4 Methods of Knowledge Representation

The field of artificial intelligence encompasses a diverse range of techniques and methodologies for encoding knowledge in a manner that enables machines to comprehend and effectively utilize it. The selection of a specific methodology frequently relies on the characteristics of the problem at hand, the nature of the data involved, and the particular demands of the artificial intelligence application. Let us explore some of the prevailing methodologies utilized in the discipline as shown in Fig. 4.1.

4.4.1 Semantic Networks

Semantic networks, a pioneering method in the realm of knowledge representation, serve as a graphical depiction of knowledge, capturing the intricate web of relationships and concepts that constitute a domain. Rooted in cognitive psychology and theories of human memory, semantic networks have evolved to become a foundational tool in artificial intelligence, especially in areas like natural language processing, expert systems, and ontology development. Let's delve deeper into their structure, functionality, and significance.

A semantic network is essentially a graph, composed of nodes and edges.

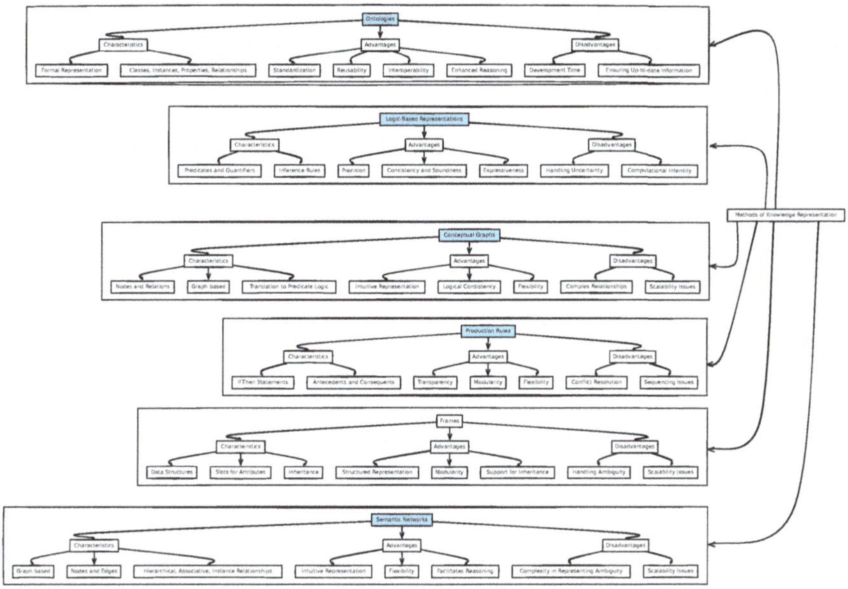

Fig. 4.1 Methodologies for knowledge representation

Nodes: These represent entities, objects, or concepts. For instance, in a semantic network about animals, nodes might represent specific animals like "Lion," "Whale," or broader categories like "Mammals."

Edges (or Links): These represent the relationships or associations between nodes. Each edge is typically labeled, indicating the nature of the relationship. For instance, an edge might indicate a hierarchical relationship like "is-a" (a lion is a mammal) or a property-based relationship like "has-color" (a lion has color yellow).

Semantic networks can represent a variety of relationships, including the following:

Hierarchical Relationships: These depict the hierarchy or taxonomy of concepts. For instance, "Canary" might be linked to "Bird" with an "is-a" relationship.

Associative Relationships: These capture associations or properties. For example, "Lion" might be linked to "Savannah" with a "lives-in" relationship.

Instance Relationships: These specify instances of a particular category. For instance, "Mango" might be linked to "Fruit" indicating that mango is an instance of fruits.

The applications and use cases of semantic networks are diverse and wide-ranging. Semantic networks have proven to be valuable tools in a myriad of domains.

Natural Language Processing (NLP): They assist in word sense disambiguation, understanding context, and semantic parsing.

Expert Systems: Semantic networks help in encoding domain-specific knowledge, enabling the system to make informed decisions.

Ontology Development: In the realm of the Semantic Web, semantic networks serve as the foundation for developing ontologies, which provide structured representations of web data.

There are several advantages associated with the use of semantic networks.

Intuitive Representation: Their graphical nature makes semantic networks easily interpretable, providing a clear visualization of relationships and concepts.

Flexibility: They can represent diverse types of knowledge, from taxonomies to properties and instances.

Facilitates Reasoning: The structured representation allows for efficient querying and reasoning, enabling AI systems to derive new insights from existing knowledge.

While semantic networks offer numerous advantages, they also come with challenges. Representing ambiguous relationships or handling intricate nuances can be complex. Moreover, as the size and complexity of the network grow, ensuring efficient querying and reasoning can become challenging.

4.4.2 Frames

Frames, a prominent method in knowledge representation, emerged from the need to represent stereotypical situations, objects, or events in a structured manner. Rooted in the work of Marvin Minsky in the 1970s, frames provided a way to encapsulate both the attributes of an entity and the relationships between them. Over time, frames have become a foundational tool in artificial intelligence, especially in areas like expert systems, cognitive robotics, and natural language understanding. Let's delve deeper into their structure, functionality, and significance.

A frame can be visualized as a data structure, akin to a record in databases or an object in object-oriented programming. It comprises various slots, each representing an attribute or a feature of the entity. These slots can contain the following:

Values: Specific data or information related to the attribute.

Facets: Meta-information about the slot, such as the type of data it can hold, default values, or constraints.

For instance, consider a frame representing a "Car." It might have slots like "Color," "Model," and "Manufacturer." The "Color" slot might have a value "Red" and a facet indicating the type of data as "String."

Inheritance: One of the powerful features of frames is the ability to support inheritance. This means that if a particular piece of information or attribute is not specified in a frame, the system can look up in a parent or a more general frame. For instance, if we have a general frame for "Vehicle" that specifies a slot "has_wheels" with a value "True," a specific frame for "Car" can inherit this attribute without explicitly stating it.

Frames have found diverse applications in the realm of AI.

Expert Systems: Frames help in organizing domain-specific knowledge in a structured manner, enabling the system to make decisions or offer recommendations.

Cognitive Robotics: In robots that need to understand and navigate human environments, frames can represent common situations or objects, aiding in perception and action.

Natural Language Understanding: Frames assist in parsing and understanding sentences by providing a structured representation of common scenarios or events.

There are several advantages associated with the use of frames.

Structured Representation: Frames offer a clear and organized way to represent entities, capturing both attributes and their relationships.

Modularity: Each frame can be seen as a modular unit, making it easier to update or expand the knowledge base.

Support for Inheritance: This allows for efficient representation, reducing redundancy and ensuring consistency.

While frames offer a structured way to represent knowledge, they also come with challenges. Handling ambiguity or representing non-stereotypical situations can be complex. Moreover, as the knowledge base grows, ensuring efficient querying and reasoning while maintaining consistency can become challenging.

4.4.3 Production Rules

Production rules, often referred to as "rules of production" or simply "rules," are a foundational method in knowledge representation, particularly in the domain of expert systems. These rules encapsulate domain-specific knowledge in a structured, conditional format, enabling systems to make decisions, draw inferences, or perform actions based on specific criteria. The simplicity and clarity of production rules have made them a preferred choice for many AI applications, especially in scenarios where decision-making processes need to be transparent and interpretable. Let's explore their structure, functionality, and significance in greater depth.

At its core, a production rule is a conditional statement, typically represented in an "if-then" format.

Antecedent (or Condition): This is the "if" part of the rule, specifying the criteria or conditions that need to be met.

Consequent (or Action): This is the "then" part, detailing the actions, decisions, or inferences that should be made if the conditions are satisfied. For instance, a production rule in a medical diagnostic system might state: "If a patient has a persistent cough and high fever, then recommend a test for pneumonia."

Production rules are the backbone of rule-based systems, especially expert systems. In such systems, knowledge is encoded as a collection of production rules. When presented with a problem or a set of data, the system evaluates the rules in a specific sequence, triggering actions or making decisions based on which rules' conditions are met.

Production rules have found widespread applications across various domains.

Medical Diagnosis: Systems can recommend tests or treatments based on symptoms and medical history.

Financial Systems: Rules can guide investment strategies, risk assessments, or fraud detection based on market conditions and transaction patterns.

Control Systems: In automated environments, rules can dictate machinery operations based on sensor data.

Advantages of Production Rules include the following:

Transparency: The "if-then" structure makes the decision-making process clear and easy to understand.

Modularity: Rules are independent units, making it easier to add, modify, or remove knowledge without disrupting the entire system.

Flexibility: They can represent a wide range of knowledge, from simple facts to complex decision-making criteria.

While production rules offer a clear and modular way to represent knowledge, they come with their own set of challenges. As the number of rules grows, ensuring that they don't conflict with each other becomes crucial. Moreover, determining the sequence in which rules are evaluated can impact the system's decisions and efficiency.

4.4.4 Conceptual Graphs

Conceptual graphs, a versatile method in knowledge representation, offer a visual and formal framework to depict and reason about information and knowledge. Introduced by John F. Sowa in the 1970s, conceptual graphs were inspired by the existential graphs of Charles Sanders Peirce and semantic networks in artificial intelligence. They provide a bridge between natural language and logical representations, making them particularly useful for natural language processing tasks and knowledge-based systems. Let's explore their structure, functionality, and significance in greater depth.

Conceptual graphs are essentially bipartite graphs, comprising two main types of nodes.

Concept Nodes: These represent entities, objects, or ideas. They are typically enclosed within rectangles and can have a type label (indicating the kind of concept) and an optional referent (providing specific details about the concept).

Relation Nodes: These capture the relationships or associations between concepts. They are typically enclosed within circles and are connected to concept nodes through edges.

For instance, a conceptual graph representing the statement "John loves Mary" would have two concept nodes for "John" and "Mary" and a relation node for "loves" connecting them.

Conceptual graphs can be translated into formal logic expressions, specifically predicate logic. This formal underpinning allows for rigorous reasoning and inferencing capabilities, making conceptual graphs suitable for tasks that require logical consistency and deductive reasoning.

Conceptual graphs have been employed in a myriad of AI domains.

Natural Language Processing (NLP): They assist in semantic parsing, information retrieval, and knowledge extraction from textual data.

Knowledge Bases and Expert Systems: Conceptual graphs help in encoding and organizing domain-specific knowledge in a structured and logical manner.

Database Querying: They can be used to translate natural language queries into formal database query languages.

Advantages of Conceptual Graphs include the following:

Intuitive Representation: Their graphical nature provides a clear and easily interpretable visualization of knowledge.

Logical Consistency: The ability to translate graphs into formal logic ensures logical consistency and rigorous reasoning capabilities.

Flexibility: They can represent a wide range of knowledge, from simple facts to complex scenarios.

While conceptual graphs offer a structured and logical way to represent knowledge, they also come with challenges. Representing intricate or ambiguous relationships can be complex. Moreover, as the knowledge base grows, ensuring efficient querying, reasoning, and maintaining consistency can become challenging.

4.4.5 Logic-Based Representations

Logic-based representations are foundational to the realm of artificial intelligence, providing a structured and formalized means to encode, reason about, and infer knowledge. Rooted in classical logic and mathematical formalisms, these representations offer a rigorous approach to capture the intricacies of various domains, from natural language semantics to expert system rules. By leveraging the principles of logic, AI systems can ensure consistency, validity, and soundness in their operations. Let's delve deeper into the essence, applications, and significance of logic-based representations.

At the heart of logic-based representations lies the desire to encode knowledge in a way that is both precise and unambiguous. This is achieved through.

Predicates: These are statements that can be evaluated as true or false. For instance, "IsRainy(Tuesday)" might be a predicate indicating whether it's rainy on Tuesday.

Quantifiers: Elements like "for all (∀)" (universal quantifier) and "there exists (∃)" (existential quantifier) allow for generalizations and specificities in statements.

Inference Rules: These are formal rules that dictate how new facts can be derived from existing ones. Modus Ponens, for instance, is a classic inference rule. It can be formally stated as

$$\frac{\begin{array}{c} P \rightarrow Q \quad \text{(If } P\text{, then } Q\text{)} \\ P \quad \text{(Statement } P \text{ is true)} \end{array}}{\therefore Q \quad \text{(Therefore, } Q \text{ is true)}}$$

Example:

$$\frac{\begin{array}{r} \text{If it is raining, then the ground will be wet.} \quad (P \rightarrow Q) \\ \text{It is raining.} \quad (P) \end{array}}{\therefore \text{The ground is wet.} \quad (Q)}$$

Logic-based representations find utility across a spectrum of AI applications.

Expert Systems: These systems use logical rules to make decisions or offer recommendations based on a given set of facts.

Semantic Web: Logic-based ontologies and reasoning engines power the semantic web, enabling richer and more meaningful web interactions.

Natural Language Processing: Logic is used to capture the semantics of language, enabling tasks like semantic parsing and question answering.

Advantages of Logic-Based Representations include the following:

Precision: Logic offers a clear and unambiguous way to represent knowledge, eliminating vagueness.

Consistency and Soundness: Logical systems ensure that derived facts are consistent with the given knowledge and that the reasoning processes are sound.

Expressiveness: Complex relationships, generalizations, and constraints can be effectively captured using logical formalisms.

While logic-based representations are powerful, they come with their set of challenges. Real-world scenarios often involve uncertainty, which classical logic is not inherently equipped to handle. Moreover, ensuring that a large set of logical statements is consistent and free from contradictions can be computationally intensive.

Recognizing the limitations of classical logic in handling uncertainties, researchers have developed extensions like fuzzy logic (which deals with degrees of truth) and probabilistic logic (which incorporates probabilities into logical statements).

4.4.6 Ontologies

In the realm of artificial intelligence and knowledge representation, ontologies play a pivotal role in structuring and formalizing domain-specific knowledge. Originating from philosophy, where ontology is the study of the nature of being and existence, in the context of AI, ontology refers to a formal representation of knowledge within a particular domain, capturing entities, their attributes, and the relationships between them. Let's delve deeper into the essence, structure, and significance of ontologies in AI.

Ontologies provide a structured framework to represent, categorize, and interlink concepts within a specific domain. They serve as a "blueprint" that defines the types of entities that exist, the properties they possess, and how they relate to each other. By doing so, ontologies facilitate a shared understanding of a domain, enabling effective communication, reasoning, and knowledge sharing.

Ontologies are typically composed of the following:

Classes: These represent general categories or types of entities. For instance, in a medical ontology, "Disease" and "Medication" might be classes.

Instances: These are specific occurrences or members of classes. For the class "Disease," instances might include "Diabetes" or "Hypertension."

Properties or Attributes: These define characteristics of classes or instances. For the instance "Diabetes," an attribute might be "symptoms."

Relationships: These capture how classes or instances relate to one another. For instance, a relationship might denote that "Insulin" is a "treatment for" "Diabetes."

Ontologies find utility in a myriad of AI and information science domains.

Semantic Web: Ontologies form the backbone of the semantic web, enabling richer, more meaningful web interactions and data interlinking.

Natural Language Processing: They assist in semantic parsing, information extraction, and knowledge-based question answering.

Expert Systems: Ontologies help in encoding domain-specific knowledge, facilitating reasoning and decision-making.

Data Integration: They enable the integration of heterogeneous data sources by providing a common framework of understanding.

Advantages of Ontologies include the following:

Standardization: They provide a standardized vocabulary and structure for a domain, ensuring consistency.

Reusability: Once developed, ontologies can be reused across various applications and systems.

Interoperability: They facilitate data exchange and integration across different systems and platforms.

Enhanced Reasoning: Ontologies, especially when combined with logic-based systems, enable powerful reasoning and inferencing capabilities.

While ontologies offer a structured way to represent knowledge, their development can be time-consuming and requires domain expertise. Ensuring that an ontology captures all nuances of a domain and remains up-to-date as the domain evolves can be challenging.

Each of these methods offers unique advantages and is suited to specific types of problems and domains. The choice of knowledge representation technique can significantly influence the efficiency, accuracy, and capability of an AI system. It's akin to choosing the right tool for the job, ensuring that the AI system can effectively understand, reason, and act upon the knowledge it possesses.

4.5 Conclusion

The field of knowledge representation, which is a fundamental aspect of artificial intelligence, has undergone a significant transformation. The journey from the initial stages of symbolic AI, characterized by the explicit encoding of knowledge in rules and logic, to the emergence of neural networks that capture knowledge implicitly through their weights and activations, has undergone a significant transformation. This chapter extensively explored the diverse methodologies of knowledge representation, encompassing semantic networks, frames, production rules, and ontologies. Each of these approaches provides distinct viewpoints and functionalities.

The investigation of knowledge representation within the framework of Neuro-Symbolic AI highlights the continuous merging of symbolic and neural approaches. As we approach the threshold of this integration, the potential for artificial intelligence systems that possess the ability to engage in logical reasoning akin to symbolic systems, while also demonstrating the adaptability characteristic of neural networks, is vast. These systems offer not only improved functionalities but also increased interpretability and transparency, effectively tackling longstanding issues in the field of artificial intelligence.

As we progress towards the forthcoming chapters, specifically the one focusing on Neuro-Symbolic Integration techniques, it is imperative to acknowledge the fundamental significance of knowledge representation. The methodologies and concepts explained in this chapter will establish a solid foundation, facilitating a more comprehensive exploration of the complexities involved in the integration of symbolic reasoning and neural learning.

4.6 Exercise Questions

1. Suppose you are developing an AI system for a university library. You decide to use a semantic network to manage the relationships between books, authors, subjects, and publishers. Draw a semantic network representing at least five books, their authors, subjects, and publishers. Explain how this network helps in querying the system to find books by a specific author or on a specific subject. What challenges might arise in scaling this semantic network as the library's collection grows, and how could you address them?
2. Suppose you are working on an AI project to analyze social media interactions. You decide to use conceptual graphs to represent the relationships between users, posts, and reactions. Design a simple conceptual graph for a scenario where User A posts a status update, User B likes it, and User C comments on it. Include nodes and relations. How can translating these conceptual graphs into predicate logic enhance the system's ability to perform semantic analysis?
3. Imagine you are developing an expert system for legal advice. The system relies on logic-based representations to handle legal rules and facts. Formulate three

predicates and two inference rules that could be used in this system. For example, predicates like "IsContractValid(Contract)" and rules like "If Contract is signed and Contract is dated, then IsContractValid(Contract)." What are the benefits and limitations of using logic-based representations in this context, especially when dealing with complex and ambiguous legal scenarios?
4. You are creating a knowledge management system for a healthcare organization. You decide to use ontologies to represent medical knowledge, including diseases, symptoms, treatments, and medications. Develop a basic ontology that includes at least three diseases, their symptoms, and possible treatments. Define classes, instances, properties, and relationships. How does using an ontology improve interoperability between different healthcare systems and enhance the decision-making process for medical professionals?
5. Propose a hybrid approach that combines semantic networks with production rules for a specific application (e.g., customer service chatbot). Describe how these two methods complement each other. What are the potential benefits and challenges of implementing this hybrid approach?

References

1. Brachman, R., Levesque, H.: Knowledge Representation and Reasoning. Morgan Kaufmann (2004)
2. Ash, R.B.: Information Theory. Courier Corporation (2012)
3. Frické, M.: The knowledge pyramid: the dikw hierarchy. Ko Knowl. Organization **46**(1), 33–46 (2019)

Chapter 5
Rule-Based Systems and Expert Systems

> *"I am putting myself to the fullest possible use, which is all I think that any conscious entity can ever hope to do."*
>
> HAL 9000, *2001: A Space Odyssey*

5.1 Introduction

Imagine a bustling hospital emergency room. Amid the cacophony of beeping monitors and hurried medical staff, a doctor sits at a computer terminal. A patient has just been rushed in with a rare tropical disease, and every second counts. The doctor, while experienced, hasn't encountered this particular ailment in years. Instead of flipping through thick medical textbooks or making frantic calls to specialists, the doctor inputs the symptoms into an expert system. Within moments, the system, built upon decades of medical knowledge and thousands of case studies, provides a diagnosis, suggests a treatment plan, and even predicts potential complications. The patient receives timely and accurate treatment, all thanks to a rule-based expert system that encapsulates vast medical expertise.

This scenario underscores the transformative power of rule-based systems and expert systems in the realm of artificial intelligence. These systems, rooted in logic and symbolic reasoning, have been pivotal in bridging the gap between vast domains of human knowledge and the computational prowess of machines. From medical diagnoses to financial forecasting, they have found applications across diverse sectors, mimicking human expertise and often surpassing it in accuracy and speed.

In this chapter, we embark on a journey to understand the intricacies of these systems. We'll delve deep into their origins, explore the mathematical logic that

B. P. Bhuyan et al., *Neuro-Symbolic Artificial Intelligence*, Studies in Computational Intelligence 1176, https://doi.org/10.1007/978-981-97-8171-3_5

powers them, and witness their evolution over the years. Through this exploration, we aim to appreciate the foundational role they have played in the AI landscape and envision their place in the future of Neuro-Symbolic AI.

5.2 The Genesis of Rule-Based Systems

During the early phases of the development of artificial intelligence, researchers encountered the significant obstacle of emulating human cognition. Consequently, a fundamental inquiry arose: How can we effectively transfer human knowledge to machines through encoding? Initially, the solution resided in the development of systems based on rules.

The origins of rule-based systems can be attributed to the initial efforts in the field of artificial intelligence, which aimed to develop systems capable of logical reasoning and problem-solving in a manner similar to human cognition. These systems were constructed based on the underlying assumption that human decision-making, particularly in specialized domains, frequently adheres to a collection of explicit rules or heuristics. The act of embedding these rules into a machine was thought to enable the machine to replicate human-like cognitive processes.

The core component of a rule-based system is its knowledge base, which serves as a repository of rules that govern the behavior of the system. The structure of each rule is commonly in the form of an "IF-THEN" statement. As an example, within the context of a medical diagnostic system, a rule could be formulated as follows: "IF a patient exhibits symptoms of fever AND cough, THEN it is plausible to consider the possibility of influenza." The system functions by iteratively assessing these rules in relation to the data it is provided, thereby generating decisions or drawing inferences based on identified correspondences.

The appeal of rule-based systems stems from their transparency and predictability. In contrast to black-box models, rule-based systems provide transparent visibility into their decision-making processes. Each decision within the system can be attributed to a particular rule, thereby rendering the system's operations comprehensible and defensible.

Nevertheless, the development of these systems encountered various difficulties. The task of encoding extensive knowledge, particularly in intricate domains, was a formidable undertaking. The translation of domain experts' expertise into machine-readable rules necessitated meticulousness, as it was a time-consuming process that was susceptible to oversight. Furthermore, it can be argued that these systems possessed an inherent inflexibility. The individual's conduct was limited exclusively to the regulations within their existing knowledge framework, rendering them unsuitable for tasks necessitating flexibility or the assimilation of novel information.

Notwithstanding these challenges, rule-based systems have achieved a noteworthy milestone in the progression of artificial intelligence.

5.3 Logic in Rule-Based Systems

At the core of rule-based systems lies the principle of logic. Logic, in its essence, provides a structured framework for reasoning, allowing for the formulation of arguments based on established premises to derive valid conclusions. In rule-based systems, logic serves as the foundational mechanism that drives decision-making and inference.

5.3.1 Propositional Logic

Propositional logic, also known as sentential logic or statement logic, is one of the most fundamental forms of logic used in rule-based systems. It operates on simple, declarative statements, or propositions, that are either true or false but not both. The primary objective of propositional logic is to evaluate the truth values of more complex statements, which are constructed using logical connectives from simpler ones.

The basic elements of propositional logic are as follows:

Propositions: These are atomic statements that declare a fact. For instance, "The sky is blue" or "It is raining" are propositions. Each proposition can be assigned a truth value: true (T) or false (F).

Variables: Propositions are often represented by variables, typically using letters like P, Q, R, etc. For example, P might represent the statement "The sky is blue."

Logical connectives are operators that combine one or more propositions to form compound propositions. The primary logical connectives are as follows:

NOT ($\neg$): This is a unary operator that negates the truth value of a proposition. If P is true, then $\neg P$ (not P) is false, and vice versa.

AND ($\wedge$): This binary operator yields true if both propositions it combines are true. For instance, $P \wedge Q$ is true only if both P and Q are true.

OR ($\vee$): Another binary operator, it yields true if at least one of the propositions it combines is true. $P \vee Q$ is true if either P, Q or both are true.

IMPLIES ($\Longrightarrow$): This operator represents implication. $P \Longrightarrow Q$ can be read as "If P, then Q." It's true unless P is true and Q is false.

BICONDITIONAL ($\Longleftrightarrow$): Represents equivalence. $P \Longleftrightarrow Q$ is true if both P and Q have the same truth value.

Truth tables are a pivotal tool in propositional logic, used to represent the truth value of compound propositions based on the truth values of their constituent propositions. For instance, for the AND operator:

P	Q	$P \land Q$
T	T	T
T	F	F
F	T	F
F	F	F

In rule-based systems, propositional logic serves as the foundational layer, enabling the system to make decisions based on predefined rules. For instance, a rule might state: "IF it is sunny AND it is a weekend, THEN go for a picnic." Here, the conditions "it is sunny" and "it is a weekend" are propositions, and the rule can be represented in propositional logic as $P \land Q \implies R$, where R is "go for a picnic."

In essence, propositional logic provides a structured way to represent and reason with declarative knowledge, making it an indispensable tool in the realm of rule-based systems.

Logical equivalences are foundational in propositional logic. They provide a set of rules that allow us to transform logical expressions into equivalent forms. These transformations are useful for simplifying expressions, proving theorems, and understanding the inherent properties of logical connectives. Let's delve into some of the primary logical equivalences:

1. **Identity Laws**:
   - Law of Identity: $P \equiv P$
   - Law of Negation: $P \lor \neg P \equiv \text{True}$ and $P \land \neg P \equiv \text{False}$
2. **Domination Laws**:
   - Law of Zero (AND Domination): $P \land \text{False} \equiv \text{False}$
   - Law of One (OR Domination): $P \lor \text{True} \equiv \text{True}$
3. **Idempotent Laws**:
   - Idempotent Law for AND: $P \land P \equiv P$
   - Idempotent Law for OR: $P \lor P \equiv P$
4. **Double Negation Law**: $\neg(\neg P) \equiv P$
5. **Commutative Laws**:
   - Commutative Law for AND: $P \land Q \equiv Q \land P$
   - Commutative Law for OR: $P \lor Q \equiv Q \lor P$
6. **Associative Laws**:
   - Associative Law for AND: $P \land (Q \land R) \equiv (P \land Q) \land R$
   - Associative Law for OR: $P \lor (Q \lor R) \equiv (P \lor Q) \lor R$
7. **Distributive Laws**:
   - Distributive Law for AND over OR: $P \land (Q \lor R) \equiv (P \land Q) \lor (P \land R)$
   - Distributive Law for OR over AND: $P \lor (Q \land R) \equiv (P \lor Q) \land (P \lor R)$

8. **De Morgan's Laws**:
   - De Morgan's Law for AND: $\neg(P \wedge Q) \equiv \neg P \vee \neg Q$
   - De Morgan's Law for OR: $\neg(P \vee Q) \equiv \neg P \wedge \neg Q$
9. **Absorption Laws**:
   - Absorption Law for AND: $P \vee (P \wedge Q) \equiv P$
   - Absorption Law for OR: $P \wedge (P \vee Q) \equiv P$
10. **Implication**: $P \rightarrow Q \equiv \neg P \vee Q$
11. **Contrapositive**: $P \rightarrow Q \equiv \neg Q \rightarrow \neg P$
12. **Biconditional**: $P \leftrightarrow Q \equiv (P \rightarrow Q) \wedge (Q \rightarrow P)$.

When we dive deeper into the structure of these propositions, we encounter concepts like clauses and literals, which play a pivotal role in the representation and manipulation of knowledge in rule-based systems.

A **literal** is a basic proposition or its negation. In other words, if P is a proposition, then P (in its positive form) and $\neg P$ (in its negated form) are both literals. For instance, in the context of a weather-based rule system, if P represents "It is sunny," then $\neg P$ would represent "It is not sunny."

A **clause** is a disjunction of literals. This means it's a statement formed by joining two or more literals using the OR ($\vee$) operator. For instance, $P \vee \neg Q$ is a clause that can be read as "It is sunny or it is not raining."

In the realm of propositional logic, a particularly important type of clause is the **Horn clause**, which contains at most one positive literal. Horn clauses are significant because they can be efficiently processed, making them valuable in rule-based systems and logic programming.

Using propositional logic, let: P be "It is sunny." Q be "The air quality is good." R be "It's a good day for a picnic." The knowledge can be represented as: $P \wedge Q \implies R$.

This is equivalent to $\neg P \vee \neg Q \vee R$ (using the implication elimination rule).

This is a Horn clause because it has at most one positive literal, which is R. P is a fact, a type of Horn clause with only one positive literal and no negative literals. Q is also a fact. Given the facts P and Q, the system can infer R using the Horn clause $\neg P \vee \neg Q \vee R$, concluding that "It's a good day for a picnic."

In rule-based systems, especially in logic programming languages like Prolog, Horn clauses are fundamental. They allow for efficient inference mechanisms and are the primary forms of representation in such systems.

A proposition is said to be in **Conjunctive Normal Form (CNF)** if it is a conjunction (AND operation) of one or more clauses. For instance, the expression $(P \vee Q) \wedge (\neg P \vee R) \wedge R$ is in CNF. CNF is especially important in computational logic because many algorithms, like those used in SAT solvers, operate on formulas expressed in this form.

In rule-based systems, clauses and literals are used to represent and evaluate rules. For example, a rule might be: "IF it is sunny OR it is not raining, THEN go for a walk." This rule can be represented as the clause $P \vee \neg Q$.

Furthermore, systems often convert complex rules into CNF to simplify processing and reasoning. By breaking down rules into combinations of literals and clauses, the system can more efficiently evaluate which actions to take based on the given conditions.

Propositional logic operates on simple, declarative propositions without internal structure. It is concerned with statements to which the values of true or false can be assigned. While it provides a foundational understanding of logical reasoning, it has several limitations:

Lack of Expressiveness: One of the most significant limitations of propositional logic is its lack of expressiveness. It cannot express statements about individual objects or specific properties of objects. For instance, while we can represent "John is a man" and "John is tall" as two separate propositions, we cannot express more complex relationships like "All men are tall."

No Quantification: Propositional logic does not have the capability to handle quantifiers like "all" or "some." This means it cannot represent general statements about classes of objects or their properties. For example, "All birds can fly" or "Some students are intelligent" cannot be directly represented.

Inefficiency in Representation: For complex scenarios, propositional logic can become cumbersome. If we want to represent information about ten different students in a class, we would need ten different propositions, even if they share common attributes.

No Modularity: In propositional logic, if a piece of knowledge changes, it often requires a change in many parts of the knowledge base. This lack of modularity makes it less flexible for dynamic domains.

No Internal Structure: Propositional logic treats each statement as an atomic entity without any internal structure. This means it cannot differentiate or analyze the internal components of a statement.

Inability to Handle Incomplete Information: Propositional logic works on definite true or false values. It doesn't have the capability to handle uncertainty or incomplete information, which is often required in real-world scenarios.

Lack of Semantic Richness: While propositional logic can tell if a compound statement is true or false based on the truth values of its components, it doesn't capture the meaning or semantics of statements. This makes it less suitable for tasks that require understanding the context or meaning of statements.

In light of these limitations, while propositional logic serves as a good starting point for understanding logical reasoning, more expressive logic like first-order logic (predicate logic) is often preferred for complex knowledge representation tasks, especially in artificial intelligence. Predicate logic extends propositional logic by introducing terms, predicates, and quantifiers, thereby addressing many of the limitations of propositional logic.

5.3.2 Predicate Logic

Predicate Logic or first-order logic (FOL) is a formal logical system that extends propositional logic by introducing terms, predicates, and quantifiers. It allows for more expressive statements about objects and their relationships. **Terms** are the fundamental entities or objects in first-order logic. They can represent specific things or general concepts. There are three primary types of terms:

Constants: These represent specific, unchanging entities. For example, "John," "Paris," or "3.14" are constants. They denote a particular object in the domain of discourse.

Variables: Variables can stand in for any object. They are often represented by lowercase letters like x, y, z. In the context of a statement, a variable might represent any object in the domain.

Functions: Functions take a set of objects as input and produce another object. For instance, "father of John" or "mother of Alice" can be considered functions. In a mathematical context, a function like $f(x)$ is also a term in first-order logic.

Predicates are akin to verbs in natural language. They describe properties of objects or relationships between objects.

Properties: A predicate can describe a property of an object. For instance, "IsTall(John)" might be true if John is tall and false otherwise.

Relationships: Predicates can also describe relationships between multiple objects. For example, "Loves(John, Mary)" might be true if John loves Mary.

Quantifiers specify how many objects in the domain of discourse satisfy a particular property or relationship.

Universal Quantifier (∀): This quantifier asserts that a statement holds for all objects in the domain. For instance, $\forall x, IsHuman(x) \implies IsMortal(x)$ means that for every object x, if x is human, then x is mortal.

Existential Quantifier (∃): This quantifier asserts that there is at least one object in the domain for which a statement holds. For example, $\exists x, IsUnicorn(x)$ means there is some object x that is a unicorn.

A well-formed formula (often abbreviated as **WFF**) in first-order logic is a string of symbols from the language of FOL that adheres to the syntactic rules of the logic. WFFs are the formulas that have a clear and unambiguous meaning in the logic. WFFs are constructed using terms, predicates, logical connectives, and quantifiers. Here's how they are built:

1. **Atomic Formulas**:
   - If P is a predicate of arity n (meaning it takes n arguments) and $t_1, t_2, \ldots, t_n$ are terms, then $P(t_1, t_2, \ldots, t_n)$ is a WFF.
2. **Logical Connectives**:
   - If ϕ and ψ are WFFs, then so are:
     - $\neg\phi$ (negation)
     - $\phi \wedge \psi$ (conjunction)

   - $\phi \vee \psi$ (disjunction)
   - $\phi \to \psi$ (implication)
   - $\phi \leftrightarrow \psi$ (biconditional)

3. **Quantifiers**:
   - If ϕ is a WFF and x is a variable, then:
     - $\forall x\, \phi$ is a WFF (universal quantification)
     - $\exists x\, \phi$ is a WFF (existential quantification)
4. **Parentheses**: Parentheses can be used to group formulas and clarify the order of operations. For instance, $\phi \wedge (\psi \vee \chi)$ is a WFF where the disjunction is evaluated before the conjunction.

Bound and Free Variables: In the context of WFFs, it's essential to distinguish between bound and free variables:

- A variable is **bound** if it's within the scope of a quantifier. For instance, in the WFF $\forall x\, P(x)$, the variable x is bound by the universal quantifier.
- A variable is **free** if it's not bound by any quantifier. In the WFF $P(x) \wedge \forall y\, Q(y)$, the variable x is free, while y is bound.

Importance: Ensuring that a formula is well-formed is crucial for several reasons:

- **Semantics**: Only WFFs have a clear and unambiguous meaning in the logic.
- **Inference**: Logical systems use WFFs to derive new truths.
- **Consistency**: By adhering to the rules for constructing WFFs, we ensure that our logical system remains consistent.

Skolemization is a process used to eliminate existential quantifiers by introducing new function symbols, known as Skolem functions. The idea is to replace existentially quantified variables with functions of the universally quantified variables that are in scope.

Example: Consider the FOL sentence: $\forall x \exists y P(x, y)$. After Skolemization, this becomes: $\forall x P(x, f(x))$. Here, f is a Skolem function that represents the object that y would have referred to, given a particular x.

Skolemization is crucial because it allows us to work with a simpler form of logic without existential quantifiers, making the resolution process more straightforward. It doesn't change the satisfiability of the formula; if the original formula was satisfiable, the Skolemized version will be too, and vice versa.

Resolution is a powerful inference rule that is used both in propositional logic and first-order logic. In FOL, resolution is a bit more complex due to the presence of variables, quantifiers, and predicates. The primary goal of the resolution is to derive a contradiction (an empty clause) from a set of clauses, thereby proving the unsatisfiability of the set.

Steps for Resolution in FOL:

1. Convert to Clausal Form: The first step is to convert all sentences in the knowledge base to conjunctive normal form (CNF). This involves expressing each sentence as a conjunction of disjunctions.
2. Skolemization: Remove existential quantifiers by introducing Skolem functions. This step ensures that the logic remains equivalent while eliminating the existential quantifiers.
3. Drop Universal Quantifiers: In the clausal form, universal quantifiers are implicit, so they can be dropped.
4. Resolve: The resolution rule is applied to pairs of clauses. If two clauses contain complementary literals (a predicate and its negation), they can be resolved together to produce a new clause. The goal is to derive an empty clause, indicating a contradiction.

The **Herbrand Universe** for a given first-order formula is the set of all ground terms (terms without variables) that can be constructed using only the constant and function symbols from the formula. If the formula has no constants, an arbitrary constant is added. If the formula also has no function symbols, then the Herbrand Universe consists of just that one constant.

Example: For the formula $P(a) \wedge \exists x Q(f(x), g(a))$, the Herbrand Universe consists of the terms: $a, f(a), g(a), f(f(a)), f(g(a)), g(f(a)), g(g(a)), \ldots$ and so on.

The **Herbrand Base** for a given first-order formula is the set of all possible ground atomic formulas (atomic formulas without variables) that can be constructed using the predicate symbols of the formula and the terms from the Herbrand Universe.

Example: Continuing with the previous formula $P(a) \wedge \exists x Q(f(x), g(a))$, the Herbrand Base would include atomic formulas like: $P(a), P(f(a)), P(g(a))$, $Q(a, a), Q(f(a), a), Q(g(a), a), Q(a, f(a)), \ldots$ and so on.

The importance of the Herbrand Universe and Base lies in the Herbrand Theorem, which states the following:

A first-order formula is satisfiable if and only if it has a model in its Herbrand Universe.

In simpler terms, if there's a way to make the formula true by picking specific ground terms from the Herbrand Universe to replace its variables, then the formula is satisfiable.

This theorem forms the basis for many automated theorem-proving techniques in FOL. Instead of searching for a model in an abstract, potentially infinite domain, theorem provers can focus on the concrete (but still potentially infinite) Herbrand Universe. In practice, this is useful for formulas with a finite Herbrand Universe. For others, it provides a systematic way to approach the problem, even if it's not always feasible to check every possible interpretation.

First-Order Logic (FOL) is a powerful formalism that has been widely used in various domains, especially in artificial intelligence, for knowledge representation and reasoning. However, like all formal systems, FOL has its limitations. Here are some of the primary limitations of First-Order Logic.

Expressiveness Limitations: FOL cannot naturally represent higher-order information, i.e., properties of properties or predicates of predicates. For instance, while you can state that "John believes that the sky is blue" in FOL, expressing more complex beliefs about beliefs becomes cumbersome. Gödel's incompleteness theorems show that there are true statements about the natural numbers that cannot be proven within first-order Peano arithmetic.

Undecidability: The general problem of determining the truth of an arbitrary FOL statement is undecidable. This means there's no algorithm that can determine the truth or falsehood of every possible FOL statement.

Infinite Models: The compactness theorem of FOL states that if a set of FOL sentences has a model, then it has a finite submodel. However, determining this finite model isn't always straightforward. Moreover, some FOL statements have only infinite models, making them challenging to handle in practice.

Complexity: Even though the general problem is undecidable, there are many decidable fragments of FOL. However, many of these problems are NP-hard or worse, making them computationally intensive to solve.

Lack of Probabilistic Reasoning: Traditional FOL doesn't handle uncertainty or probabilistic reasoning well. While there are probabilistic extensions to logic, basic FOL is deterministic in nature.

Commonsense Reasoning: FOL, being a formal system, often struggles with capturing the nuances of commonsense reasoning that humans find intuitive. Representing and reasoning about vague or ambiguous concepts can be challenging.

Over-Specification: FOL requires precise specification of all details, which can be cumbersome and might not always align with the way humans think or communicate.

Non-Monotonicity: Real-world reasoning often requires the ability to retract conclusions as new information becomes available. Traditional FOL is monotonic, meaning once a conclusion is drawn, it cannot be undone. Non-monotonic logics have been developed to address this, but they are not part of standard FOL.

Despite these limitations, First-Order Logic remains a foundational tool in formal reasoning, mathematics, and computer science. Its strength lies in its rigorous formalism and wide applicability, but for certain domains or problems, other tools or extensions to FOL might be more appropriate.

5.3.3 Second-Order Logic

Second-Order Logic (SOL) extends FOL by allowing quantification over sets, relations, and functions, not just individual entities. This added expressiveness allows SOL to capture properties and structures that FOL cannot.

In SOL, the basic building blocks are as follows:

- **Individual variables**: $x, y, z, \ldots$
- **Predicate variables**: $P, Q, R, \ldots$

Terms:

1. Individual variable: x
2. Function symbol: $f(x)$
3. Variable function symbol: $F(x, y)$

Formulas:

1. Atomic: $t_1 = t_2, \quad P(t)$
2. Compound: $\neg\phi, \quad \phi \wedge \psi, \quad \phi \vee \psi$
3. Quantification: $\forall x\phi, \quad \exists x\phi, \quad \forall P\phi, \quad \exists P\phi$

In the realm of logic, a model provides a concrete realization of a formal system. It offers a specific interpretation of the symbols and formulas of that system. For Second-Order Logic, models play a crucial role in understanding the semantics of the language, especially given its enhanced expressive power over First-Order Logic. A model $\mathcal{M}$ for SOL consists of the following:

- A domain D.
- An interpretation function I.
- An assignment g.

Domain of Discourse (D): The domain, often denoted as D, is a non-empty set of objects that the variables in our logic can refer to. In SOL, this domain is similar to that in First-Order Logic, representing the basic entities of interest. For instance, if we're discussing numbers, the domain might be the set of natural numbers. If we're discussing people, the domain might consist of individual persons.

Interpretation Function (I): I maps the non-logical symbols of our language to specific entities within our model. This function ensures that each symbol has a clear, unambiguous meaning. In SOL, the interpretation function has the following duties:

Individual Constants: For every individual constant in the language, I assigns it a specific object from the domain D. For instance, if "Socrates" is an individual constant and our domain consists of people, I might map "Socrates" to the actual person Socrates.

Predicate Constants: For each n-ary predicate constant, I assigns it an n-ary relation on D. For example, if "loves" is a binary predicate constant in our language and our domain is people, I might map "loves" to the relation of romantic affection between pairs of people.

Function Constants: For each n-ary function constant, I assigns it an n-ary function on D. For instance, if "+" is a binary function constant and our domain is numbers, I might map "+" to the addition operation on numbers.

Variable Assignment (g): While the interpretation function deals with the constants of our language, the variable assignment function, g, deals with its variables. This function assigns values from the domain D to the individual variables and relations over D to the predicate variables. In SOL, this assignment extends to second-order variables:

Individual Variables: Each individual variable is assigned an object from the domain D.

Predicate Variables: Each n-ary predicate variable is assigned an n-ary relation on D. This is where SOL distinguishes itself from First-Order Logic, as it allows for quantification over relations, not just individual entities.

In essence, a model in SOL provides a structured way to interpret the symbols and formulas of the language. It offers a bridge between the abstract, formal world of logic and the concrete entities and relations of a specific domain of interest.

Truth in SOL is determined recursively:

1. Atomic formulas: based on I and g.
2. Logical connectives: standard truth-functional semantics.
3. $\forall x \phi$: true if ϕ is true for all objects in D.
4. $\forall P \phi$: true if ϕ is true for all relations over D.

Second-order quantification is a distinguishing feature of Second-Order Logic (SOL) that sets it apart from its first-order counterpart. At its core, while First-Order Logic (FOL) allows for quantification over individual objects within a domain, SOL takes a step further. It permits the quantification of sets of these objects and even the relations between them. This nuanced capability allows SOL to articulate more complex and higher-order properties that FOL might struggle with.

In the realm of FOL, quantifiers typically range over individual elements of a domain. For instance, when we encounter a statement like $\forall x P(x)$, the universal quantifier $\forall$ is understood to range over individual elements. It asserts that the property P is true for every individual x within the domain.

However, SOL broadens this scope. It introduces the idea of talking about sets of individuals and making quantified statements about these sets. A statement in SOL, such as $\forall X \exists y Y(y)$, can be interpreted as saying, "For every set of individuals X, there exists an individual y such that y possesses the property Y." This kind of expressiveness allows us to convey intricate properties of sets, such as the notion that every set has at least one member or even the idea that a set exists that encompasses all individuals.

Beyond just sets, SOL's prowess extends to quantification over relations among these individuals. Consider a binary relation, symbolized as R. In SOL, a statement like $\forall R \exists x, y R(x, y)$ can be deciphered as, "For every conceivable binary relation R, there are individuals x and y such that the relation R is true between them." This capability to quantify relations offers a powerful avenue to express statements about all possible relations of a certain kind.

Yet, the true might of SOL is unveiled when we touch upon higher-order quantification. This allows us to quantify over properties of sets or relations themselves. It's a profound ability, enabling us to express intricate statements like the existence of a relation that's both transitive and symmetric but not reflexive.

However, with great power comes great responsibility. The enhanced expressiveness of SOL, especially with its second-order quantifiers, introduces increased computational complexity. Determining the veracity of SOL statements can be a more intricate task than their FOL equivalents. But, in return, SOL offers a richer tapestry of expression, making it invaluable for advanced applications across mathematics, philosophy, and computer science.

5.3.4 Monadic SOL

In the vast landscape of logical systems, Monadic Second-Order Logic carves out a unique niche. While it's a subset of Second-Order Logic, it doesn't compromise much on the expressive might. The term "monadic" in its name is indicative of its restriction: it only deals with unary predicates or monadic predicates. This means that while MSO can talk about sets of individuals, it doesn't venture into the realm of relations between these individuals.

To understand this better, consider a domain of discourse comprising individuals. In First-Order Logic, we can make statements about these individuals. For instance, we might say "Every individual has property P," or in formal terms, $\forall x P(x)$. In full-fledged Second-Order Logic, we can make statements about sets of these individuals and the relations between them. For example, "There exists a set S such that every individual in S has property P," represented as $\exists S \forall x (x \in S \implies P(x))$.

Monadic Second-Order Logic operates in the middle ground. It allows us to talk about sets of individuals but doesn't extend to relations between them. So, while we can express the aforementioned statement about the existence of set S, we wouldn't be able to express a statement that involves a binary relation between individuals.

Despite this seeming limitation, MSO is incredibly powerful. One of its most celebrated properties is its ability to define many properties of finite structures, especially in the realm of graph theory. For instance, properties like "being connected" or "having a Hamiltonian cycle" can be succinctly expressed in MSO. This expressiveness has profound implications, especially in the world of formal language theory, where MSO can be used to define certain classes of formal languages.

5.4 Decidability and Completeness of Logic

The study of logic, especially in the context of artificial intelligence and formal systems, often leads to questions about the properties of the logical systems themselves. Two of the most fundamental and intriguing properties are decidability and completeness. These properties address the capabilities and limitations of logical systems, providing insights into what can be achieved with them and where they might fall short.

5.4.1 Decidability

Decidability is a foundational concept in logic, computer science, and the philosophy of mathematics. It delves into the question of determinability, asking whether there exists a definitive procedure or algorithm that can ascertain the truth or falsity of a statement within a given system. The implications of decidability are profound, as they touch upon the very nature of computation, reasoning, and the limits of knowledge.

The roots of decidability trace back to the early twentieth century, during a period of intense investigation into the foundations of mathematics. Mathematicians and logicians sought to determine if there was a mechanical method to solve all mathematical problems.

A problem or a set is said to be **decidable** if there exists an algorithm that can determine, in a finite amount of time, whether any given input belongs to that set (or satisfies the problem). If no such algorithm exists, the problem or set is termed **undecidable**.

One of the most famous undecidable problems is the Halting Problem, proposed by Alan Turing in 1936 [1]. The problem asks whether, given a description of an arbitrary computer program and an input, there exists an algorithm that determines whether the program will eventually halt (stop running) or continue to run indefinitely. Turing proved that a general algorithm to solve the Halting Problem for all possible program-input pairs cannot exist. This was a groundbreaking result, showing inherent limitations in algorithmic determinability.

In the realm of formal logic, decidability has significant implications. For instance, propositional logic is decidable. This means that there exists an algorithm (like the truth table method) that can determine the truth value of any propositional formula.

However, when we move to more expressive systems like first-order logic, the waters become murkier. While certain subsets of first-order logic are decidable, the general system is not. This means that there isn't an algorithm that can determine the truth value of every possible first-order statement.

Gödel's incompleteness theorems [2] further underscored the limitations of formal systems. He showed that in any consistent, recursive axiomatic system powerful enough to describe the arithmetic of the natural numbers, there are true propositions that cannot be proved within the system. This was a monumental result, indicating that there are inherent limits to what can be known or proven within formal logical systems.

While SOL is generally undecidable, the decidability of MSO varies depending on the structures over which it is interpreted. One of the most notable results concerning MSO's decidability is its relationship with finite structures, particularly finite trees and words [3]. Its decidability is a topic of significant interest in mathematical logic and theoretical computer science due to its expressive power and its applicability in various domains.

The realm of finite words offers a rich tapestry of results connecting MSO with automata theory. Any property of finite words that can be expressed in MSO can be recognized by a finite automaton. This equivalence is encapsulated in the Büchi-Elgot-Trakhtenbrot theorem [4], which states that a language of finite words is MSO-definable if and only if it is regular. This theorem provides a bridge between the logical world of MSO and the computational world of automata, enabling techniques from one domain to be applied to problems in the other.

Trees, as hierarchical structures, introduce additional complexity. However, for finite trees, MSO retains its decidability. Specifically, for ordered binary trees, MSO corresponds to the class of regular tree languages. This decidability is captured in Rabin's Tree Theorem [5], which asserts that a set of finite trees is MSO-definable if and only if it is recognizable by a tree automaton.

The landscape changes dramatically when we move to infinite structures. MSO over the infinite binary tree, for instance, is undecidable. This is a stark contrast to its decidability over finite structures and showcases the limitations of MSO's expressive power.

The decidability of MSO, especially over finite structures, has practical implications. In formal verification, for instance, systems can be modeled as transition systems, and properties to be verified can be expressed in MSO. The decidability of MSO provides a theoretical guarantee that algorithms exist to check these properties, leading to the development of model-checking tools and techniques.

5.4.2 Completeness

Completeness, in the context of formal logic, is a fundamental concept that speaks to the expressive power and reliability of a logical system. It ensures that if a statement is semantically true, then there exists a proof of its truth within the system. In other words, a logical system is complete if every truth in its model can be derived using its axioms and inference rules.

The notion of completeness finds its roots in the early twentieth century, most notably in the work of David Hilbert, a German mathematician. Hilbert posed the question of whether the axioms of arithmetic were complete [6]. This question was later famously addressed by Kurt Gödel in his incompleteness theorems.

Kurt Gödel's Incompleteness Theorems [7], presented in 1931, are two of the most significant results in the history of mathematical logic. The first theorem states that in any consistent, formal mathematical system that is capable of expressing basic arithmetic (like Peano arithmetic), there exist true propositions that cannot be proven within that system. This directly implies that such systems are inherently incomplete.

The second theorem goes further, stating that if a formal system is consistent, it cannot prove its own consistency.

These theorems were groundbreaking because they showed that our quest for a complete and self-contained set of axioms for all of mathematics was unattainable.

Propositional Logic: The completeness of propositional logic was established early on. For this logic, if a formula is logically valid, then it can be derived from the axioms of propositional logic using its inference rules.

First-Order Logic: Gödel also proved the completeness of first-order logic, which means that if a statement is universally valid, then it has a proof in first-order logic. This result is in contrast to his incompleteness theorems, which apply to specific formal systems like Peano arithmetic, rather than the entirety of first-order logic.

Higher-Order Logics: As we move to second-order and higher-order logics, completeness becomes elusive. Second-order logic, for instance, is not complete. This means there are semantically true statements in second-order logic that cannot be derived from any set of axioms. On structures like finite trees and words, where MSO is decidable, the logic is also complete in the sense that every MSO-definable property can be recognized by an automaton (tree automaton for trees and finite automaton for words).

The decidability and expressiveness of MSO on specific structures make it a powerful tool in formal language theory, automata theory, and database theory. Its ability to quantify over sets allows for the expression of complex properties that are beyond the reach of first-order logic.

In the context of artificial intelligence and knowledge representation, completeness ensures that a logical system can capture all the truths of its domain. This is crucial for tasks like theorem proving, where we rely on the logical system to derive new truths from given information.

5.5 Expert Systems: Mimicking Human Expertise

In the realm of artificial intelligence, expert systems stand out as a testament to the early aspirations of emulating human expertise within a computer program. These systems, designed to replicate the decision-making abilities of a human expert, have been pivotal in various domains, from medical diagnosis to financial forecasting. Let's delve deeper into the intricacies of expert systems and understand their significance.

The 1970s and 1980s marked the golden era of expert systems [8]. As the limitations of generalized AI became evident, there was a shift towards building specialized systems that could exhibit expert-level competence in specific domains. The idea was simple yet profound: if a computer could be fed with the knowledge of experts and could reason with that knowledge, it could make decisions or give advice at an expert level.

5.5.1 Anatomy of an Expert System

Expert systems, often hailed as the early triumphs of artificial intelligence, are designed to replicate the decision-making prowess of human experts in specific domains. Their architecture is meticulously crafted to ensure that they not only store domain-specific knowledge but also reason with it to derive conclusions or recommendations. Let's dissect the anatomy of an expert system to understand its core components and their interplay.

5.5.1.1 Knowledge Base: The Repository of Expertise

The knowledge base stands as the central pillar of an expert system, embodying its core intelligence and expertise. This repository is not merely a collection of facts but a meticulously organized structure that mirrors the inherent hierarchies and relationships of its domain, ensuring efficient retrieval and application of knowledge.

Within this structured organization, facts serve as the foundational building blocks. These facts can be broadly categorized into two types: static and dynamic. Static facts represent the unchanging truths about the domain. For instance, in a chemistry-oriented expert system, a static fact might state that "Water boils at 100 °C at standard atmospheric pressure." On the other hand, dynamic facts are mutable, changing based on the situation or context. In the realm of a stock market expert system, the current price of a particular stock would be considered a dynamic fact.

Beyond facts, the knowledge base houses rules, which are essentially conditional directives encapsulating the domain's inherent logic and relationships. Typically expressed in an "IF condition THEN conclusion" format, these rules might also carry certainty factors, indicating the confidence level associated with them. For instance, a rule might suggest, "IF symptoms A and B are present THEN there's an 80% chance of Disease X." Some rules, known as meta-rules, even dictate the application of other rules, especially in situations where conflicts arise.

Heuristics add another layer of depth to the knowledge base. These aren't just about hard facts but encapsulate the strategies, guidelines, or best practices that experts in the domain have derived from their experiences. They offer the system a semblance of intuition, enabling it to make educated guesses or decisions when faced with inconclusive data. Some heuristics might carry more weight than others, reflecting their reliability or relevance in specific contexts.

However, the knowledge base isn't a static entity. As domains evolve and new insights come to the fore, the knowledge base requires regular updates to maintain its relevance. This involves processes of knowledge acquisition, where new knowledge is gathered from human experts or automated sources, and knowledge verification, ensuring the accuracy and consistency of new information. Over time, some knowledge might become outdated or less pertinent, necessitating a refinement process to keep the knowledge base streamlined and current.

In essence, the knowledge base of an expert system is a dynamic and evolving entity. It not only stores information but also captures the very essence of domain expertise, allowing the system to reason, adapt, and evolve in a manner reminiscent of human experts.

5.5.1.2 Inference Engine: The Reasoning Powerhouse

The inference engine is the cognitive heart of an expert system, responsible for simulating the decision-making prowess of human experts. It doesn't merely access the knowledge stored within the knowledge base but actively reasons with it, drawing inferences, making deductions, and arriving at conclusions or recommendations based on the given inputs.

At its core, the inference engine operates by applying rules from the knowledge base to the facts or data presented to it. This process might seem straightforward, but given the vast and intricate web of rules within a typical expert system, the engine's task is anything but simple. It must navigate this web, determining which rules are applicable in the current context, resolving conflicts between competing rules, and chaining rules together in a logical sequence to derive meaningful outcomes.

There are primarily two modes of reasoning that an inference engine might employ: forward chaining and backward chaining. Forward chaining starts with the known facts and systematically applies rules to infer additional facts or conclusions. It's a data-driven approach, ideal for situations where the system needs to discover what outcomes can be derived from a given set of data. Conversely, backward chaining begins with a hypothetical conclusion and works backward to identify the facts or conditions that would support this conclusion. This goal-driven approach is particularly useful when the system has a specific outcome in mind and needs to determine if the current data supports it.

The inference engine also grapples with the challenge of uncertainty. In many real-world scenarios, data might be incomplete, ambiguous, or even contradictory. To navigate this uncertainty, the engine often employs techniques like fuzzy logic, which allows for degrees of truth rather than binary true/false evaluations, or probabilistic reasoning, which assigns likelihood values to different outcomes.

Furthermore, the engine's efficiency is paramount. With potentially thousands of rules to consider, the engine must be adept at pruning irrelevant paths and focusing its computational resources on the most promising lines of reasoning. Techniques like heuristic search, which guide the engine based on best-guess estimates, play a crucial role in ensuring timely and accurate conclusions.

In essence, the inference engine is more than just a rule applicator. It embodies the expert system's analytical prowess, mirroring the logical, strategic, and sometimes even intuitive reasoning processes of human experts. Through its intricate algorithms and methodologies, the inference engine ensures that the system's responses are not just accurate but also contextually relevant and timely, truly reflecting the depth and nuance of human expertise.

5.5.1.3 User Interface: Bridging Humans and the System

The user interface (UI) of an expert system serves as the primary point of interaction between the human user and the underlying computational machinery. It's the gateway through which users present their queries, concerns, or problems and receive solutions, recommendations, or insights from the system. But beyond its functional role, the UI plays a crucial part in determining the overall user experience, influencing how users perceive the system's utility, reliability, and even its "intelligence."

At a fundamental level, the UI must be intuitive and user-friendly. Users, especially those unfamiliar with the intricacies of the domain the expert system caters to, should be able to navigate the interface with ease, inputting data, asking questions, and understanding the system's responses without undue difficulty. This requires a design that is both visually appealing and logically structured, with clear prompts, feedback mechanisms, and guidance at every step.

However, the UI of an expert system goes beyond mere aesthetics and usability. Given the complex nature of the problems these systems address, the interface must be adept at facilitating intricate dialogues. It should be capable of guiding users in framing their questions correctly, offering clarifications when inputs are ambiguous, and even proactively seeking additional information when necessary to refine its conclusions.

Moreover, the UI plays a pivotal role in building trust. Expert systems often operate in domains where stakes are high, such as medical diagnosis, financial forecasting, or legal advisement. Users must believe in the system's competence and reliability. To this end, the interface should not only provide answers but also offer justifications or explanations for its conclusions. By making its reasoning transparent, the system can assuage doubts, address skepticism, and foster confidence in its capabilities.

Another essential aspect of the UI in expert systems is its adaptability. Given the diverse user base, from novices to domain experts, the interface should be customizable to cater to varying levels of expertise. While a novice might prefer a more guided, step-by-step interaction, an expert might seek shortcuts, advanced querying capabilities, or deeper dives into the system's reasoning processes.

Lastly, in the age of multi-device accessibility, the UI should be responsive, ensuring seamless interaction whether accessed via a desktop, tablet, or smartphone. This flexibility ensures that users can consult the expert system whenever the need arises, irrespective of their device or location.

5.5.2 The Power of Expert Systems

Expert systems, a prominent branch of artificial intelligence, have been hailed for their transformative potential across various domains. Their power lies not just in automating routine tasks but in emulating the decision-making abilities of human experts. Let's delve into the multifaceted strengths of expert systems and understand what makes them so influential.

1. Emulating Human Expertise: At the heart of every expert system is its ability to mimic the decision-making prowess of a human expert. By leveraging a comprehensive knowledge base and a robust inference engine, these systems can draw conclusions, make predictions, and offer solutions that are on par with, or in some cases, even surpass the recommendations of seasoned professionals.

2. Consistency and Reliability: Unlike humans, expert systems are immune to fatigue, emotions, or cognitive biases. This means they can consistently provide accurate and reliable recommendations, ensuring that the quality of their output remains unaffected by external factors.

3. Scalability: One of the significant advantages of expert systems is their scalability. While training human experts can be time-consuming and resource-intensive, expert systems can be easily replicated, allowing organizations to scale their operations without compromising on the quality of decision-making.

4. Continuous Learning: Expert systems can be designed to continuously update their knowledge base. As new information, research, or data becomes available, the system can integrate these insights, ensuring that its recommendations are always based on the most recent and relevant information.

5. Accessibility: Expert systems democratize access to expertise. In domains like healthcare or legal advisement, where expert consultation can be expensive or hard to come by, these systems can offer valuable insights to a broader audience, bridging the gap between experts and those in need.

6. Speed: In situations where time is of the essence, expert systems can process vast amounts of data and draw conclusions in a fraction of the time a human expert might take. This rapid decision-making can be crucial in domains like financial trading or emergency medical diagnosis.

7. Cost-Efficiency: Over the long term, expert systems can prove to be cost-effective. While the initial investment in developing and training the system can be substantial, the subsequent operational costs are minimal, especially when compared to the expenses associated with hiring and training human experts.

8. Knowledge Preservation: Expert systems play a pivotal role in preserving and disseminating expertise. As seasoned professionals retire or move on, their knowledge, accumulated over years of experience, can be integrated into these systems, ensuring that their insights continue to benefit future generations.

9. Customizability: Expert systems can be tailored to cater to specific industries, domains, or even individual organizations. This customizability ensures that the solutions they offer are relevant, actionable, and aligned with the unique challenges and objectives of the user.

10. Collaborative Potential: Expert systems don't aim to replace human experts but to augment their capabilities. By taking over routine tasks or providing a second opinion, these systems allow professionals to focus on more complex, nuanced aspects of their work, fostering a collaborative synergy between man and machine.

5.5.3 Challenges and Evolution

The journey of expert systems, from their inception to their current state, has been marked by both groundbreaking achievements and formidable challenges. As with any pioneering technology, expert systems have had to navigate a landscape of evolving expectations, technical limitations, and the ever-present quest for relevance. Let's delve into the challenges faced by expert systems and how they set the stage for the emergence of Knowledge Graphs and the Semantic Web [9].

1. Static Knowledge Base: Traditional expert systems relied on a static knowledge base, which meant that any new information or change in the domain knowledge required manual updating. This limitation made it challenging to keep the system current and relevant, especially in rapidly evolving fields.

2. Lack of Contextual Understanding: While expert systems were adept at drawing conclusions based on predefined rules, they often lacked the ability to understand the context. This meant that nuances, which human experts would naturally consider, were sometimes overlooked.

3. Scalability Issues: As the complexity and size of the knowledge base grew, ensuring the efficient performance of the system became a challenge. The sheer volume of rules could lead to slower inference times and reduced responsiveness.

4. Difficulty in Knowledge Acquisition: Extracting expertise from human experts and translating it into a machine-readable format was a significant challenge. This process, known as knowledge engineering, was time-consuming and often required the collaboration of domain experts, knowledge engineers, and system designers.

5. Over-reliance on Explicit Knowledge: Expert systems primarily dealt with explicit knowledge, often missing out on the tacit knowledge that human experts possess. This tacit knowledge, which includes intuition, experience-based insights, and heuristics, is challenging to codify.

5.5.3.1 Evolution Towards Knowledge Graphs and the Semantic Web

The challenges faced by expert systems highlighted the need for a more dynamic, context-aware, and scalable approach to knowledge representation and reasoning. This realization paved the way for the emergence of Knowledge Graphs and the Semantic Web.

1. Dynamic Knowledge Representation: Knowledge Graphs offer a more flexible and dynamic way to represent knowledge. They capture relationships between entities, allowing for a richer and more interconnected representation of information.

2. Contextual Understanding: The Semantic Web, with its emphasis on linking data and providing context, addresses one of the significant limitations of traditional expert systems. By understanding the relationships between different pieces of information, systems can make more informed and nuanced decisions.

3. Scalability and Performance: Modern architectures, combined with the graph-based representation of Knowledge Graphs, ensure that even vast and complex datasets can be queried and analyzed efficiently.

4. Integration of Tacit Knowledge: The Semantic Web's emphasis on ontologies—structured frameworks that define relationships between concepts—allows for a more in-depth capture of both explicit and tacit knowledge.

5. Continuous Learning: With the integration of machine learning techniques, modern systems can continuously update and refine their knowledge bases, ensuring they remain current and relevant.

5.6 Conclusion

The journey through the realm of rule-based systems and expert systems has been both enlightening and intricate. From the rudimentary beginnings of rule-based systems to the sophisticated architectures of expert systems, the evolution of knowledge representation and reasoning in AI has been nothing short of revolutionary.

We commenced our exploration with the genesis of rule-based systems, understanding their foundational principles and their significance in the early days of AI. The dive into the various logical systems, from propositional to predicate and then to the more complex second-order logic, provided a comprehensive view of the mathematical underpinnings that drive these systems. Each logical system, with its unique characteristics, serves as a testament to the versatility and depth of human thought processes and our attempts to replicate them in machines.

The discussion on decidability and completeness further accentuated the challenges and intricacies involved in ensuring that our logical systems are both sound and capable of representing the vastness of human knowledge.

Expert systems, as a pinnacle of rule-based AI, showcased the potential of machines to mimic human expertise. We delved deep into their anatomy, understanding the critical components like the knowledge base and the inference engine. The power of expert systems, their ability to revolutionize industries and make informed decisions based on vast amounts of data, was evident. However, like any technology, they too had their set of challenges, which in turn paved the way for newer paradigms like Knowledge Graphs and the Semantic Web.

As we conclude this chapter, it's essential to appreciate the symbiotic relationship between rule-based systems and expert systems. While rule-based systems provide the structure and logic, expert systems bring in the depth of domain-specific knowledge. Together, they represent a significant chapter in the story of AI, a chapter that laid the groundwork for the sophisticated AI systems of today.

As we transition to the next phase of our exploration, the integration of these traditional systems with modern AI paradigms, we carry forward the lessons, insights, and rich legacy of rule-based and expert systems.

References

1. Lucas, S.: The origins of the halting problem. J. Logical Algebraic Methods in Program. **121**, 100687 (2021)
2. Smullyan, R.M.: Godel's Incompleteness Theorems. Oxford University Press (1992)
3. Hliněný, P., Seese, D.: Trees, grids, and MSO decidability: from graphs to matroids. Theoretical Comput. Sci. **351**(3), 372–393 (2006)
4. Engelfriet, J., Vogler, H.: A Buchi-Elgot-Trakhtenbrot theorem for automata with MSO graph storage. Discrete Math. Theoretical Comput. Sci. **22**(4), 1d–1d (2020)
5. Rabin, M.O.: Decidability of second-order theories and automata on infinite trees. Trans. Am. Math. Soc. **141**, 1–35 (1969)
6. Hill, C.O.: Husserl and hilbert on completeness. In: From Dedekind to Gödel: Essays on the Development of the Foundations of Mathematics, pp. 143–163. Springer (1995)
7. Gödel, K.: Kurt Gödel: collected works: volume I: publications 1929–1936, vol. 1. Oxford University Press, USA (1986)
8. Liao, S.H.: Expert system methodologies and applications—a decade review from 1995 to 2004. Expert Syst. Appl. **28**(1), 93–103 (2005)
9. Zhou, H., Shen, T., Liu, X., Zhang, Y., Guo, P., Zhang, J.: Survey of knowledge graph approaches and applications. J. Artif. Intell. **2**(2), 89–101 (2020)

Chapter 6
Knowledge Graphs: Representation and Reasoning

> *"We at Google have made tremendous advances in understanding language. Our knowledge graph has been fundamental to that."*
>
> Amit Singhal, *Google*

6.1 Introduction

Consider a hypothetical scenario in which a youthful innovator named Sarah is engaged in the development of a pioneering software application that possesses the capability to anticipate and alleviate potential ecological catastrophes. The individual conceives of a system capable of analyzing extensive quantities of data derived from diverse sources, identifying patterns, and rendering intelligent decisions instantaneously. However, the primary obstacle she encounters pertains to the intricate nature and interdependencies of the data. Conventional databases and algorithms are limited in their ability to effectively capture the complex relationships and multidimensional characteristics of the data sources in question.

Knowledge graphs are a type of data structure that represents knowledge in a structured and interconnected manner.

In the current era characterized by rapid technological advancements and the increasing importance of data-driven decision-making, Knowledge Graphs have emerged as a promising solution that holds great potential for individuals like Sarah who are actively seeking innovative opportunities. Structured yet flexible methods are employed to represent and engage in logical reasoning with intricate data. Knowledge Graphs provide a comprehensive perspective of data by capturing entities, their

B. P. Bhuyan et al., *Neuro-Symbolic Artificial Intelligence*, Studies in Computational Intelligence 1176, https://doi.org/10.1007/978-981-97-8171-3_6

attributes, and the relationships that exist among them. This holistic representation enables the identification of concealed patterns, insights, and knowledge that would otherwise remain undisclosed.

This chapter will provide an in-depth exploration of Knowledge Graphs. This analysis will examine the fundamental principles of the subject under investigation, elucidate their modes of representation, and reveal the inherent capacity for logical reasoning that they possess. This discourse will explore the complexities involved in the construction of Knowledge Graphs, as well as delve into the sophisticated methods of induction. By doing so, we aim to gain a comprehensive understanding of the multifaceted nature of Knowledge Graphs and their profound relevance in the contemporary data-driven society.

6.2 Knowledge Graphs

Knowledge Graphs (KGs) have become a prominent and effective tool for representing, organizing, and querying extensive quantities of structured and semi-structured data [1]. Applications of these systems have been discovered in various domains, ranging from search engines to recommendation systems. They have played a crucial role in connecting structured databases with unstructured textual data.

The concept of a "knowledge graph" has been a topic of discussion, leading to the emergence of different definitions over the course of its evolution [2]. Nevertheless, in the context of our discourse, we choose to embrace an all-encompassing viewpoint. The fundamental essence of a knowledge graph lies in its purpose as a data structure meticulously crafted to amass and communicate comprehensive knowledge pertaining to the tangible realm. Within this particular framework, the nodes serve as representations of the entities that are of significance, while the edges symbolize the connections or relationships that exist between these entities. The data model that serves as the foundation for this type of graph can manifest in different ways, with two common forms being a directed edge-labeled graph or a property graph.

In the domain of a knowledge graph, knowledge encompasses more than a mere aggregation of factual information. The statements can be classified into two distinct categories: **simple statements**, exemplified by the statement "Santiago is the capital of Chile," and **quantified statements**, which provide generalized facts, such as the statement "All capitals are cities." The former can be easily represented as edges in the graph, whereas the latter requires more sophisticated means, such as ontologies or rules, to capture its complexity. The delineation between these two concepts holds significant importance as it dictates the manner in which additional knowledge is obtained. Deductive methodologies, such as the one employed in this scenario, have the capacity to derive the conclusion that "Santiago is a city" by drawing logical inferences from the given information. Conversely, inductive approaches can be utilized to derive and amass more intricate knowledge derived from these fundamental assertions.

The diversity of knowledge graphs presents an inherent challenge. Due to their frequent integration of data from diverse sources, they can display a diverse array of structures and varying levels of granularity. In order to effectively navigate this complexity, three key components emerge as crucial:

The **schema** provides a comprehensive framework for organizing the knowledge graph. The representation can be likened to a schematic plan, elucidating the systematic arrangement and interconnectedness of the data within the graph.

The issue of redundancy arises in knowledge graphs due to their reliance on various sources, which may result in multiple nodes potentially representing the same real-world entity. The notion of **identity** guarantees the unification of these nodes, establishing their recognition as representing a singular entity.

In certain instances, knowledge may not always be considered as an absolute certainty. The validity of a particular piece of information can vary across different cultural or temporal contexts. The role of **context** becomes significant in facilitating the precise interpretation and application of a fact or relationship by providing a specific backdrop.

6.2.1 Graph Data Models in Knowledge Graphs

Knowledge graphs inherently possess a graph-structured nature. The specific type of graph data model they employ can vary, and we discuss several of these models, emphasizing their characteristics and use-cases.

Directed Edge-Labeled Graphs (DELGs): In the DELG model, the graph consists of nodes and directed edges. Each edge carries a label, indicating the type of relationship it represents. Mathematically, a DELG can be represented as $G = (V, E, L)$, where V is the set of nodes, $E \subseteq V \times V$ is the set of directed edges, and L is the set of labels associated with the edges. In this model, nodes typically symbolize entities, while the directed edges capture the relationships between these entities. The directionality of the edges is crucial, as it signifies the source and target of the relationship.

Property Graphs: Property graphs enhance the basic structure of DELGs by allowing both nodes and edges to possess properties. These properties are represented as key-value pairs. A node or edge in a property graph can be represented as $n = (id, P)$, where id is a unique identifier and P is a set of properties. This model is especially beneficial when capturing additional attributes about entities or relationships. For instance, a node representing a person might have properties like "age" or "nationality." Modern graph databases, such as Neo4j, often utilize the property graph model due to its adaptability.

RDF (Resource Description Framework): RDF is a foundational model for data interchange on the web. It's built upon making statements about resources in the form of subject-predicate-object expressions, known as triples. In RDF, a triple can be represented as $t = (s, p, o)$, where s is the subject, p is the predicate, and o is the object. In the context of knowledge graphs, the subject denotes the entity, the predicate specifies the type of relationship, and the object indicates the entity or value the subject is related to. RDF offers a standardized way to represent information and is foundational to many semantic web technologies.

Hypergraphs: Hypergraphs [3] differ from traditional graphs in that an edge can connect any number of nodes, not just two. Mathematically, a hyperedge e in a hypergraph H can connect a subset of nodes $V_e \subseteq V$, where V is the set of all nodes in H. This model is beneficial when representing n-ary relationships, where a relationship might involve more than two entities. However, due to their complexity, hypergraphs are less commonly used in knowledge graph contexts compared to the other models.

6.2.2 Schema, Identity, and Context in Knowledge Graphs

6.2.2.1 Schema

The concept of a schema in the realm of knowledge graphs is analogous to the architectural blueprint of a building. Just as a blueprint provides a detailed layout, specifying where each room, door, or window should be, a schema delineates how data within a knowledge graph is organized and how relationships between different pieces of data are managed.

At its core, a schema provides a structured framework that defines both the entities that can exist within the knowledge graph and the possible relationships or connections between these entities. This structure ensures that the data within the graph adheres to a consistent format, making it easier to query, analyze, and interpret. For instance, in a knowledge graph dedicated to literature, the schema might specify entities such as "Author," "Book," "Genre," and "Publisher." It would also define potential relationships like "wrote" (linking "Author" to "Book") or "belongs to" (connecting "Book" to "Genre").

The term "**semantic**" in the context of data and knowledge representation refers to the meaning or context associated with data elements. When we talk about a semantic schema, we're delving into a more advanced and nuanced layer of schema design that goes beyond just defining entities and their relationships. A semantic schema aims to capture the inherent meaning, context, and nuances of the data it represents.

Meaningful Representation: Unlike traditional database schemas that focus primarily on the structure and relationships of data, a semantic schema emphasizes the meaning of data. For instance, while a traditional schema might simply link an "Author" to a "Book," a semantic schema would capture deeper nuances, such as the genre of the book, the era in which the author wrote, or the cultural significance of the work.

Ontologies and Taxonomies: Central to the concept of semantic schemas are ontologies and taxonomies. An ontology defines a set of concepts and categories in a subject area or domain, along with the relationships between them. It provides a framework for comprehensively representing knowledge within that domain. Taxonomies, on the other hand, are hierarchical classifications that categorize entities based on shared characteristics. In a semantic schema, ontologies and taxonomies work hand in hand to provide a rich, multidimensional representation of data.

Interoperability: One of the significant advantages of semantic schemas is their ability to promote interoperability. Given that they capture the meaning and context of data, semantic schemas can facilitate smoother data integration from diverse sources. They can help in understanding how a particular data element in one system corresponds to an element in another system, not just based on names or labels, but based on inherent meaning.

Enhanced Querying: With a semantic schema in place, querying becomes more intuitive and powerful. Users can query the knowledge graph based on concepts and meanings rather than just data labels. This allows for more complex and context-aware queries, fetching results that are more aligned with the user's intent.

Contextual Understanding: Semantic schemas [4] provide a framework for understanding the context in which data exists. For instance, understanding that a "bat" can be both a flying mammal and a piece of sports equipment requires a semantic understanding of the term based on the context in which it's used. A semantic schema can capture and represent such nuances.

Schema **validation** stands as a cornerstone in the realm of knowledge graphs, ensuring the integrity, consistency, and accuracy of the data they encompass. As these graphs burgeon in complexity, often drawing from a myriad of sources, the imperative of ensuring data's adherence to a predefined schema becomes even more pronounced. Let's delve deeper into the nuances of schema validation within knowledge graphs.

The very essence of schema validation lies in its ability to uphold data integrity. By ensuring that data entering the knowledge graph adheres to a certain standard and structure, the overall health and reliability of the graph are preserved. This becomes especially vital when one considers the diverse origins from which data might be sourced, bringing with it the looming shadow of inconsistency. Schema validation acts as a beacon, ensuring that regardless of where the data originates, it aligns seamlessly with the established structure. Moreover, it serves as an early warning system, detecting anomalies or potential errors during the data ingestion process, thereby saving significant downstream resources.

Several methodologies have been developed to facilitate schema validation. One of the most prevalent is **rule-based validation**, where specific rules are established that data must conform to before being integrated into the knowledge graph. For instance, such a rule might mandate that every "Author" entity is accompanied by a "Date of Birth" attribute. **Automated testing** has also emerged as a potent tool, mirroring the testing processes in software development to catch discrepancies in data. **Constraints**, too, play a pivotal role, defining the boundaries within which entities and relationships must operate. For example, a constraint could dictate that a "Book" entity's "Date of Publication" cannot be projected into the future. Another nuanced approach is **ontological validation**, which cross-references data against established ontologies to ensure semantic consistency and accuracy.

However, the path of schema validation is not without its challenges. Schemas, by their very nature, are dynamic, evolving in tandem with the domains of knowledge they represent. This constant flux can make validation a moving target, with the challenge exacerbated by the granularity of the data. Striking a balance between thoroughness and resource efficiency becomes a tightrope walk. Furthermore, not all data anomalies can be painted with the broad brush of errors. Some represent genuine exceptions, outliers that defy the norm but are nonetheless valid. These necessitate a more flexible approach to validation, one that can accommodate such deviations.

To round off, an effective schema validation process is incomplete without a robust feedback mechanism. When data fails to pass the validation muster, a system should be in place to alert the relevant stakeholders, be they data providers or system overseers. This feedback loop not only ensures prompt rectification but also continually refines the validation process, adapting it to the ever-shifting landscape of data and knowledge.

6.2.2.2 Identity

In the intricate web of knowledge representation, the concept of identity holds paramount importance. It serves as the foundation upon which data is recognized, differentiated, and connected. Within this realm, persistent identifiers emerge as a crucial tool, ensuring that entities maintain a consistent and unique identity across diverse datasets and over time. Let's delve deeper into the significance and mechanics of persistent identifiers within knowledge graphs.

At its core, a **persistent identifier** is a long-lasting reference to a document, file, web page, or other digital resource. Unlike transient identifiers, which might change over time due to various reasons like system updates or data migration, persistent identifiers remain unchanged, ensuring that the entity they represent can always be located, accessed, and cited. This stability is especially vital in knowledge graphs, where the interconnectedness of data necessitates reliable reference points.

The adoption of persistent identifiers addresses several challenges inherent to digital data management. Firstly, it combats the issue of link rot, where URLs or other references become obsolete or broken over time. Ensuring that an entity's identifier remains constant guarantees that the data remains accessible, even if its location or the underlying infrastructure changes. This is particularly crucial for scholarly references, datasets, or any digital resource that requires long-term citation and access.

Furthermore, persistent identifiers play a pivotal role in disambiguating data. In the vast expanse of digital information, it's not uncommon for different entities to share similar or even identical names. By assigning a unique, unchanging identifier to each entity, knowledge graphs can ensure that data points are accurately linked and that there's no confusion regarding their references. This becomes especially crucial when integrating data from multiple sources, where the risk of duplication or misidentification is high.

Several systems and standards have been developed to implement persistent identifiers. Digital Object Identifiers (DOIs) [5] are perhaps the most well-known and widely used for academic and research publications. Other systems like Archival Resource Keys (ARKs) [6] or Persistent Uniform Resource Locators (PURLs) [7] serve similar purposes in different contexts. These systems not only provide the identifier itself but also the infrastructure to resolve these identifiers to the actual digital resources they represent.

However, the implementation of persistent identifiers is not without challenges. It requires meticulous planning, consistent data management practices, and, often, collaboration between different stakeholders. Ensuring that identifiers remain persistent demands a commitment to long-term data stewardship, which can be resource-intensive.

Within the domain of knowledge graphs, the notion of identity surpasses the confines of an individual dataset or system. The graphs in question strive to offer a comprehensive perspective on knowledge by incorporating data from various sources. The process of integration requires the establishment of **external identity** links, which function as connectors linking entities across various datasets or knowledge bases.

External identity links play a pivotal role due to a multitude of reasons.

(a) The contemporary era of digital technology has resulted in the dispersion of data pertaining to a singular entity across multiple databases, each providing distinct viewpoints or fragments of the overall picture. External identity links play a crucial role in consolidating fragmented information, thereby offering a comprehensive perspective of the entity.
(b) The process of disambiguation involves addressing the potential discrepancies in naming conventions or identifiers used by different sources to refer to the same entity. The inclusion of external links serves to establish a connection between various representations, thereby acknowledging their reference to a singular real-world entity or concept. This practice effectively mitigates the occurrence of duplicity and ambiguity.
(c) The enrichment of data in knowledge graphs can be achieved through the establishment of connections with external sources. For example, a graphical representation that encompasses data pertaining to authors and their publications could establish external connections to a database housing comprehensive author biographies, thereby enhancing the richness and complexity of the existing dataset.
(d) Interoperability plays a crucial role within the expansive framework of data and knowledge bases. External identity links facilitate seamless interaction, sharing, and leveraging of data between diverse systems, thereby mitigating compatibility challenges.
(e) The inclusion of external, reputable sources can be utilized as a method to validate the information contained within the knowledge graph. The provision of a reference point serves to guarantee the accuracy and reliability of the data.

In the domain of knowledge graphs, the term "**lexicalization**" pertains to the act of assigning human-readable labels or descriptions to entities and relations [8]. These textual annotations, frequently presented as words or phrases, aid in the comprehension and interpretation of the information depicted in the graph for its users. In essence, lexicalization serves as a means of connecting the structured and formal representation of data in a graph with the natural language that humans use to communicate and understand information.

The significance of lexicalization lies in its ability to facilitate user navigation within a knowledge graph, particularly for individuals lacking technical expertise. The utilization of abstract identifiers or codes can often present a daunting challenge for such users. The utilization of lexicalized labels offers a more intuitive approach to the exploration and interaction with the graph.

The primary objective of knowledge graphs is to depict real-world entities and their interconnectedness. Lexical labels play a crucial role in ensuring that the meaning and significance of these entities and relationships are unambiguous and well-defined. An example can be observed in the case of a relation denoted by the code "R123," which may not possess significant meaning on its own. However, when this relation is expressed linguistically as "is the capital of," its intended significance becomes apparent.

Knowledge graphs frequently serve a diverse range of international users. Lexicalization facilitates the incorporation of linguistic designations in various languages, thereby promoting broader accessibility and comprehension. When users make a request to access a knowledge graph, they commonly employ natural language expressions. The utilization of lexicalized labels enhances the effectiveness and precision of search and query procedures, as they closely correspond with the input provided by the user.

The process of integrating data from multiple sources into a knowledge graph is facilitated by the use of lexicalization. This technique assists in the identification and mapping of similar entities and relations by analyzing their labels.

6.2.2.3 Context

Within the realm of knowledge representation, the concept of context holds significant importance, extending beyond a mere supplementary component and instead serving as a fundamental cornerstone. The determination of relevance, interpretation, and fundamental nature of information is a crucial aspect. The significance of context becomes particularly evident when discussing knowledge graphs, which strive to accurately represent the complex and diverse characteristics of the real world.

In the realm of knowledge graphs, the term "context" pertains to the particular environment or framework in which a given piece of information is situated. The context in which the information is presented plays a crucial role in shaping its interpretation and comprehension. One example to consider is the term "Paris." In the absence of contextual information, the given entity merely represents a nominal designation. However, when considering the geographical context, it is possible that the term could pertain to either a city located in France or a town situated in Texas. The interpretation of a given data point can vary significantly depending on the specific context in which it is situated.

Time is an additional dimension of context. The temporal relevance of information frequently constrains its scope. Please consider the available data pertaining to the Chief Executive Officer (CEO) of a company. A knowledge graph has the capacity to store information pertaining to the Chief Executive Officer (CEO) from both a historical perspective, five years ago, as well as the contemporary period. In the absence of temporal context, the presented data may engender confusion. However, when a specific temporal context is established, the knowledge graph has the capability to offer accurate and pertinent information within that particular timeframe.

The concept of context encompasses not only the information or data at hand but also takes into consideration the characteristics and needs of the user. A knowledge graph that possesses the ability to comprehend the contextual environment in which a user is functioning has the potential to greatly augment the user's experience. As an illustration, when a user situated in Paris, France conducts a search query for "renowned landmarks" using a mobile device, a contextually aware system could assign higher importance to search results pertaining to the Eiffel Tower or the Louvre.

Nevertheless, the representation of context poses certain challenges. The context is subject to change and is influenced by various factors. The context is subject to change as situations and settings evolve, rendering it a dynamic entity. Additionally, one must consider the issue of granularity. Certain situations necessitate a comprehensive framework, whereas others call for a precise and limited one. Furthermore, one must also consider the issue of ambiguity. The existence of an entity in various contexts poses a challenge in effectively representing its multifaceted nature without causing excessive complexity for both the system and the user. Achieving this balance requires careful consideration.

6.3 Deduction in Knowledge Graphs

The role of deductive knowledge is of great significance within the domain of knowledge graphs. It refers to the process of deriving new knowledge from the existing information stored in the graph. The derivation presented herein is grounded in logical reasoning, thereby ensuring the coherence and compatibility of the newly acquired knowledge with the pre-existing data.

Within the domain of knowledge graphs, deductive knowledge is frequently linked to the utilization of rules and ontologies. Rules are predetermined logical constructs that enable the system to deduce new relationships or attributes by utilizing the available data. As an illustration, in the context of a knowledge graph, if there exists the information that "Person A is the parent of Person B" and "Person B is the parent of Person C," a logical rule could be applied to infer that "Person A is the grandparent of Person C."

In contrast, ontologies offer a systematic framework that establishes the connections among various entities and their characteristics within the knowledge graph. They function as a schematic, providing direction for the process of deductive reasoning. The knowledge graph can ensure the relevance and accuracy of deduced information by following the structure and constraints specified by the ontology. Among the ontology languages that have gained prominence in recent years, the Web Ontology Language (OWL) stands out. Endorsed by the World Wide Web Consortium (W3C), OWL is designed to be compatible with RDF graphs, making it a preferred choice for many knowledge graph applications.

However, given OWL's broader adoption across various domains, our focus will be on its features and how it interprets graphs. It's worth noting that many features of OWL find parallels in OBOF, underscoring the foundational principles shared by ontology languages.

6.3.1 Interpreting Graphs with OWL

The Web Ontology Language (OWL) stands as a beacon in the realm of semantic web technologies, offering a robust framework for interpreting Resource Description Framework (RDF) graphs. Its design philosophy and features are tailored to ensure that knowledge graphs not only represent data but also encapsulate the underlying semantics of that data. Here's a more detailed exploration of how OWL interprets graphs:

Classes and Hierarchies: At the heart of OWL's interpretative power are its classes. These classes serve as categories or types, encapsulating entities that share common characteristics. For instance, a class named "Bird" might encompass entities like "Sparrow," "Eagle," and "Penguin." But OWL goes beyond mere categorization. It allows for the creation of class hierarchies, where subclasses can inherit properties from their parent classes. This hierarchical structure mirrors the taxonomies often found in the natural world and other domains, enabling a more nuanced representation of entities and their relationships.

Properties and Their Nuances: In OWL, relationships between entities are articulated through properties. These properties can be of two main types: object properties and datatype properties. While object properties describe relationships between two individuals (e.g., "hasParent" or "worksWith"), datatype properties link an individual to a data value (e.g., "hasAge" or "hasName"). Furthermore, OWL allows for the specification of property characteristics, such as symmetry, transitivity, and functionality, adding depth to the relationships being represented.

Axiomatic Foundations: Axioms in OWL serve as foundational statements or truths about the ontology. They define the constraints and relationships that hold within the represented knowledge domain. Whether it's specifying that a particular class is a subclass of another or asserting specific characteristics of properties, axioms provide the logical bedrock upon which the ontology stands. Their presence ensures that the knowledge graph remains consistent and valid in its assertions.

Reasoning and Inference: One of OWL's standout features is its support for logical reasoning. Given a set of axioms and facts, OWL-based reasoners can deduce new knowledge that wasn't explicitly stated. This inferencing capability allows knowledge graphs to be dynamic and continually enriched. For instance, if the ontology states that "All men are mortal" and "Socrates is a man," a reasoner can infer that "Socrates is mortal."

Annotations and Metadata: Beyond the logical content of the ontology, OWL provides mechanisms for annotations. These annotations, while not directly influencing the logical assertions, play a crucial role in offering context, documentation, and additional metadata about the ontology's elements. They enhance the understandability and traceability of the ontology, especially when collaborating with others or when revisiting the ontology after a time.

6.3.2 Semantics and Entailment

Understanding the intricate web of relationships and assertions within knowledge graphs necessitates a deep dive into the concepts of semantics and entailment. The conditions highlighted in our discussions provide a roadmap for interpreting each feature within the graph. These conditions subsequently lead to entailments. For instance, if we consider a hypothetical feature, say "Symmetric," and have an assertion that "City A is adjacent to City B," the symmetric nature would entail that "City B is also adjacent to City A." Let's further explore how these conditions lead to such entailments.

Every axiom, when incorporated into a graph, imposes certain conditions on the interpretations that satisfy the graph. Consider a simple example: We have a graph that asserts "Tom is a friend of Jerry." The models of this graph would be any interpretation where this relationship holds true. Now, if we introduce another axiom, say "All friends trust each other," it restricts the models of the graph. Now, not only should "Tom be a friend of Jerry," but also "Tom should trust Jerry" for any interpretation to be a valid model of the graph.

Entailment is a foundational concept in knowledge graphs. One graph is said to entail another if every model of the first graph is also a model of the second graph. For instance, if we have a graph asserting "All men are mortal" and another graph asserting "Socrates is a man," the combination of these two graphs would entail a third graph that asserts "Socrates is mortal." To elucidate the distinction between "if-then" and "if-and-only-if" semantics, consider the following example: Let's say we have a graph that asserts "All roses are flowers." Under "if-then" semantics, if we find a rose in our data, we can conclude it's a flower. However, finding a flower doesn't necessarily mean it's a rose. On the other hand, under "if-and-only-if" semantics, a rose in our data would mean it's a flower, and vice versa. The distinction is subtle but crucial for understanding the entailment relationships between graphs. Notably, the more comprehensive OWL ontology language predominantly employs an "if-and-only-if" semantics.

6.3.3 *Reasoning*

The realm of knowledge graphs is vast and intricate, and reasoning plays a pivotal role in navigating this landscape. Given two graphs, determining if the first entails the second, based on the definitions and features we've discussed, is a complex task. In fact, it's undecidable. This means there's no finite algorithm that can, with certainty, decide entailment for all inputs. However, there are practical reasoning algorithms that can:

(a) Halt on any input but might miss some entailments.
(b) Always halt with the correct answer but only for input ontologies with specific features.
(c) Return correct answers for any input but might not halt for certain inputs.

Among these, the third option has been explored using theorem provers for First Order Logic. However, the first two options, which use rules and/or Description Logic, are more prevalent. The first is efficient and scalable, suitable for scenarios with incomplete data. The second is ideal for domains like medical ontologies, where missing an entailment could have significant repercussions.

Rules provide a structured way to access deductive knowledge. They are essentially "if-then" statements. For instance, a rule might state: "If X is a mammal, then X is an animal." Here, the "if" part is the body of the rule, and the "then" part is the head. Both parts are represented as graph patterns. Rules can capture entailments under ontological conditions. For instance, if we have a rule that states "All mammals are animals," and our data graph has "Dogs are mammals," we can entail that "Dogs are animals."

Rules can be applied in various ways. One approach is "materialization," where rules are applied recursively to a graph, adding the conclusions back to the graph until no more additions can be made. Another strategy is "query rewriting," which automatically extends a query to find solutions entailed by a set of rules.

While rules can capture ontological entailments, they can also be defined for specific domains. Some rules might not be captured by ontology features, especially if they infer relations from cyclical graph patterns. Various languages, like Notation3 (N3) [9], Rule Interchange Format (RIF) [10], Semantic Web Rule Language (SWRL) [11], and SPARQL Inferencing Notation (SPIN) [12], allow for expressing rules over graphs.

In essence, reasoning within knowledge graphs, through the use of rules and entailments, provides a structured way to derive new knowledge from existing data. It's a powerful tool that enhances the utility and depth of knowledge graphs, making them invaluable assets in the realm of AI and data science.

6.3.4 Description Logic

Description Logics (DLs) have a storied lineage, originating from the need to formalize the meanings of frames and semantic networks. Given that semantic networks are often seen as the forerunners of knowledge graphs, and considering the significant influence DLs have exerted on the Web Ontology Language (OWL), it's clear that DLs occupy a central position in the logical formalization of knowledge graphs.

DLs aren't a monolithic logic but rather a family of logic. Initially, they were envisioned as decidable fragments of First Order Logic (FOL), ensuring that reasoning tasks, such as entailment checking, were achievable. Over time, DLs have matured, striking a balance between their expressive might and the computational complexity of reasoning. They've also integrated features beyond FOL, like transitive closure and datatypes, which are especially beneficial for modeling graph data.

At their essence, Description Logics are constructed on three primary elements:

Individuals: Specific entities like "Paris."

Classes (or Concepts): General categories such as "City."

Properties (or Roles): Relations like "capitalOf."

Using these foundational elements, DLs facilitate the creation of axioms. For instance, one can assert that Paris is a city (City(Paris)) or that Paris is the capital of France (capitalOf(Paris, France)). These assertional axioms form what is known as the Assertional Box (A-Box). Beyond these, DLs introduce logical symbols to craft class axioms (Terminology Box or T-Box) and property axioms (Role Box or R-Box). For example, stating that "City" is a subclass of "Place" would be a class axiom.

DLs offer a rich array of logical symbols, enabling the definition of new classes based on pre-existing terms. For instance, one could define a class of individuals that are renowned tourist destinations. Using DL symbols, it's possible to state that any city that's a capital must also be a significant tourist attraction. Furthermore, DLs can specify that an entity, like an airport, belongs to one of several classes, such as being either a domestic or international airport. Mathematically,

Sets:

$$I = \{\text{Paris, France}\}$$
$$C = \{\text{City, Country}\}$$
$$P = \{\text{capitalOf}\}$$

Relationships:

$$\text{capitalOf} : I \times I \rightarrow \{\text{true, false}\}$$

Examples:

$$\text{capitalOf(Paris, France)} = \text{true}$$
$$\text{capitalOf(France, Paris)} = \text{false}$$

Class Axioms:

$$\text{If capitalOf}(x, y) = \text{true and } x \in \text{City}, \text{ then } y \in \text{Country}.$$

Subclass-Superclass Relationship:

$$\text{City} \sqsubseteq \text{Place}$$

Subproperty Relationship:

$$\text{flight} \sqsubseteq \text{connectsTo}$$

Defining New Classes Based on Existing Terms:

$$\exists\text{nearby.Airport}$$

Universal Class:

$$\exists\text{flight}.\top \sqsubseteq \exists\text{nearby.Airport}$$

Union of Classes:

$$\text{Airport} \sqsubseteq \text{DomesticAirport} \sqcup \text{InternationalAirport}$$

The parallels between DL features and the OWL features previously discussed aren't coincidental. OWL was profoundly influenced by DLs. For instance, OWL 2 DL is a subset of OWL tailored to ensure that entailment remains decidable. However, certain features, like counting chains, are often restricted in DLs to preserve this decidability.

Expressive DLs can support intricate entailments involving a range of logical constructs. A common strategy to determine such entailments is to reduce the problem of entailment to satisfiability, which checks if an ontology is consistent. Techniques like tableau can then be employed to verify satisfiability. While these methods are potent, they can be computationally demanding, making them less apt for large-scale data. However, in complex domains where detailed modeling is paramount, they prove invaluable.

6.4 Induction in Knowledge Graphs

Inductive knowledge stands in contrast to deductive knowledge. While the latter is rooted in deriving exact logical consequences, the former is about discerning patterns from observed data and generalizing them. This generalization can then be employed to make new predictions, albeit with a degree of uncertainty.

Imagine a vast data graph filled with geographical data and flight information. From this, one might discern a pattern: most capital cities have international airports. Using this pattern, one could predict that if Paris is a capital city, it probably has an international airport. However, this prediction isn't foolproof. Vaduz, the capital of Liechtenstein, for instance, doesn't have an international airport. Predictions based on inductive knowledge often come with a confidence level. In this case, if 187 out of 195 capitals have international airports, the confidence level of such a prediction would be 0.959. This kind of knowledge, termed inductive knowledge, encompasses both the models that encapsulate these patterns and the predictions these models make. Despite its inherent uncertainty, inductive knowledge can be invaluable.

Inductive Techniques for Knowledge Graphs:

1. Unsupervised Methods: A significant portion of work in graph analytics falls under this category. These methods employ well-established algorithms to perform tasks like detecting communities or clusters within a graph, identifying central nodes and edges, and more.

2. Knowledge Graph Embeddings: This technique uses self-supervision to create a low-dimensional numerical representation of a knowledge graph. The primary aim is to map input edges to an output score that indicates the likelihood of the edge being true.

3. Graph Neural Networks (GNNs): GNNs leverage the inherent structure of graphs for supervised learning. They are particularly adept at tasks that involve understanding the relationships and patterns within graph data.

4. Symbolic Learning: Unlike the previously mentioned techniques that focus on numerical models, symbolic learning is about deriving symbolic models from a graph. These models are logical formulas, often in the form of rules or axioms, and are learned in a self-supervised manner.

Each of these techniques offers unique advantages and can be chosen based on the specific requirements of the task at hand.

6.4.1 Graph Analytics

Graph Analytics is the process of applying analytical techniques to graph data to uncover meaningful patterns and insights. The unique structure of graphs, which consists of nodes (entities) and edges (relationships), makes them particularly suited for certain types of analyses. These analyses often revolve around understanding the topology of the graph, i.e., how its nodes are interconnected. Techniques from graph theory and network analysis, which have been applied to diverse domains such as social networks, the Web, transportation systems, and biological networks, form the foundation of graph analytics.

Consider the domain of tourism. A tourism board might employ graph analytics to glean insights about key transportation hubs that cater to numerous tourist attractions (using centrality measures), or to identify clusters of attractions frequented by the same set of tourists (via community detection). They might also be interested in understanding the robustness of their transportation network, especially in the face of disruptions like strikes (using connectivity measures). Or, they might want to identify attractions that are similar in nature or appeal (through node similarity measures). For illustrative purposes, imagine a graph depicting transportation connections in France, focusing on popular tourist destinations like Paris, Nice, and Bordeaux.

A plethora of techniques are available for graph analytics.

Centrality: This technique identifies the most crucial nodes or edges in a graph. There are various measures of centrality, including degree, betweenness, closeness, Eigenvector, and PageRank. For instance, in our hypothetical French transportation graph, centrality measures could pinpoint Paris as a major transport hub.

Community Detection: The goal here is to find communities or clusters within the graph. These are sub-graphs that have denser internal connections compared to their connections with the rest of the graph. Algorithms like the Louvain modularity [13] method can be employed for this purpose. In our example, community detection might reveal clusters of attractions in Northern, Southern, and Central France.

Connectivity: This assesses how well the graph is interconnected. It can provide insights into the resilience of the network and potential points of failure. Measures include graph density, k-connectivity, and the identification of strongly or weakly connected components. For instance, if there are only a couple of train routes connecting Paris to Bordeaux, those routes become critical for connectivity.

Node Similarity: This technique identifies nodes that are similar based on their connections within the graph. Various metrics, such as structural equivalence or diffusion kernels [14], can be employed. In our French transportation example, Nice and Bordeaux might be deemed similar if both cities have direct flights and train routes to Paris.

Additionally, while the above techniques primarily require just the graph as input, there are other forms of graph analytics that might need additional inputs like specific nodes or pairs of nodes. One such technique is Path Finding.

Path Finding: This aims to find paths within the graph, especially between specified pairs of nodes. There are various definitions of valid paths, including simple paths, shortest paths, and paths that satisfy certain conditions. In our example, pathfinding algorithms could be used to determine the shortest route from Paris to Nice. It's worth noting that many traditional graph algorithms are designed for simpler graphs without edge labels. However, as we delve deeper into more complex graph models, we'll see how these techniques can be integrated with other methods like reasoning and querying.

For large-scale graph analytics, especially in distributed settings, several frameworks have been proposed. Notable among these are Apache Spark (GraphX) [15], GraphLab [16], Pregel [17], Signal Collect [18], and others. These frameworks employ a systolic abstraction [19] based on directed graphs. In this model, nodes act as processors that can send messages to other nodes via edges. The computation is iterative: in each cycle, nodes read incoming messages, perform computations, and then send messages based on the results. The systolic computational abstraction is superimposed on the data graph being analyzed.

Consider the task of determining the most accessible places based on the routes in a given graph. A metric like PageRank can be useful. PageRank calculates the probability of a user randomly navigating the routes and ending up at a specific location after a certain number of steps. Implementing PageRank on large graphs can be achieved using a graph parallel framework.

Mathematically, the PageRank algorithm can be described as follows:

$$R_{i+1}(v) = \frac{(1-d)}{|V|} + d \sum_{u \in M(v)} \frac{R_i(u)}{|E(u)|}$$

where

- $R_i(v)$ is the rank of node v at iteration i .
- d is the damping factor, typically set to 0.85.
- $M(v)$ are the nodes that link to v.
- $|V|$ is the total number of nodes.
- $|E(u)|$ is the number of edges from node u.

This iterative process continues until a termination criterion, such as a maximum number of iterations or a residual threshold, is met.

However, these frameworks, which rely on message passing between neighboring nodes, have their limitations. Not all analytics can be expressed within such frameworks. To address this, some frameworks introduce additional features, like a global step for computations across all nodes or a mutation step allowing for dynamic changes to the graph during processing.

While many analytics are designed for simpler graphs, data graphs often come with additional metadata in the form of edge labels or property-value pairs. Several strategies can be employed to adapt these analytics to data graphs.

Projection: This involves selecting a sub-graph and removing all edge metadata. For instance, a graph might be reduced to only consider specific edge labels, with those labels then being discarded.

Weighting: This strategy converts edge metadata into numerical values. Many techniques can be adapted to weighted graphs. For instance, edge labels might be mapped to weights based on certain criteria.

Transformation: This involves converting the graph to a simpler model. Transformations can be either lossy (some information is discarded) or lossless (the original graph can be fully reconstructed). Both types of transformations aim to preserve the graph's inherent structure.

Customization: This involves modifying the analytical procedure to account for edge metadata. For instance, path-finding might be adapted to consider edge labels or centrality measures might be aggregated by edge labels.

The choice of strategy can significantly influence the results of the analysis. Deciding on the best approach often requires empirical validation, and more research is needed to understand the broader implications of these strategies on various analytical techniques.

Graph query languages offer a plethora of options for querying graphs [20]. The interplay between query languages and analytics can be viewed from multiple angles.

1. Projection or Transformation using Queries: Query languages can be employed to project or transform a graph, making it suitable for specific analytical tasks.

2. Analytics Enhancing Queries: Analytics can aid the querying process. For instance, an analysis of connectivity can suggest optimal ways to distribute a large graph across multiple machines for efficient querying. Furthermore, analytics can be used to rank query results, especially in large graphs, prioritizing the most relevant results for the user.

3. Interleaving Querying and Analytics: In certain scenarios, there's a need to combine querying and analytical processes. For instance, in the event of an airline strike, one might want to determine events in cities that are inaccessible from a major hub. This would involve querying to extract specific transport networks, using analytics to determine connectivity, and then querying again to pinpoint events in isolated cities. While imperative languages like Gremlin, GraphX, or R can handle such tasks, there's a growing interest in more declarative languages that can seamlessly integrate querying and analytics.

Knowledge graphs often come with an associated semantic schema or ontology. This schema defines the semantics of domain-specific terms, leading to entailments. The decision to apply analytics before or after these entailments can drastically alter the results. For instance, while an edge indicating "Bikram Bhuyan hosts EID24" might be semantically equivalent to "EID24 venue Bikram Bhuyan" when considering the inverse axiom "hosts inv. of venue," from an analytical perspective that considers edge direction, these are distinct. Including one type of edge, its inverse, or both can significantly influence analytical outcomes.

To date, the intersection of analytics and entailment remains relatively uncharted, presenting a plethora of intriguing research questions. One potential avenue of exploration is the development of semantically invariant analytics. These would produce consistent results across semantically equivalent graphs (graphs that entail one another). Such analytics would focus on the semantic content of the knowledge graph rather than just the topological features of the data graph. For instance, semantically invariant analytics would treat a graph containing the inverse axiom "hosts inv. of venue" and a set of "hosts" edges the same as a graph where each "hosts" edge is replaced by an inverse "venue" edge, or the union of both graphs.

6.4.2 Knowledge Graph Embeddings

Machine learning methods, especially in the realm of knowledge graphs, have been the focal point of significant research and development. These methods can be employed either to directly refine a knowledge graph or for downstream tasks using the knowledge graph, such as recommendation, information extraction, question answering, and more.

However, a challenge arises: traditional machine learning techniques often expect dense numeric input in vector form, which contrasts with the typical representation of graphs. So, the question is, how can we convert graphs, or their components (nodes, edges, etc.), into numeric vectors?

An initial approach might involve a one-hot encoding, producing a vector for each node with dimensions $|L| \cdot |V|$ with $|L|$ being the number of edge labels and $|V|$ being the total number of nodes. But this often results in vast, sparse vectors, which aren't ideal for most machine learning models. This is where knowledge graph embedding techniques come into play. Their primary objective is to generate a dense representation of the graph in a continuous, low-dimensional vector space suitable for machine learning applications.

The embedding typically consists of an entity embedding for each node and a relation embedding for each edge label. The overarching aim of these vectors is to encapsulate and retain the latent structures within the graph. There are various ways to achieve this, but most commonly, a specific embedding method will employ a scoring function to compute the plausibility of an edge, i.e., its likelihood of being true.

These embeddings can then be utilized for several tasks related to the nodes and edge labels of the original graph. For instance.

Edge Confidence: The plausibility scoring function can be used to assign confidence levels to edges, especially those extracted from external sources.

Link Prediction: The scoring function can also be employed to complete edges with missing nodes or edge labels.

Similarity Measures: Since embedding models often assign similar vectors to analogous nodes and edge labels, they can be used to measure similarity. This is particularly useful for identifying duplicate nodes or for recommendation purposes.

There is a plethora of knowledge graph embedding techniques that have been proposed [21]. In this chapter, we aim to mention some of the most prominent techniques which will be explored in later chapters.

Translational Models: These models adopt a geometric perspective, where relation embeddings translate subject entities to object entities in the low-dimensional space.

Tensor Decomposition Models: These models focus on extracting latent factors that approximate the structure of the graph.

Neural Models: These models employ neural networks to train embeddings, ensuring they provide accurate plausibility scores.

Language Models: These models leverage existing word embedding techniques and propose methods to generate graph-like analogs for their expected textual inputs.

In the subsequent chapters, we will deep dive into each of these techniques, exploring their intricacies, advantages, and potential applications in the realm of knowledge graphs.

6.4.3 Graph Neural Networks (GNNs)

Graph Neural Networks (GNNs) represent a fusion of traditional neural networks and graph structures, aiming to harness the power of both worlds. While embeddings focus on converting graph data into dense numerical representations suitable for conventional machine learning models, GNNs take a more direct approach. They construct custom machine learning models specifically tailored for graph-structured data.

At a fundamental level, there's an inherent similarity between neural networks and graphs. A traditional neural network can be visualized as a weighted, directed graph where nodes act as artificial neurons and edges represent weighted connections or axons. However, the structure of a standard neural network, especially a fully connected feed-forward one, is quite uniform. It's typically organized in sequential layers, with each node in one layer connected to every node in the subsequent layer. In contrast, data graphs have a more diverse and intricate structure, shaped by the relationships between the entities they represent.

GNNs bridge this gap by constructing a neural network that mirrors the topology of the data graph. In essence, nodes in a GNN are linked to their neighboring nodes as dictated by the data graph. The primary objective is to learn a model that maps input features of nodes to output features. This mapping can be supervised, where output features for certain nodes might be manually labeled or derived from the knowledge graph.

What sets GNNs apart from knowledge graph embeddings is their capability for end-to-end supervised learning tailored to specific tasks. With a set of labeled examples, GNNs can be employed for various tasks like classifying elements within the graph or even the entire graph. Their applications are vast, ranging from classifying compounds, and objects in images, and documents to predicting traffic patterns, building recommendation systems, and even software verification.

Interestingly, given the right set of labeled examples, GNNs can even serve as substitutes for traditional graph algorithms. For instance, instead of using conventional methods to identify central nodes in knowledge graphs, GNNs can achieve this through supervised learning [22].

In the upcoming chapters, we'll delve deeper into the core concepts behind GNNs. We'll explore two primary variants of GNNs: recursive GNNs and convolutional GNNs, shedding light on their mechanisms, advantages, and potential applications in the ever-evolving domain of graph-based machine learning.

6.4.4 Symbolic Learning in Knowledge Graphs

Symbolic learning offers an alternative to the numerical models like knowledge graph embeddings and graph neural networks. While numerical models provide powerful predictive capabilities, they often lack interpretability. This means that while they can predict relationships or patterns within a graph, they don't necessarily explain the "why" behind those predictions.

For instance, consider a scenario where a knowledge graph embedding predicts a high likelihood of a flight connection between two airports. While the model might be accurate, it doesn't provide insights into why this connection is deemed likely. Moreover, these models face challenges when encountering new, unseen data. If a new airport is added to the graph, the model would need retraining to make predictions involving this new entity.

Symbolic learning addresses these challenges by focusing on learning logical, interpretable rules from the data. Instead of dense numerical representations, symbolic learning generates hypotheses in a logical language that provides explanations for observed relationships in the graph. For example, from observing flight data, symbolic learning might deduce that if there's a flight from Airport A to Airport B, there's likely a return flight from Airport B to Airport A. Such rules are not only interpretable but also offer a framework for deductive reasoning, allowing domain experts to validate the derived models.

Furthermore, symbolic rules are generalizable. They can be applied to new, previously unseen data without the need for retraining. Using the earlier example, even if a new airport is introduced to the graph, the rule about return flights still holds.

Two primary forms of symbolic learning are as follows:

Rule Mining: This involves discovering meaningful patterns in the form of rules from vast amounts of data. In the context of knowledge graphs, the aim is to identify rules that can explain a high proportion of observed relationships while contradicting a minimal number of them. Various metrics, like support and confidence, are used to evaluate the quality of these rules. Techniques like AMIE have been developed to efficiently mine rules from large knowledge graphs.

Axiom Mining: Beyond simple rules, axioms provide more complex logical structures that can explain patterns in the data. For instance, an axiom might state that all airports are either domestic, international, or both. Such axioms can be used for more advanced deductive reasoning.

Recent advancements in symbolic learning have also explored the integration of neural networks. Differentiable rule mining [23], for instance, represents rule-based joins as matrix multiplications, allowing for end-to-end learning of rules using neural architectures. This approach combines the strengths of symbolic reasoning with the power of neural networks, although it's currently limited to specific rule types.

In summary, symbolic learning offers a more interpretable and generalizable approach to understanding knowledge graphs. Focusing on logical rules and axioms provides insights into the underlying structure and relationships within the graph, making it a valuable tool for experts and researchers.

6.5 Conclusion

In this chapter, we embarked on a comprehensive exploration of Knowledge Graphs, delving into their representation and the intricate mechanisms of reasoning that they support. Beginning with an introduction that set the stage for our discussions, we delved into the foundational aspects of Knowledge Graphs, understanding their graph data models and the pivotal roles of schema, identity, and context.

Our journey into deduction in Knowledge Graphs illuminated the interpretative power of OWL, the semantics and entailment intricacies, and the reasoning capabilities that these graphs possess. We also touched upon Description Logic, shedding light on its significance in the realm of Knowledge Graphs.

Transitioning from deduction to induction, we explored the analytical prowess of graph analytics, the potential of Knowledge Graph Embeddings, and the innovative capabilities of Graph Neural Networks (GNNs). Our discussions on symbolic learning underscored the importance of extracting logical structures from Knowledge Graphs, offering insights into both rule and axiom mining.

As we conclude this chapter, it's evident that Knowledge Graphs are not just repositories of interconnected data but are dynamic entities capable of reasoning, learning, and evolving. Their multifaceted nature, combined with the tools and techniques we've discussed, makes them invaluable assets in the world of data science and artificial intelligence.

As we look ahead to the next unit on Neural Networks and Deep Learning, we anticipate diving deeper into the neural architectures that power many modern AI applications. These architectures, when combined with the foundational knowledge of Knowledge Graphs, promise to unlock even more potent capabilities and insights.

References

1. Fensel, D., Şimşek, U., Angele, K., Huaman, E., Kärle, E., Panasiuk, O., Toma, I., Umbrich, J., Wahler, A., Fensel, D., et al.: Introduction: what is a knowledge graph? Knowledge graphs: methodology, tools and selected use cases, pp. 1–10 (2020)
2. Zou, X.: A survey on application of knowledge graph. J. Phys.: Conf. Ser. **1487**, 012016. IOP Publishing (2020)
3. Bhuyan, B.P., Singh, T.P., Tomar, R., Ramdane-Cherif, A.: Nesykhg: neuro-symbolic knowledge hypergraphs. Procedia Comput. Sci. **235**, 1278–1288 (2024)
4. Sikos, L.F., Philp, D.: Provenance-aware knowledge representation: a survey of data models and contextualized knowledge graphs. Data Sci. Eng. **5**, 293–316 (2020)
5. Paskin, N.: Digital object identifier (doi®) system. Encyclopedia of Library Inf. Sci. **3**, 1586–1592 (2010)
6. Hilse, H.W.: Implementing persistent identifiers. Consortium of European Research Libraries (2006)
7. Lyons, S.: Persistent identification of electronic documents and the future of footnotes. Law Libr. J. **97**, 681 (2005)
8. Song, D., Schilder, F., Hertz, S., Saltini, G., Smiley, C., Nivarthi, P., Hazai, O., Landau, D., Zaharkin, M., Zielund, T., et al.: Building and querying an enterprise knowledge graph. IEEE Trans. Serv. Comput. **12**(3), 356–369 (2017)

9. Arndt, D., Mennicke, S.: Notation3 as an existential rule language. In: International Joint Conference on Rules and Reasoning, pp. 70–85. Springer (2023)
10. Kifer, M.: Rule interchange format: the framework. In: International Conference on Web Reasoning and Rule Systems, pp. 1–11. Springer (2008)
11. Horrocks, I., Patel-Schneider, P.F., Boley, H., Tabet, S., Grosof, B., Dean, M., et al.: Swrl: a semantic web rule language combining owl and ruleml. W3C Member submission, vol. 21, no. 79, pp. 1–31 (2004)
12. Fürber, C., Hepp, M.: Using sparql and spin for data quality management on the semantic web. In: Business Information Systems: 13th International Conference, BIS 2010, Berlin, Germany, May 3–5, 2010. Proceedings 13, pp. 35–46. Springer (2010)
13. Ghosh, S., Halappanavar, M., Tumeo, A., Kalyanaraman, A., Lu, H., Chavarria-Miranda, D., Khan, A., Gebremedhin, A.: Distributed louvain algorithm for graph community detection. In: 2018 IEEE international parallel and distributed processing symposium (IPDPS), pp. 885–895. IEEE (2018)
14. Kondor, R.I., Lafferty, J.: Diffusion kernels on graphs and other discrete structures. In: Proceedings of the 19th International Conference on Machine Learning, vol. 2002, pp. 315–322 (2002)
15. Malak, M., East, R.: Spark GraphX in Action. Simon and Schuster (2016)
16. Low, Y., Gonzalez, J.E., Kyrola, A., Bickson, D., Guestrin, C.E., Hellerstein, J.: Graphlab: A new framework for parallel machine learning. arXiv preprint arXiv:1408.2041 (2014)
17. Malewicz, G., Austern, M.H., Bik, A.J., Dehnert, J.C., Horn, I., Leiser, N., Czajkowski, G.: Pregel: a system for large-scale graph processing. In: Proceedings of the 2010 ACM SIGMOD International Conference on Management of data, pp. 135–146 (2010)
18. Stutz, P., Bernstein, A., Cohen, W.: Signal/collect: graph algorithms for the (semantic) web. In: The Semantic Web–ISWC 2010: 9th International Semantic Web Conference, ISWC 2010, Shanghai, China, November 7–11, 2010, Revised Selected Papers, Part I 9, pp. 764–780. Springer (2010)
19. Li, H.F., Jayakumar, R.: Systolic structures: a notion and characterization. J. Parallel Distrib. Comput. **3**(3), 373–397 (1986)
20. Angles, R., Arenas, M., Barceló, P., Hogan, A., Reutter, J., Vrgoč, D.: Foundations of modern query languages for graph databases. ACM Comput. Surv. (CSUR) **50**(5), 1–40 (2017)
21. Wang, Q., Mao, Z., Wang, B., Guo, L.: Knowledge graph embedding: a survey of approaches and applications. IEEE Trans. Knowl. Data Eng. **29**(12), 2724–2743 (2017)
22. Xie, Y., Xu, Z., Zhang, J., Wang, Z., Ji, S.: Self-supervised learning of graph neural networks: a unified review. IEEE Trans. Pattern Anal. Machine intell. **45**(2), 2412–2429 (2022)
23. Sadeghian, A., Armandpour, M., Ding, P., Wang, D.Z.: Drum: End-to-end differentiable rule mining on knowledge graphs. Adv. Neural Inf. Proces. Syst. **32** (2019)

Part III
Neural Networks and Deep Learning

Chapter 7
Feedforward Neural Networks and Backpropagation

"It's not a bug, it's a feature!"

–Douglas Adams, *The Hitchhiker's Guide to the Galaxy*

7.1 Introduction

Imagine yourself in a bustling farmer's market, surrounded by an array of colorful fruits and vegetables. You focus on apples and notice their varying hues: vibrant reds, tart greens, and sunny yellows. Based on your past experiences, you predict that red apples are sweet, green apples are tart, and yellow apples have a balanced flavor. You've made these predictions without tasting a single apple, relying solely on their colors. This intuitive process of making predictions based on observed features is something we do effortlessly. But how can we teach an artificial intelligence system to make similar predictions?

Feedforward neural networks [1] is a type of artificial neural network that propagates information in a unidirectional manner, from the input layer to the output layer, without any feedback connections. The aforementioned computational models draw inspiration from the cognitive processes employed by the human brain in information processing. Similar to how one may utilize the color of an apple as a means of anticipating its flavor, a neural network operates by receiving inputs (such as color) and generating an output (in this case, a predicted taste) following the processing of information through a network of interconnected nodes, resembling the functionality of neurons within the human brain.

B. P. Bhuyan et al., *Neuro-Symbolic Artificial Intelligence*, Studies in Computational Intelligence 1176, https://doi.org/10.1007/978-981-97-8171-3_7

The efficacy of refining these predictions is attributed to a computational technique known as backpropagation. Consider it a mechanism for providing feedback. In the event that our neural network makes an initial prediction that a green apple possesses a sweet taste but subsequently discovers that it is in fact tart, the process of backpropagation serves to facilitate the adjustment of the network's internal parameters. This adjustment is crucial in order to enhance the network's predictive capabilities for future instances.

This chapter will provide a comprehensive analysis of the architectural design of feedforward neural networks, elucidating the distinct functions of each constituent element. Next, we will delve into the concept of backpropagation, which is a fundamental algorithm that allows neural networks to learn from their errors. This process is supported by essential mathematical principles such as the chain rule of calculus.

Upon completion of this endeavor, you will acquire a thorough comprehension of the predictive and refining mechanisms employed by feedforward neural networks. This advancement will contribute to the ongoing pursuit of emulating the predictive capabilities exhibited by the human brain within artificial intelligence systems.

7.2 Understanding Feedforward Neural Networks

Consider a scenario where one aims to forecast the value of a residential property by considering multiple factors such as its size, location, number of bedrooms, age, and proximity to amenities like schools and parks. Each of these features serves as an input to the system. When modeled as a Feedforward Neural Network (FNN), the system processes these multiple inputs through a series of layers. Eventually, the network outputs a predicted price for the house.

Fundamentally, this is how an FNN operates. The system takes in multiple inputs, processes them sequentially through interconnected layers of nodes or neurons, and generates an output. The term "feedforward" denotes the unidirectional flow of data from input to output without any feedback loops or cycles.

7.2.1 Basic Architecture

The structure of an FNN can be likened to a multi-stage computational pipeline, wherein data undergoes successive transformations and refinements at each stage, ultimately resulting in a final output. Let us further examine the intricacies of its structure.

7.2.1.1 The Input Layer: Gateway to the Neural Network

The input layer serves as the neural network's entry point, where raw data is initially introduced into the system. This layer is crucial as it establishes the foundation for all subsequent processing and transformations within the network.

In the input layer, each neuron corresponds to a specific feature or attribute of the input data. For instance, if the task is to predict housing prices, features such as size, location, number of bedrooms, age, and proximity to amenities would each be represented by a neuron. The number of neurons in the input layer typically matches the dimensionality of the input data.

For example, consider an image with dimensions of 28×28 pixels. Each pixel's intensity would be represented by a neuron, resulting in an input layer comprising 784 neurons. This structure ensures that the raw input data is effectively communicated to the network for further processing.

Before data enters the input layer, it is often normalized or standardized. This preprocessing step ensures that all input features are on a similar scale, which facilitates faster and more efficient training. For example, house sizes may vary from several hundred to several thousand square feet. Normalizing these values ensures that no single feature disproportionately influences the learning process.

Normalization techniques [2] such as min-max scaling or z-score standardization are commonly employed. Min-max scaling adjusts the range of the data to a specific interval, usually [0, 1]. Z-score standardization, on the other hand, adjusts the data to have a mean of zero and a standard deviation of one.

Neural networks require numerical input values. When the dataset includes categorical data, such as the type of house (e.g., "detached," "semi-detached," "terraced"), it must be converted into a numerical format. One common approach is one-hot encoding, where each category is represented by a binary vector. For example, if there are three house types, "detached" could be represented as [1, 0, 0], "semi-detached" as [0, 1, 0], and "terraced" as [0, 0, 1].

In real-world datasets, missing values are often encountered. The input layer, or rather the preprocessing steps preceding it, must handle these missing values effectively. Techniques such as imputation [3] can be employed to fill in missing values based on certain criteria. For example, missing house sizes could be replaced with the mean or median size from the dataset.

7.2.1.2 The Hidden Layers: The Crux of Deep Learning

The hidden layers of a neural network are where the magic of deep learning truly happens. These layers are responsible for transforming the input data into increasingly abstract and useful representations, enabling the network to perform complex tasks with high accuracy.

The depth, referring to the number of hidden layers, and the width, referring to the number of neurons in each layer, determine the network's capacity to model complex relationships. A deeper network can capture intricate patterns, but it also risks overfitting, necessitating more data and computational resources.

Neurons within a hidden layer, unlike input neurons that receive raw data, process information that has already been transformed by preceding layers. Each connection between neurons has an associated weight, adjusted during training to minimize prediction errors. Additionally, each neuron has a bias, which modifies the output of its activation function. Hidden layers perform various critical operations.

Feature Extraction: In tasks like image recognition, initial hidden layers detect basic visual elements such as edges and textures. Middle layers combine these into shapes and patterns, while deeper layers identify high-level structures like faces or objects.

Data compression: In autoencoders [4], initial hidden layers reduce the input data's dimensionality, creating a condensed representation. Subsequent layers decompress this data to reconstruct the original input, with the bottleneck layer capturing the most essential features.

As the complexity of a model increases with more hidden layers and neurons, it becomes prone to overfitting—performing well on training data but poorly on new, unseen data. Regularization techniques, such as dropout, help mitigate this issue by randomly setting a fraction of input units to zero during each update, thus preventing neurons from becoming too specialized.

Hidden layers are the primary locus for learning and refining the network's understanding of the data. Through backpropagation, the network adjusts weights and biases in response to discrepancies between its predictions and actual values. This iterative process enhances the network's ability to capture the underlying patterns in the data.

7.2.1.3 Connections and Weights: The Synapses of Neural Networks

Connections in a neural network facilitate the transmission of data, directing the output of one neuron to become the input for another. This seamless flow of information from the input layer through hidden layers to the output layer ensures that each connection fulfills its role in a well-coordinated manner. However, the true complexity of this process is introduced by the weights attached to these connections.

Weights, represented as numerical values, determine the strength and direction of the influence that one neuron exerts on another. This modulation of data as it traverses a connection can be compared to adjusting the volume of a musical note, ensuring that the information is amplified or diminished appropriately.

Consider an urban metropolis where roads facilitate travel between various points of interest. In this analogy, connections within a neural network are like these roads, serving as pathways for data transmission. Weights, on the other hand, act like traffic signals, managing the flow of information and ensuring it reaches the most relevant destinations effectively.

Initially, weights are often assigned small random values or determined through techniques like Xavier initialization [5]. This initial setup is crucial as it significantly impacts the network's learning efficiency. During training, the network iteratively adjusts these weights to minimize the discrepancy between its predictions and the actual target values. This adjustment process is governed by algorithms such as back-propagation, where the learning rate—a hyperparameter—determines the magnitude of these adjustments.

7.2.1.4 Activation Functions: The Heartbeat of Neurons

Fundamentally, an activation function is a mathematical expression that determines the output of a neuron by transforming the weighted sum of its inputs. This transformation allows the network to capture non-linear relationships within the data, enabling it to perform more complex tasks than a linear model could.

Consider the baking process: combining ingredients like flour, sugar, and butter is essential, but it is the application of heat in the oven that transforms these ingredients into a cake. Similarly, while weights and inputs are the fundamental components of a neuron, the activation function acts as the catalyst that enables the neural network to function effectively.

Each activation function has distinct characteristics, making it suitable for specific tasks or network architectures. Over Table 7.1, we note several notable activation functions, examining their properties.

Table 7.1 Activation functions and their properties

Activation function	Equation	Parametric	Monotonic	Smooth	Bounded	Output range
Binary step	$f(x) = \begin{cases} 1 & \text{if } x > 0 \\ 0 & \text{otherwise} \end{cases}$	No	Yes	No	Yes	0–1
Linear	$f(x) = x$	No	Yes	Yes	No	$-\infty$ to $+\infty$
ReLU	$f(x) = \max(0, x)$	No	Yes	No	No	0 to $+\infty$
Leaky ReLU	$f(x) = \begin{cases} x & \text{if } x \geq 0 \\ \alpha x & \text{otherwise} \end{cases}$	Yes	Yes	No	No	$-\infty$ to $+\infty$
Sigmoid	$f(x) = \frac{1}{1+e^{-x}}$	No	Yes	Yes	Yes	0–1
Tanh	$f(x) = \frac{e^x - e^{-x}}{e^x + e^{-x}}$	No	Yes	Yes	Yes	-1 to 1
Softmax	$f(z_i) = \frac{e^{z_i}}{\sum_j e^{z_j}}$	No	Yes	Yes	Yes	0–1
Swish	$f(x) = x \cdot \text{sigmoid}(x)$	No	No	Yes	No	$-\infty$ to $+\infty$
ELU	$f(x) = \begin{cases} x & \text{if } x \geq 0 \\ \alpha(e^x - 1) & \text{otherwise} \end{cases}$	Yes	Yes	Yes	No	$-\alpha$ to $+\infty$

7.2.1.5 The Output Layer: The Neural Network's Grand Finale

The output layer of a neural network is where the transformed data from the hidden layers is finalized into a form that meets the network's specific task. This stage is crucial as it determines the type and range of predictions or classifications the network can make.

The output layer receives processed information from the preceding hidden layer and transforms it according to the network's goals. The choice of activation function here is pivotal, as it directly influences the nature of the output.

For example, in a neural network designed to predict house prices based on factors such as location, size, and age, the output layer typically consists of a single neuron. This neuron outputs a continuous value representing the price, often using a linear activation function to cover a wide range of values.

In contrast, for classification tasks such as identifying whether an image contains a cat, dog, or bird, the output layer contains multiple neurons—one for each class. Each neuron's output represents the probability of the input belonging to its respective class. The softmax activation function is frequently employed in such cases, converting the raw output scores into a probability distribution.

After producing the outputs, the network's predictions are compared to the actual target values using a loss function. This comparison generates an error value, which is then propagated back through the network during training. This backpropagation process adjusts the weights, refining the model's predictions over successive iterations.

7.2.2 Loss Functions: Gauging the Neural Network's Performance

Loss functions are critical in training neural networks as they quantify the difference between the predicted outputs and the actual target values. By providing a metric for this discrepancy, loss functions guide the optimization process to adjust the network's parameters to improve its performance. Different types of loss functions are used depending on the nature of the task, such as regression, binary classification, or multi-class classification. Some of the popular loss functions are shown in Table 7.2.

7.3 Backpropagation: The Backbone of Neural Network Training

Given the potentially vast number of parameters in a network, determining the precise adjustments for each one is a formidable challenge. We require a systematic approach to understand how each weight influences the overall loss. Essentially, we

Table 7.2 Loss functions and their key characteristics

Loss function	Equation	Key characteristics and symbols
Mean Squared Error (MSE)/L2 loss	$\text{MSE} = \frac{1}{N}\sum_{i=1}^{N}(y_i - \hat{y}_i)^2$	• Sensitive to outliers due to squaring errors • Provides higher penalties for larger errors • Commonly used for regression tasks • N: Number of data points • y_i: Actual value • $\hat{y}_i$: Predicted value
Mean Absolute Error (MAE)/L1 loss	$\text{MAE} = \frac{1}{N}\sum_{i=1}^{N}\|y_i - \hat{y}_i\|$	• Less sensitive to outliers compared to MSE • Treats all errors equally • Suitable for regression tasks where outliers should not be heavily penalized • N: Number of data points • y_i: Actual value • $\hat{y}_i$: Predicted value
Huber loss	$L_\delta(a) = \begin{cases} \frac{1}{2}a^2 & \text{for } \|a\| \leq \delta \\ \delta(\|a\| - \frac{1}{2}\delta) & \text{for } \|a\| > \delta \end{cases}$	• Combines properties of MAE and MSE • Robust to outliers while still penalizing large errors more than MAE • Adjustable sensitivity through parameter δ • a: Residual error $(y_i - \hat{y}_i)$ • δ: Threshold parameter
Cross-entropy loss/log loss	$L(y, \hat{y}) = -\sum_{i=1}^{N}[y_i \log(\hat{y}_i) + (1 - y_i)\log(1 - \hat{y}_i)]$	• Measures the performance of a classification model whose output is a probability value between 0 and 1 • Heavily penalizes incorrect predictions with high confidence • Commonly used in binary and multi-class classification • N: Number of data points • y_i: Actual binary label • $\hat{y}_i$: Predicted probability

(continued)

Table 7.2 (continued)

Loss function	Equation	Key characteristics and symbols
Hinge loss	$L(y, f(x)) = \max(0, 1 - y \cdot f(x))$	• Used for training classifiers, especially Support Vector Machines (SVMs) • Focuses on maximizing the margin between classes • Suitable for binary classification tasks • y: Actual label (+1 or −1) • $f(x)$: Raw output of the classifier for input x
KL divergence	$D_{KL}(P \parallel Q) = \sum_i P(i) \log \frac{P(i)}{Q(i)}$	• Measures how one probability distribution diverges from a second, expected probability distribution • Useful for tasks like Variational Autoencoders (VAEs) and other generative models • $P(i)$: True probability distribution • $Q(i)$: Approximate probability distribution
Mean Bias Deviation (MBD)	$\text{MBD} = \frac{1}{N} \sum_{i=1}^{N} (y_i - \hat{y}_i)$	• Measures the average bias in predictions • Can be used to detect systematic errors in a model • N: Number of data points • y_i: Actual value • $\hat{y}_i$: Predicted value
Poisson loss	$L(y, \hat{y}) = \hat{y} - y \log(\hat{y})$	• Used for modeling count data • Assumes the data follows a Poisson distribution • y: Actual count • $\hat{y}$: Predicted count

(continued)

Table 7.2 (continued)

Loss function	Equation	Key characteristics and symbols
Focal loss	$L(p_t) = -\alpha_t(1 - p_t)^\gamma \log(p_t)$	• Designed to address class imbalance by focusing more on hard-to-classify examples • Adjustable focusing parameter γ to balance easy and hard examples • α_t: Balancing factor for class imbalance • p_t: Model's estimated probability for the true class • γ: Focusing parameter
Contrastive loss	$L = (1 - Y)\frac{1}{2}(D_W)^2 + (Y)\frac{1}{2}\{\max(0, m - D_W)\}^2$	• Used in metric learning tasks • Encourages the network to distinguish between similar and dissimilar pairs of inputs • Y: Binary label indicating if pairs are similar or dissimilar • D_W: Distance metric between pair of inputs • m: Margin parameter

seek to determine the "sensitivity" of the loss with respect to each weight. With this information, we can fine-tune the weights to reduce the loss effectively.

Backpropagation is the cornerstone algorithm for training feedforward neural networks. By applying the chain rule of calculus, backpropagation calculates the gradient of the loss function with respect to each weight. This gradient indicates both the direction and magnitude of weight adjustments needed to minimize the loss. To fully appreciate this process, a deeper dive into its mathematical foundation is necessary.

7.3.1 The Core Idea: Chain Rule of Calculus

To grasp backpropagation, one must first understand the chain rule in calculus, a fundamental theorem underpinning the method. The chain rule provides a technique for differentiating composite functions, which is crucial for determining how weights and biases affect changes in the loss.

Consider two functions, $y = g(u)$ and $u = f(x)$. The composite function $y = g(f(x))$ expresses y with respect to x. The chain rule states that the derivative of

y with respect to x is the product of the derivative of y with respect to u and the derivative of u with respect to x. Mathematically, this is represented as follows:

$$\frac{\partial y}{\partial x} = \frac{\partial y}{\partial u} \times \frac{\partial u}{\partial x} \tag{7.1}$$

In a neural network, the final output (prediction) results from a composite function involving inputs, weights, biases, and activation functions. The loss function, measuring the discrepancy between the network's predictions and the actual targets, depends on these parameters.

The goal during neural network training is to optimize weights and biases to minimize the loss function. Understanding how slight changes in individual weights or biases impact the overall loss is crucial. The chain rule facilitates this by breaking down the derivative of the loss concerning a specific weight into a series of products of derivatives. This involves tracing a path from the output layer back to the weight in question, iterating this procedure for every weight and bias in the network.

7.3.1.1 Illustrative Example: A Simple Neural Unit

Let's consider a single neuron, often referred to as a perceptron, to illustrate the chain rule's application in the context of neural networks.

Imagine you're trying to predict the price of a house based on its size (in square feet). You decide to use a single neuron for this task.

- Model Setup:
  - Input: Size of the house (x)
  - Weight: w
  - Bias: b
  - Activation function: Linear function (i.e., $f(z) = z$)
  - Output: Predicted price (Y)
  - Actual price: (y)
  - Loss function: Mean Squared Error (MSE)
- Forward Pass:
  - Compute the weighted sum: $z = w \times x + b$
  - Compute the prediction: $y = f(z)$
  - Compute the loss: $L = \frac{1}{2}(Y - y)^2$
- Backward Pass using Chain Rule:
  - We want to know how the loss L changes with respect to the weight w. This is given by $\frac{\partial L}{\partial w}$
  - Using the chain rule: $\frac{\partial L}{\partial w} = \left(\frac{\partial L}{\partial Y} \times \frac{\partial Y}{\partial z} \times \frac{\partial z}{\partial w}\right)$
  - To break it down:

$\frac{\partial L}{\partial Y}$ is the change in loss with respect to the predicted price. This is $(Y - y)$.
$\frac{\partial Y}{\partial z}$ is the change in predicted price with respect to the weighted sum. Since f is a linear function, this value is 1.
$\frac{\partial z}{\partial w}$ is the change in the weighted sum with respect to the weight. This is the input x.

In this simple example, we've seen how the chain rule allows us to compute the gradient of the loss concerning a weight in our neural unit. This gradient tells us how to adjust w to minimize the loss and make better predictions. The same principle, when expanded, applies to complex neural networks with multiple layers and neurons.

7.4 Conclusion

As we close this chapter, we reflect on the marvel that is the feedforward neural network. These architectures, inspired by the intricate workings of our brain, have paved the way for numerous advancements in the realm of artificial intelligence. By processing inputs through layers of interconnected nodes, they can make predictions, classify data, and even generate new content.

The true strength of these networks, however, lies not just in their structure but in their ability to learn and adapt. Backpropagation, with its roots in the chain rule of calculus, provides a systematic mechanism for these networks to refine their predictions. Through iterative adjustments of weights and biases, informed by the gradients computed during backpropagation, a neural network gradually improves its performance, inching closer to the desired output with each iteration.

While the mathematics behind these concepts can be intricate, the core idea remains intuitive: make a prediction, assess the error, learn from it, and improve. It's a testament to the adaptability and resilience inherent in both biological and artificial systems. They provide the foundational mechanisms for continuous learning, setting the stage for their integration with symbolic systems.

References

1. Bebis, G., Georgiopoulos, M.: Feed-forward neural networks. IEEE. Potentials **13**(4), 27–31 (1994)
2. Singh, D., Singh, B.: Investigating the impact of data normalization on classification performance. Appl. Soft Comput. **97**, 105524 (2020)
3. Donders, A.R.T., Van Der Heijden, G.J., Stijnen, T., Moons, K.G.: A gentle introduction to imputation of missing values. J. Clin. Epidemiol. **59**(10), 1087–1091 (2006)
4. Pinaya, W.H.L., Vieira, S., Garcia-Dias, R., Mechelli, A.: Autoencoders. In: Machine learning, pp. 193–208. Elsevier (2020)
5. Kumar, S.K.: On weight initialization in deep neural networks. arXiv preprint arXiv:1704.08863 (2017)

Chapter 8
Convolution in Neural Networks

> *"Training a Convolutional Neural Network is like teaching a computer to see patterns and objects in a Where's Waldo book, but much faster and without ever getting frustrated."*

8.1 Introduction

Envision traversing an art exhibition, wherein every canvas showcases a vivid amalgamation of hues, forms, and motifs. As the observer transitions between artworks, their visual perception becomes attuned to nuanced elements, such as the meticulous application of brushstrokes, the interplay of illumination and obscurity, and the intricate interlacing of various textures. The human brain demonstrates remarkable efficiency in processing visual stimuli, effortlessly engaging in tasks such as object recognition, emotional discernment, and even anticipation of future events within a series of paintings. The remarkable capacity to analyze and comprehend visual information is a consequence of intricate neural mechanisms, which researchers and technologists have endeavored to replicate in artificial systems.

Explore the realm of convolutional operations and neural networks, where the underlying mathematical and computational principles strive to emulate the remarkable cognitive abilities of the human brain in comprehending visual and structured information. The convolution operation, which is fundamental in this field, is a mathematical technique that effectively captures localized patterns within data, analogous to the way our visual system selectively attends to specific details in a painting.

B. P. Bhuyan et al., *Neuro-Symbolic Artificial Intelligence*, Studies in Computational Intelligence 1176, https://doi.org/10.1007/978-981-97-8171-3_8

When incorporated into neural networks, particularly Convolutional Neural Networks (CNNs) [1], it enables machines to visually perceive and comprehend images, discerning features and patterns in a manner akin to human perception.

However, the domain of data extends beyond solely visual representations. Consider the potential integration of interactive elements within our art gallery, such as the incorporation of three-dimensional holographic displays, dynamic videos depicting motion and transformation, or even virtual reality environments that transcend the limitations of a two-dimensional grid and instead embrace a network-like interconnectedness. In order to effectively analyze a wide range of data, we expand upon our fundamental concepts. Graph Neural Networks (GNNs) are utilized to address the challenge of processing non-grid data by effectively capturing the intricate relationships and structures present in interconnected data [2]. In contrast, convolutions of higher dimensions, such as 3D convolutions [3], are capable of handling volumetric and temporal data, thereby incorporating the dimensions of depth and time.

This chapter extensively explores the mathematical foundations of these concepts, elucidating the complexities of convolution operations, the efficacy of CNNs, the adaptability of GNNs, and the vastness of convolutions in higher dimensions. Throughout our exploration, we will elucidate the impact of these tools and techniques on the cognitive processes of machines, thereby transforming our hypothetical art gallery scenario into a palpable actuality.

8.2 Mathematical Foundation of Convolution

Convolution is a fundamental mathematical operation used in neural networks to extract features from input data. It operates by sliding a small filter (also known as a kernel) over the input data and performing element-wise multiplication and summation. Mathematically, the convolution of a filter Λ on input $\mathbf{X}$ at location (i, j) is defined as follows:

$$\Lambda \star \mathbf{X} = \sum_m \sum_n \Lambda(m, n) \cdot \mathbf{X}(i - m, j - n) \tag{8.1}$$

Convolution possesses several key properties that make it a powerful tool in neural networks:

- **Associativity**: Associativity means that the order of convolving two functions doesn't affect the result. Mathematically, for three functions f, g, and h, associativity can be expressed as follows:

$$(f \star g) \star h = f \star (g \star h) \tag{8.2}$$

  Associativity ensures that when we apply multiple convolutions in a sequence, the order doesn't affect the final result. This is crucial for maintaining consistent outcomes when convolutions are stacked in deep neural network architectures.

- **Commutativity**: Commutativity means that the order of the functions being convolved doesn't affect the result. Mathematically, for two functions f and g, commutativity can be expressed as follows:

$$f \star g = g \star f \tag{8.3}$$

  This symmetry is valuable in various contexts, such as when designing filters for specific feature detection tasks.
- **Distributive**: Convolution obeys the distributive property, similar to multiplication. For functions f, g, and h, the distributive property can be expressed as follows:

$$f \star (g + h) = (f \star g) + (f \star h) \tag{8.4}$$

  The distributive property is akin to how multiplication distributes over addition. In convolution, it signifies that convolving a function with the sum of two other functions is equivalent to convolving the function with each term separately and then adding the results.
- **Fourier Transform Multiplication**: In the frequency domain, convolution corresponds to the element-wise multiplication of the Fourier Transforms of two functions. Mathematically, this is expressed as follows:

$$F \star (f \star h) = F(f) \cdot F(g) \tag{8.5}$$

  where $F(\cdot)$ represents the Fourier Transform. This property enables the use of Fast Fourier Transform (FFT) techniques for efficient convolution calculations as multiplication in the frequency domain is faster than convolution in the spatial domain. In neural networks, the frequency-domain perspective of convolution underscores its role in capturing patterns and features. Convolutional layers emulate frequency filters, extracting different features by convolving the input with various filters.
- **Convolution and Discrete Summation**: Convolution can be intuitively understood as a discrete summation of the product of two functions. It involves overlaying one function onto another, multiplying corresponding elements, and then summing up the results. This process captures the interactions between the two functions and extracts relevant information. Mathematically, the convolution of two discrete functions f and g at a specific point t is expressed as follows:

$$(f \star g)(t) = \sum_{\tau=-\infty}^{\infty} f(\tau) \cdot g(t - \tau) \tag{8.6}$$

  where $f(\tau)$ is one of the input functions, which is flipped and shifted. $g(t - \tau)$ is the other input function, shifted to align with the flipped function. Imagine f and g as sequences of values. Convolution involves placing g on top of f, shifting it across f, multiplying corresponding elements at each shift, and then summing up the results.

8.3 CNNs: Understanding Convolution with Image Data

Embedded within every visual representation, whether it be a captivating depiction of nature's beauty or a humble portrayal of a cherished individual, resides a fundamental structure comprised of numerical elements. The numerical values depicted in this context serve as a representation of pixel data, encapsulating the fundamental characteristics of the image in relation to color intensities and patterns. Convolutional Neural Networks (CNNs) are specifically engineered to comprehend and interpret matrices, thereby extracting significant patterns that can be effectively utilized for a range of tasks, encompassing image classification and object detection.

The design of a CNN is intricately constructed to effectively handle data structures that exhibit a grid-like nature, with a particular emphasis on image processing. In contrast to conventional feedforward neural networks, CNN exhibits the ability to capture spatial hierarchies within images, thereby incorporating the relative positions of pixels into their computations. Let us explore the fundamental elements that constitute a CNN.

8.3.1 Image Data and Pixels

Within the domain of digital technology, visual representations are encoded as arrays composed of individual pixels. Each pixel within this array corresponds to a distinct color or intensity value.

A pixel, which is an abbreviation for "picture element," represents the most fundamental component of an image. Every individual pixel within an image grid is associated with a distinct location and is responsible for storing pertinent data pertaining to either color or grayscale intensity. In the context of a color image, it is important to note that each individual pixel comprises three distinct color channels, namely red, green, and blue (RGB). These channels are represented by intensity values that span a numerical range of 0–255.

In the context of grayscale images, it can be observed that each individual pixel possesses a solitary intensity value, which serves as a representation of the overall brightness of said pixel. In the context of image processing, the intensity value is commonly expressed as an integer within the range of 0–255. This scale allows for the representation of different shades of gray, with 0 denoting black and 255 denoting white. The following matrix is an example of the same:

$$\begin{pmatrix} 50 & 100 & 150 & 200 & 250 & 50 \\ 75 & 120 & 200 & 240 & 180 & 25 \\ 90 & 110 & 170 & 220 & 170 & 10 \\ 120 & 160 & 220 & 160 & 100 & 12 \\ 140 & 180 & 140 & 180 & 0 & 120 \end{pmatrix}$$

Images can be conceptualized as arrays with two dimensions. The rows of the array correspond to the height of the image, while the columns represent the width. The resolution of an image is determined by its dimensions. Images with higher resolution contain a greater number of pixels, thereby enabling the representation of finer details and enhancing the overall visual quality.

The spatial relationships between objects and features are maintained through the arrangement of pixels in an image. The proximity of pixels in an image frequently corresponds to adjacent regions in the scene, facilitating the perception of shapes, textures, and patterns.

Within the realm of digital systems, the process of discretization is employed to convert images into a limited quantity of pixels. The level of detail in an image is determined by its resolution, which indicates the degree to which the scene has been sampled. High-resolution images offer enhanced fidelity in capturing the original scene, albeit at the cost of increased memory and processing requirements.

8.3.2 Convolutional Filters (Kernels)

Convolutional filters, commonly known as kernels, are a fundamental component of CNNs. These small matrices are of utmost importance in the process of extracting features, patterns, and information from images. Their utilization allows CNNs to effectively identify and classify objects, textures, and intricate visual structures. Convolutional filters have been purposefully designed to identify and extract distinct features present in an image, including but not limited to edges, corners, textures, and more complex structures. The filters function as templates, capturing specific attributes through the process of convolving over the pixel values of an image. The convolution operation involves sliding the filter Λ with dimensions $h \times w$, where h is the height and w is the width of the filter, over an input image matrix $\mathbf{I}$. At each position, the filter's weights are element-wise multiplied with the corresponding pixel values in the region. The results are then summed to produce a single value in the output feature map. Mathematically, the convolution operation at position (x, y) is defined as follows:

$$(\Lambda \star \mathbf{I})(x, y) = \sum_{i=1}^{h} \sum_{j=1}^{w} \Lambda(i, j) \cdot \mathbf{I}(x + i - 1, y + j - 1) \tag{8.7}$$

where $(\Lambda \star \mathbf{I})(x, y)$ is the value of the output feature map at position (x, y), $\Lambda(i, j)$ is the weight of the filter at position (i, j), and $\mathbf{I}(x + i - 1, y + j - 1)$ is the pixel value at the corresponding position in the input image.

A convolutional filter is a matrix of small dimensions relative to the input image. The numerical coefficients assigned to each element within the filter are commonly referred to as kernel weights in academic literature. The weights assigned to the filter determine the specific type of feature it is intended to detect. The convolution operation entails the process of systematically moving the filter across the image, performing element-wise multiplication between the filter weights and the corresponding pixel values within the current region, and subsequently aggregating the outcomes through summation. The aforementioned procedure yields a singular value within the output feature map, thereby emphasizing the existence of the identified feature. Let's consider an example.

Consider a grayscale image of size 6×6 given as follows:

$$\mathbf{I} = \begin{pmatrix} 50 & 100 & 150 & 200 & 250 & 50 \\ 75 & 120 & 200 & 240 & 180 & 25 \\ 90 & 110 & 170 & 220 & 170 & 10 \\ 120 & 160 & 220 & 160 & 100 & 12 \\ 140 & 180 & 140 & 180 & 0 & 120 \end{pmatrix}$$

Now, let's apply the filter Λ:

$$\Lambda = \begin{pmatrix} -1 & -1 & -1 \\ -1 & 8 & -1 \\ -1 & -1 & -1 \end{pmatrix}$$

Next, we'll perform convolution at each position using the convolution operation formula:

$$(\Lambda \star \mathbf{I})(x, y) = \sum_{i=1}^{3} \sum_{j=1}^{3} \Lambda(i, j) \cdot \mathbf{I}(x + i - 1, y + j - 1) \tag{8.8}$$

For instance, calculating the value at position (1, 1):

$$\begin{aligned} (\Lambda \star \mathbf{I})(1, 1) &= (-1 \cdot 50) + (-1 \cdot 100) + (-1 \cdot 150) + (-1 \cdot 75) \\ +(8 \cdot 120) &+ (-1 \cdot 200) + (-1 \cdot 90) + (-1 \cdot 110) + (-1 \cdot 170) = 15. \end{aligned}$$

8.3.3 *Feature Detection Using Filters and Sliding Window Technique*

CNNs demonstrate their efficacy in discerning complex patterns within image data through the utilization of a fundamental procedure referred to as feature detection. This crucial capability is attained through the combination of convolutional filters and the sliding window technique, allowing the network to reveal significant characteristics in images. Common filters used for feature detection include the following:

- **Sobel Filter** [4]
  - **Purpose**: Primarily used for edge detection.
  - **Description**: The Sobel filter uses two 3×3 kernels, one for detecting edges in the horizontal direction and the other for the vertical direction. The magnitude of the gradient can be found using these two filtered images.
  - **Horizontal** Λ:

$$\begin{pmatrix} -1 & 0 & 1 \\ -2 & 0 & 2 \\ -1 & 0 & 1 \end{pmatrix}$$

  - **Vertical** Λ:

$$\begin{pmatrix} -1 & -2 & -1 \\ 0 & 0 & 0 \\ 1 & 2 & 1 \end{pmatrix}$$

- **Prewitt Filter** [5]
  - **Purpose**: Primarily used for edge detection.
  - **Description**: Similar to the Sobel filter but uses different convolution kernels. It's less sensitive to diagonal edges compared to Sobel.
  - **Horizontal** Λ:

$$\begin{pmatrix} -1 & 0 & 1 \\ -1 & 0 & 1 \\ -1 & 0 & 1 \end{pmatrix}$$

  - **Vertical** Λ:

$$\begin{pmatrix} -1 & -1 & -1 \\ 0 & 0 & 0 \\ 1 & 1 & 1 \end{pmatrix}$$

- **Laplacian Filter** [6]
  - **Purpose**: Used for edge detection by identifying zero crossings after filtering.
  - **Description**: The Laplacian filter is a second-order derivative filter that detects areas of rapid intensity change in an image. It's rotationally symmetric.
  - Λ:

$$\begin{pmatrix} -1 & -1 & -1 \\ -1 & 8 & -1 \\ -1 & -1 & -1 \end{pmatrix}$$

The underlying concept of the sliding window technique involves the systematic examination of an image in order to identify particular features or objects. This is achieved by iteratively shifting a smaller region, referred to as a "window" or "patch" across the image. The window functions as a localized region of interest, wherein the features or patterns contained within said region are subjected to analysis. The

Algorithm 8.1 presents a fundamental framework for the sliding window methodology. In practical applications, it may be necessary to incorporate additional measures or refinements, such as conducting a comprehensive analysis of the window content at various scales using image pyramids or implementing post-processing techniques to mitigate the occurrence of false positives, such as non-maximum suppression.

Algorithm 8.1 Sliding Window Technique for Image Analysis

Require: Image I of size $W \times H$
Window size $w \times h$
Stride s
Ensure: List of features or objects detected at various window positions
Initialization:
$x \leftarrow 0$
$y \leftarrow 0$
while $y + h \leq H$ **do**
  while $x + w \leq W$ **do**
    Extract the sub-image I_{sub} from I defined by the top-left corner (x, y) and the window size $w \times h$
    Analyze I_{sub} (e.g., feature extraction, classification, etc.)
    Store results if any feature or object of interest is detected
    $x \leftarrow x + s$
  end while
  $x \leftarrow 0$
  $y \leftarrow y + s$
end while
return list of detected features or objects

8.3.4 *Hierarchical Representations and Role of Padding and Stride*

Feature maps are the output of filters (or kernels) applied to an input image or a previous layer in a neural network. As we progress deeper into a network, especially in Convolutional Neural Networks (CNNs), these feature maps represent increasingly abstract features of the input. The initial layers might capture edges and textures, while deeper layers might capture more complex patterns or even objects. This progression forms a hierarchical representation of features in the network.

- **Padding**: It refers to adding extra pixels around the border of the input image. Padding is primarily used to control the spatial dimensions of the output feature maps. It allows a convolutional layer to produce an output of the same size as the input, which is especially useful when designing architectures where spatial dimensions should be preserved. Given:

- Input size: $W \times H$
- Filter size: $f \times f$
- Padding: P
- Stride: S

The spatial dimensions of the output volume $W' \times H'$ are given by

$$W' = \frac{W - f + 2P}{S} + 1 \tag{8.9}$$

$$H' = \frac{H - f + 2P}{S} + 1 \tag{8.10}$$

- **Stride**: It defines the step size or the number of pixels by which the filter slides over the input image. A stride of 1 means the filter moves one pixel at a time, while a stride of 2 means it jumps two pixels per step. Increasing the stride reduces the spatial dimensions of the output feature map, making the computation faster but potentially losing some spatial information.

By adjusting padding and stride, one can control the spatial dimensions of the output feature maps, which can influence both the performance and computational efficiency of a neural network.

8.4 Graph Neural Network: A Gentle Introduction

Graph Neural Networks (GNNs) are a specific class of neural networks that have been specifically developed to effectively handle and analyze data that is organized in the form of graphs. Conventional neural networks, such as CNNs, are particularly suitable for data organized in a grid-like structure, such as images. Nevertheless, numerous practical issues entail data that inherently exhibits a graph structure, encompassing domains such as social networks, molecular structures, and recommendation systems. GNNs are specifically designed to process non-grid data effectively.

8.4.1 Extending Convolution to Non-grid Data with Graph Structures

In the context of visual data, the arrangement of information follows a structured, two-dimensional grid. Convolution operations involve the deliberate movement of a filter across this grid in a methodical manner. However, in the case of non-grid data such as graphs, the aforementioned regular structure is absent. In a graph, nodes possess diverse quantities of neighbors, and the arrangement of these neighbors does not possess any inherent ordering.

In order to generalize the concept of convolution to graphs, it becomes necessary to redefine the convolution operation in a manner that accommodates the inherent irregularity of graph structures.

8.4.1.1 Graph Convolution

Given a graph $G = (V, E)$ with nodes V and edges E, each node v has an associated feature vector h_v. The goal of graph convolution is to update this feature vector by aggregating information from its neighbors.

The graph convolution operation can be represented as follows:

$$h_v^{(l+1)} = \sigma\left(W^{(l)} \cdot \text{AGGREGATE}^{(l)}\left(\{h_u^{(l)} : u \in \text{Neighbors}(v)\}\right)\right) \tag{8.11}$$

where

- $h_v^{(l+1)}$ is the updated feature vector of node v at layer $l + 1$.
- $W^{(l)}$ is a learnable weight matrix at layer l.
- σ is a non-linear activation function, such as ReLU.
- $\text{AGGREGATE}^{(l)}$ is an aggregation function that combines the features of neighboring nodes.

8.4.1.2 Aggregation Functions

The choice of aggregation function determines how information from neighboring nodes is combined. Some common aggregation functions include the following:
Mean Aggregation:

$$\text{AGGREGATE}_{\text{MEAN}}^{(l)}\left(\{h_u^{(l)}\}\right) = \frac{1}{|\text{Neighbors}(v)|} \sum_{u \in \text{Neighbors}(v)} h_u^{(l)} \tag{8.12}$$

Weighted Aggregation:

If edges in the graph have weights, these weights can be used to compute a weighted average of neighbor features:

$$\text{AGGREGATE}_{\text{WEIGHTED}}^{(l)}\left(\{h_u^{(l)}, w_{uv}\}\right) = \frac{\sum_{u \in \text{Neighbors}(v)} w_{uv} \cdot h_u^{(l)}}{\sum_{u \in \text{Neighbors}(v)} w_{uv}} \tag{8.13}$$

where w_{uv} is the weight of the edge between nodes u and v.

To illustrate the same, we produce a graph with a central node v and three neighboring nodes u_1, u_2, and u_3. Edges connecting v to its neighbors are labeled with weights $w_{u,v}$. Below the graph, there's a representation of the graph convolution operation for node v.

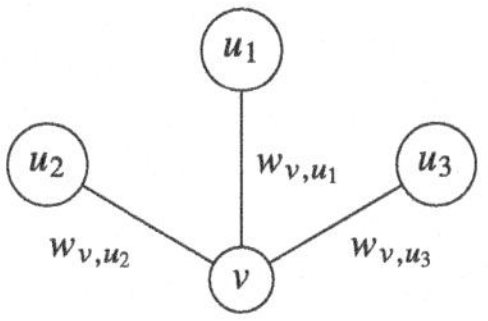

$$h_v^{(l+1)} = \sigma\left(W^{(l)} \cdot AGGREGATE^{(l)}\left(\{h_{u_1}^{(l)}, h_{u_2}^{(l)}, h_{u_3}^{(l)}\}\right)\right)$$

8.4.2 Utilizing Graph Convolution for Relational Reasoning

Relational reasoning encompasses the cognitive ability to comprehend and analyze the connections and associations among distinct entities, and subsequently employ this comprehension to formulate predictions or draw logical deductions. Within the domain of graph-structured data, the concept of relational reasoning can be conceptualized as the comprehension of the connections (edges) and associated attributes between nodes (entities), thereby facilitating an understanding of the relationships between said nodes.

GNNs are specifically designed to capture and reason about relationships by employing graph convolution. Each vertex within a graph possesses a collection of adjacent vertices. The neighboring nodes possess significant information regarding the context of the given node and its position within the broader structure of the graph.

In the process of graph convolution, individual nodes gather and combine feature information from their adjacent nodes. The process of aggregation can encompass basic operations such as averaging, as well as more intricate operations that take into account the significance or weighting of each neighboring element. Through the process of aggregating information from neighboring nodes, a single node is able to effectively capture and analyze local relational patterns. In the context of a social network, it is observed that when an individual's friends exhibit a collective interest in a specific subject matter, the process of aggregation is likely to emphasize the relevance of that particular topic to the individual as well.

The utilization of deep layers enables the capture of broader relationships. The concept of expanding the "receptive field" of each node is achieved by stacking multiple layers of graph convolution. This implies that nodes have the capability to indirectly consolidate information from nodes that are located at greater distances within the graph.

The mechanism employed in a two-layer GNN involves two distinct layers. In this context, the first layer is responsible for aggregating the immediate neighbors of a given node, while the second layer focuses on aggregating the neighbors of those neighbors. The aforementioned cascading effect persists as additional layers are added, thereby enabling nodes to capture more expansive relational patterns.

It has been observed that GNNs with increased depth possess the ability to comprehend more extensive contextual information. In the context of a family tree diagram, a deep GNN has the potential to effectively represent connections that extend to great-grandparents or second cousins, thereby facilitating a more comprehensive comprehension of an individual's extended familial relationships.

In numerous graphs, the edges, which represent the connections between nodes, contain significant information. The provided information has the capacity to elucidate the characteristics, intensity, or categorization of the connection between two entities. Contemporary GNNs have the capability to integrate edge features into the process of graph convolution. This implies that when a node assimilates data from a neighboring node, it takes into account the attributes of the edge that connects them.

GNNs possess the capability to discern different categories of relationships by taking into account the distinctive characteristics of the edges. In the context of a transportation graph, the edges can be interpreted as representations of various modes of transportation, such as roads, railways, or flight paths. By taking into account these various types of edges, a GNN can engage in logical thinking and analysis regarding diverse modes of transportation and their respective consequences.

In the subsequent chapters, we will delve into a more comprehensive examination of Graph Neural Networks.

8.5 Higher-Dimensional Convolutions

As the complexity of data increases, there is a need for the development and advancement of methods to effectively process and analyze it. An example of notable progress in the field of neural networks involves the utilization of higher-dimensional convolutions, specifically 3D convolutions, for the purpose of processing data that possesses more than two dimensions.

Consider a video clip of a bouncing ball. Each frame captures the ball's position at a specific point in time. If the clip has 10 frames, and each frame is a 100×100 pixel image, the video can be represented as a $100 \times 100 \times 10$ tensor. To capture the ball's motion over three consecutive frames, we might use a filter of size $5 \times 5 \times 3$, where the first two dimensions capture spatial patterns (like the shape of the ball) and the third dimension captures temporal patterns (like the ball's movement). Formally, the 3D convolution operation at a specific location in the output tensor is given by

$$O(x, y, z) = \sum_{i=0}^{h-1} \sum_{j=0}^{w-1} \sum_{k=0}^{t-1} I(x+i, y+j, z+k) \times F(i, j, k) \tag{8.14}$$

where O is the output tensor, I represents a segment of the input tensor (e.g., a video clip), and F denotes the 3D filter. A visual representation will look like this:

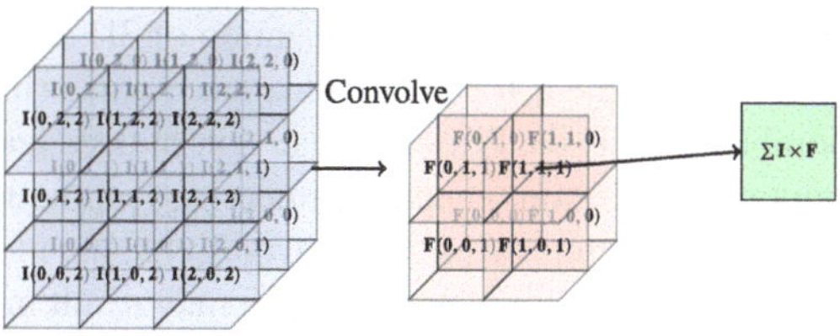

The utilization of traditional two-dimensional convolutions has served as a fundamental technique for extracting spatial features from images. Nevertheless, as the utilization of intricate data structures such as videos or volumetric data becomes more prevalent, it becomes imperative to acknowledge the additional dimension of time or depth. The efficacy of 3D convolutions is particularly notable in this context. These entities possess a high level of proficiency in extracting spatial and temporal features concurrently, rendering them exceptionally effective for analyzing such data. As an illustration, in the context of a video depicting a bird in flight, a 2D convolutional operation is capable of identifying the bird's shape. However, a 3D convolutional operation has the ability to capture the bird's flapping motion across successive frames, thereby facilitating a more comprehensive comprehension of the overall scene.

One notable benefit of 3D convolutions is their capacity to offer temporal context. Videos and other forms of time-series data frequently contain sequential information. The comprehension of the contextual information across a series of frames or data points is of utmost importance for numerous applications. In the context of a video depicting a lively traffic scenario, the ability to comprehend the sequential patterns of car movements and pedestrian crossings assumes significant importance. This understanding plays a crucial role in anticipating future actions and potentially identifying any irregular or abnormal behaviors.

The field of action recognition in videos has experienced significant advancements through the utilization of 3D convolutions. The nature of actions depicted in videos is inherently spatio-temporal in character. Let us contemplate two hypothetical situations: one in which an individual remains stationary while extending their hand upwards, and another in which an individual initiates the action of raising their hand. Although the individual frames of these scenarios may exhibit similarities, the differentiation of actions lies in the temporal progression of motion. The utilization of 3D convolutions enables the effective discrimination of various actions due to their capacity to capture temporal dynamics.

Within the field of medical imaging, where data is frequently volumetric in nature, such as with MRI or CT scans, the utilization of 3D convolutions has been seamlessly integrated. The scans possess the ability to capture not only superficial characteristics, but also to explore the intricate details of tissues and organs at a profound level. The identification of anomalies such as tumors may necessitate the comprehension of characteristics that extend across multiple sections of the scan. The inherent design of 3D convolutions enables them to effectively process and extract features that are specific to the depth dimension.

In addition to basic feature extraction, the utilization of 3D convolutions has significantly mitigated the requirement for manual feature engineering, particularly in the context of temporal data analysis. Prior to the advent of deep learning, tasks such as optical flow computation relied on manual engineering techniques to extract temporal features from videos. By employing 3D convolutions, the model is capable of acquiring these features directly from the data, which may result in the development of more resilient and adaptable models.

Furthermore, 3D convolutions have been utilized in various domains beyond mere analysis. They have a significant impact on various tasks such as video compression and generation. The utilization of convolutions can contribute to improved compression of video data and the generation of visually realistic and smooth video sequences by comprehending the inherent spatio-temporal patterns present in videos.

8.6 Conclusion

Throughout our investigation into the mathematical underpinnings of convolution, we have traversed the complex web of procedures and systems that facilitate machines' ability to analyze and comprehend data in manners akin to human cognitive processes. The fundamental convolution operation, akin to the human visual system's ability to perceive intricate details in a painting, has paved the way for the development of advanced techniques such as Convolutional Neural Networks, Graph Neural Networks, and higher-dimensional convolutions. These tools have significantly enhanced machines' capacity to analyze a wide range of data, including static images, dynamic videos, and interconnected data structures.

Nevertheless, as we have ventured into the domain of spatial and structural data processing, there lies another aspect that demands our attention: the realm of sequential data. CNNs and GNNs have proven to be effective in spatial comprehension. However, the inherent temporal characteristics of time-series data, language, and sequences necessitate the adoption of alternative methodologies. Recurrent Neural Networks (RNNs) are employed in this context. RNNs are specifically engineered to effectively capture and model temporal dependencies and sequences. As a result, they play a crucial role in various applications such as language processing, time-series forecasting, and other related tasks.

In the forthcoming chapter, an in-depth examination will be conducted on RNNs, encompassing their structural design, functional capacities, and their distinctive ability to tackle specific challenges. As we progress from spatial comprehension to sequential comprehension, we will establish the foundation for our subsequent exploration into the domain of Neuro-Symbolic computing. In this particular field, it is evident that the objective is to integrate fundamental neural computations with symbolic reasoning, resulting in a comprehensive framework of comprehension that surpasses the limited scope of pattern recognition.

References

1. O'shea, K., Nash, R.: An Introduction to Convolutional Neural Networks. arXiv preprint arXiv:1511.08458 (2015)
2. Khemani, B., Patil, S., Kotecha, K., Tanwar, S.: A review of graph neural networks: concepts, architectures, techniques, challenges, datasets, applications, and future directions. J. Big Data **11**(1), 18 (2024)
3. Tran, D., Wang, H., Torresani, L., Ray, J., LeCun, Y., Paluri, M.: A closer look at spatiotemporal convolutions for action recognition. In: Proceedings of the IEEE Conference on Computer Vision and Pattern Recognition, pp. 6450–6459 (2018)
4. Kanopoulos, N., Vasanthavada, N., Baker, R.L.: Design of an image edge detection filter using the Sobel operator. IEEE J. Solid-State Circuits **23**(2), 358–367 (1988)
5. Yang, L., Wu, X., Zhao, D., Li, H., Zhai, J.: An improved prewitt algorithm for edge detection based on noised image. In: 2011 4th International Congress on Image and Signal Processing, vol. 3, pp. 1197–1200. IEEE (2011)
6. Paris, S., Hasinoff, S.W., Kautz, J.: Local Laplacian filters: edge-aware image processing with a Laplacian pyramid. ACM Trans. Graph. **30**(4), 68 (2011)

Chapter 9
Recurrent Neural Networks (RNNs): Capturing the Dynamics of Sequences

"Wait a minute, Doc. Are you telling me you built a time machine... out of a DeLorean?"

–Marty McFly, *Back to the Future*

9.1 Introduction

Imagine a vibrant urban plaza teeming with animated dialogues, the synchronized strides of individuals on foot, and the faint blare of vehicular horns in the distance. In the midst of the surrounding noise, a street musician commences performing a well-known melody on his violin. As the melodic progression ensues, certain individuals exhibit an ability to discern the sequential arrangement of musical tones, thereby eliciting recollections tied to the composition, potentially evoking affective responses or even prompting them to reproduce the tune audibly. Individuals who are not acquainted with the melodic composition make attempts to anticipate the subsequent pitch or the increase in volume, drawing upon the patterns they have discerned thus far. The depicted scene bears a resemblance to the complex interplay of sequences and patterns that Recurrent Neural Networks (RNNs) traverse [1].

Within the expansive domain of neural networks, RNNs distinguish themselves due to their distinctive capacity to analyze and produce sequences. This capability extends to various domains, encompassing musical notes, linguistic constructs within sentences, and the temporal evolution of stock prices [2]. In contrast to conventional neural networks, which may encounter difficulties in retaining information from earlier segments of a sequence while processing subsequent segments, RNNs

B. P. Bhuyan et al., *Neuro-Symbolic Artificial Intelligence*, Studies in Computational Intelligence 1176, https://doi.org/10.1007/978-981-97-8171-3_9

excel in such situations by effectively preserving a memory of the preceding information. This characteristic renders them particularly well-suited for tasks that require consideration of context and sequencing.

This chapter extensively examines the realm of RNNs, encompassing their structural design, the problems they tackle, and the subsequent development of more sophisticated iterations. Starting with an exploration of the constraints inherent in traditional neural networks when it comes to processing sequences, we delve into the emergence of novel architectures such as Long Short-Term Memory (LSTM) [3] and Gated Recurrent Units (GRU) [4] that have significantly transformed the field of sequence modeling. This endeavor aims to comprehensively comprehend the intricacies of Recurrent Neural Networks. Throughout the course of our discussion, we will also explore their significance within the burgeoning domain of Neuro-Symbolic AI, wherein they serve as a conduit between the processing of unprocessed data and the realm of symbolic reasoning.

9.2 Sequential Data and Limitations of CNNs

9.2.1 The Nature of Sequential Data

In the course of our everyday experiences, we are consistently exposed to data that exhibit temporal dynamics, thereby exhibiting a discernible structure and sequence. Sequential data, also referred to as temporal data, is distinguished by its inherent temporal structure. The fundamental characteristic of sequential data resides in its inherent order, encompassing various forms such as the rhythmic patterns in music, the cadence of spoken sentences, and the dynamic fluctuations observed in stock market prices. The meaning of each data point, be it a word within a sentence or a note within a melody, is not solely determined by its inherent value but also by its placement within the sequence and its connections to preceding and subsequent data points.

Sequential data is widely prevalent in various domains. In the field of natural language processing, sentences can be defined as ordered sequences of words, with the interpretation of the sentence often relying on the specific arrangement of these words. Within the field of finance, time-series data plays a crucial role as it captures the evolution of variables, such as stock prices, over a specific period. This type of data exhibits a temporal dimension, where previous values frequently exert an influence on subsequent ones. In the field of biology, DNA sequences consist primarily of elongated chains of nucleotides, wherein the arrangement of these nucleotides plays a crucial role in determining the genetic information encoded within.

9.2.2 Why CNNs Fall Short for Temporal Patterns?

Convolutional Neural Networks (CNNs) have demonstrated remarkable efficacy in processing spatial hierarchies within data, particularly in the domain of images, owing to their utilization of layered filters and pooling operations. The network demonstrates high proficiency in discerning spatial patterns within a given locality, including but not limited to edges, textures, and shapes. Furthermore, as the network progresses, it exhibits the ability to combine these patterns into more intricate features.

Nevertheless, CNNs face difficulties when dealing with sequential data. There are several reasons why.

Lack of Temporal Memory: CNNs operate by processing input data in a feed-forward manner, devoid of any retention of past input information. On the other hand, it is often necessary to employ a type of memory in order to comprehend the context and interconnections among data points in sequential data. In the context of a sentence, the semantic interpretation of a word can be subject to modification based on the preceding words. Convolutional neural networks, due to their inherent design, lack the capability to capture and model complex dependencies effectively.

Fixed-size Filters: CNNs employ filters that possess a predetermined size, thereby enabling them to capture patterns of a particular length solely. In the context of sequential data, it is possible for significant patterns or dependencies to extend across different lengths. Utilizing a filter with a fixed size may result in the omission or truncation of these patterns.

Spatial versus Temporal Features: Although CNNs demonstrate proficiency in capturing spatial hierarchies, they lack inherent capability in dealing with temporal hierarchies. For instance, although a CNN may successfully identify the characteristics of a face in a stationary image, it may encounter difficulties in comprehending the evolution of a facial expression across a series of frames within a video.

Invariance versus Equivariance: The concept of invariance pertains to the characteristic whereby the output of a function or model remains unaltered even when subjected to alterations in its input. Spatial invariance is a highly sought-after characteristic in various tasks within the realm of neural networks, particularly CNNs.

For example, let us consider an image classification task wherein the objective is to determine whether an image contains a feline subject. Ideally, the classification of the image should remain consistent as “cat,” regardless of whether the cat is positioned at the center, in a corner, or even in an inverted orientation. Spatial invariance refers to a property in which the output, specifically the classification, remains unaffected by spatial transformations such as translations and rotations applied to the input, which in this context is an image.

CNNs employ pooling layers to achieve spatial invariance by reducing the spatial dimensions of feature maps. Additionally, weight sharing in convolutional layers enables the detection of features irrespective of their position in the input.

In contrast, equivariance refers to the property wherein a modification applied to the input results in a corresponding modification in the output that is equivalent in nature. In more accessible language, the correlation between the input and output remains constant even when subjected to various transformations.

In order to comprehend the concept of equivariance, it is instructive to examine the process of edge detection in an image through the utilization of a filter. When the input image is horizontally shifted to the right and subsequently subjected to edge detection, the resulting output, which represents the detected edges, will also be horizontally shifted to the right by an equivalent amount. The shift to the right in the input resulted in a corresponding shift to the right in the output, thereby establishing the property of equivariance to translation in the edge detection operation.

The pursuit of invariance is frequently sought after in tasks such as image classification, wherein the classification of an object should remain consistent regardless of its position or orientation. However, this principle may not always be appropriate for sequential data. The relative position or temporal occurrence of an event or data point holds significant importance in sequences.

In the context of time-series forecasting, the act of shifting the data points by a single time step has the potential to yield significantly divergent predictions. Likewise, in the realm of language, the arrangement of words holds the utmost significance in determining meaning. The sentences "The cat chased the dog" and "The dog chased the cat" possess distinct meanings as a result of the word order variation. Therefore, in the context of sequential data, it is frequently desirable to have models that exhibit equivariance to transformations. This property ensures that any alterations made to the input sequence will result in corresponding modifications to the output.

9.3 Foundations of Recurrent Neural Networks

9.3.1 Basic Architecture of RNNs

Recurrent Neural Networks signify a fundamental departure from conventional feed-forward neural networks. Feed-forward networks process inputs in a unidirectional manner, while RNNs incorporate loops that enable the retention of information, rendering them intrinsically well-suited for sequential data. In this discourse, we shall undertake a comprehensive analysis of the fundamental constituents and organizational framework of an RNN.

9.3.1.1 Input Node in RNNs

The input node in an RNN serves as the gateway for sequential data to enter the network. It's the point where the RNN interfaces with the external world, receiving data points from a sequence one at a time. The nature and processing of this data at the input node can vary based on the application and the representation of the data.

Natural Language Processing (NLP): In NLP tasks, the input is often a word or a character. This raw input is typically transformed into a more usable format, like a word vector using embeddings (e.g., Word2Vec, GloVe) or a one-hot encoded vector for characters.

Time-Series Analysis: Here, the input could be a scalar value (e.g., the stock price at a particular time) or a vector (e.g., multiple features like stock price, trading volume, and sentiment score).

Music Generation: The input might represent a musical note, chord, or even a measure, often encoded in a format suitable for computational processing.

Symbolic Representation

The fundamental concept underlying symbolic logic is the representation of information in a structured and discrete fashion. In the realm of symbolic logic, the concept of the input node pertains to the examination of how unprocessed data can be symbolically encoded in order to encapsulate intrinsic significance or connections.

In the field of NLP, it is important to note that a word is not merely a sequence of characters, but rather possesses inherent semantic significance. The symbolic representation may take the form of a distinct identifier or token assigned to each individual word. Nevertheless, neural networks, such as RNNs, necessitate numerical input. Embeddings play a crucial role in transforming symbolic tokens into dense vectors that effectively capture semantic relationships.

Symbolically, if we consider a word w from a vocabulary V, its symbolic representation could be a function $\delta(w)$ that maps w to a unique token in a set of tokens T. This token can then be transformed into a dense vector v using an embedding matrix E. Formally,

$$v = E \times \delta(w) \tag{9.1}$$

Once the input data is in a suitable format, it's passed to the hidden node of the RNN. However, before this transfer, there might be initial processing steps.

Normalization: Especially for numerical data, normalization (scaling values to a standard range) can help in stabilizing and speeding up training.

Feature Extraction: For some applications, raw data might be transformed into a set of features that are more informative for the task at hand.

Embedding Lookup: For tasks like NLP, the symbolic token is used to look up its dense vector representation from an embedding matrix.

The input node, although appearing to be simple, holds significant importance in facilitating the appropriate formatting and processing of data as well as capturing the inherent relationships within an RNN. The interaction between symbolic representations and neural processing, as alluded to in this statement, establishes the foundation for a more comprehensive investigation in the Neuro-Symbolic chapters.

9.3.1.2 Hidden Node (or Recurrent Node)

The hidden node, often termed the recurrent node, is the core component of an RNN. It's where the network's memory and processing capabilities reside, enabling the RNN to handle sequential data with temporal dependencies.

At each time step, the hidden node performs a series of operations.

Receiving Inputs: The node takes in two primary inputs:

(a) The processed data point from the input node.
(b) The hidden state from the previous time step, which carries information about previous data points in the sequence.

Combining Inputs: These inputs are combined, typically using a weighted sum. The weights here are learnable parameters that the network adjusts during training.

Activation Function: The combined input is then passed through an activation function to introduce non-linearity. This non-linearity allows the RNN to model complex relationships in the data. Common activation functions used in RNNs include the hyperbolic tangent ($tanh$) and the sigmoid function.

Producing Outputs: The output of the activation function serves two purposes:

(a) It's used as the RNN's output for the current time step, especially if there's no dedicated output node.
(b) It becomes the hidden state for the next time step, feeding back into the hidden node and carrying forward information.

Let's represent the operations within the hidden node mathematically.

Given:

$$x_t : \text{Input at time step } t. \tag{9.2}$$

$$h_{t-1} : \text{Hidden state from the previous time step.} \tag{9.3}$$

$$W_x : \text{Weight matrix for the input.} \tag{9.4}$$

$$W_h : \text{Weight matrix for the hidden state.} \tag{9.5}$$

$$b : \text{Bias vector.} \tag{9.6}$$

$$\phi : \text{Activation function (e.g., tanh).}$$

The hidden state h_t for the current time step is calculated as follows:

$$h_t = \phi(W_x \cdot x_t + W_h \cdot h_{t-1} + b). \tag{9.7}$$

The hidden or latent state h_t serves as a representation of the memory within the RNN. The state of the network is updated at each time step as it processes a sequence, thereby preserving information pertaining to the sequence's previous occurrences. The inclusion of memory in RNN enables the model to take into account contextual information during the processing of individual data points. This characteristic renders the RNN well-suited for tasks that heavily rely on temporal dependencies.

The distinctive capabilities of RNNs are attributed to the presence of recurrent connections. The hidden state, which is the output of the hidden node, is subsequently inputted into the node at the subsequent time step. The utilization of this loop in the network facilitates the preservation of coherence and consistency across different time steps. This mechanism guarantees that the information obtained from one time step significantly impacts the subsequent steps, thereby influencing the overall processing of the RNN.

Nevertheless, it is important to acknowledge that in vanilla recurrent neural networks, the memory component is relatively limited in its capacity for retaining information over extended periods of time. The network exhibits a bias towards assigning greater significance to recent data points within the sequence, potentially leading to difficulties in retaining information from earlier data points, particularly in instances of lengthy sequences. Advanced recurrent neural network variants, such as Long Short-Term Memory (LSTM) and Gated Recurrent Unit (GRU), incorporate mechanisms to mitigate this constraint.

9.3.1.3 Output Node in RNNs

The output node within a recurrent neural network functions as the terminal point where the network generates an outcome or forecast by leveraging the processed sequential data. The hidden state in the RNN effectively encodes the memory and contextual information of the input sequence. In contrast, the output node of the network serves the purpose of transforming this encoded information into a format that is specifically tailored to the task at hand.

Processing at the Output Node:

The output node's primary role is to transform the information from the hidden state into a format suitable for the task at hand. This transformation involves several steps, each contributing to refining the output.

a. Linear Transformation

The first step usually involves a linear transformation of the hidden state. This transformation scales and shifts the information from the hidden state, preparing it for further processing.

Mathematically, if h_t represents the hidden state at time step t, the raw output o_t can be represented as follows:

$$o_t = W_o \cdot h_t + b_o \tag{9.8}$$

where

- W_o is the weight matrix associated with the output node.
- b_o is the bias vector for the output node.

b. Activation Function

The raw output o_t is then passed through an activation function. The choice of activation function depends on the nature of the task:

- **Regression Tasks**: Often, no activation function is applied, or a linear activation is used, allowing the network to predict a range of values.
- **Binary Classification**: A sigmoid activation function is used, which constrains the output between 0 and 1, representing probabilities.

$$\sigma(x) = \frac{1}{1 + e^{-x}} \tag{9.9}$$

- **Multi-class Classification**: A softmax activation function is applied, converting raw scores into a probability distribution over multiple classes.

$$\text{softmax}(x)_i = \frac{e^{x_i}}{\sum_j e^{x_j}} \tag{9.10}$$

c. Decoding (if applicable)

In some tasks, especially those involving discrete outputs like word predictions, the processed output needs to be decoded:

- **Word Prediction**: The output, after softmax, is a probability distribution over a vocabulary. The word corresponding to the highest probability can be selected as the predicted next word.

$$w^* = \arg\max_i (o_t)_i \tag{9.11}$$

where $(o_t)_i$ is the probability of the ith word in the vocabulary, and arg max returns the index with the maximum value.

The output node of a recurrent neural network functions as the final stage of the network's computational process. It transforms the comprehensive and contextual information derived from the hidden state into predictions that are relevant and applicable to the specific task at hand. Within the domain of Neuro-Symbolic AI, the output node assumes a crucial function in facilitating the connection between the continuous and distributed representations of neural networks and the discrete and structured representations of symbolic systems.

Symbolic Logic in the Output Node:

Symbolic logic is all about structured, rule-based reasoning. In the context of the output node:

Rule-based Decoding: The output node can incorporate rule-based logic to refine its predictions. For instance, in a language translation task, the output node might use grammar rules to ensure that the predicted sentence is syntactically correct.

Constraints and Priors: Symbolic systems often operate with a set of constraints or priors. The output node can leverage these constraints to guide its predictions. For example, in a music generation task, the output node might use musical theory rules to ensure that the generated sequence is harmonically pleasing.

Interfacing with Symbolic Systems: The predictions from the output node can be used as inputs to other symbolic systems. For instance, in a robotic control application, the RNN might predict a series of actions, which are then fed into a symbolic planner to generate a plan for the robot.

For example, let us contemplate a multifaceted undertaking such as comprehending and responding to inquiries pertaining to a scientific passage. The recurrent neural network has the ability to analyze the text by capturing its subtleties and contextual information within its hidden states. However, providing precise responses to inquiries regarding the text may necessitate the utilization of symbolic reasoning, such as comprehending the interconnections between causes and effects or the hierarchical arrangement of ideas. The role of the output node is crucial in this context as it facilitates the conversion of the neural comprehension of the text into symbolic entities and relationships that are amenable to querying and logical reasoning.

Moreover, with the increasing complexity of AI systems and their anticipated interaction with various domains, the necessity for such integration becomes even more evident. A system may be required to undertake the processing of visual data, comprehend natural language instructions, and subsequently execute a sequence of actions within either a physical or virtual setting. The successful execution of each of these steps necessitates a combination of neural processing and symbolic reasoning. In the given scenarios, the output node functions as a mediator, guaranteeing that the outputs from one step are appropriately formatted to serve as inputs for the subsequent step.

In summary, the function of the output node within a Neuro-Symbolic AI framework extends beyond the mere generation of predictions or outcomes. The purpose of this is to guarantee that the outcomes are presented in a manner that is suitable for both the neural and symbolic elements of the system, thereby enabling smooth integration and comprehensive processing. The advancement of Neuro-Symbolic AI research is expected to result in a greater emphasis on the role and significance of the output node in achieving comprehensive integration of artificial intelligence.

9.3.2 Temporal Unfolding in RNNs

The fundamental concept of temporal unfolding revolves around the representation of the repetitive nature of recurrent neural networks in an extended format. Instead of perceiving the network as a singular entity that sequentially processes data in a loop, we opt to "nfold" it temporally, thereby representing each iteration of the loop as an individual step. The aforementioned structure exhibits a sequential arrangement, wherein each element within the sequence symbolizes the operations performed by the RNN at a particular moment in time.

Algorithm 9.1 Temporal Unfolding in RNNs

procedure RNNUNFOLD($X, W_x, W_h, W_o, b_h, b_o$)
  Initialize h_0 to zeros or a predefined state
  for $t = 1$ to T **do** ▷ For each time step in sequence
    $x_t \leftarrow X[t]$ ▷ Input at time step t
    $h_t = \phi(W_x \cdot x_t + W_h \cdot h_{t-1} + b_h)$ ▷ Compute new hidden state
    $y_t = W_o \cdot h_t + b_o$ ▷ Compute output for time step t
  end for
  return $y_1, y_2, \ldots, y_T$ ▷ Return outputs for all time steps
end procedure

Algorithm 9.2 Backpropagation Through Time (BPTT)

procedure BPTT($X, Y, W_x, W_h, W_o, b_h, b_o$)
  Initialize h_0 to zeros or a predefined state
  for $t = 1$ to T **do** ▷ Forward Pass
    $x_t \leftarrow X[t]$
    $h_t = \phi(W_x \cdot x_t + W_h \cdot h_{t-1} + b_h)$
    $y_t = W_o \cdot h_t + b_o$
  end for
  Compute loss L using y_t and target Y
  for $t = T$ to 1 **do** ▷ Backward Pass
    Compute gradient of L w.r.t. y_t
    Compute gradient of L w.r.t. h_t
    Accumulate gradients for W_x, W_h, W_o, b_h, b_o
    Propagate error to h_{t-1}
  end for
  Update weights using accumulated gradients
end procedure

Instead of viewing the RNN as a looped structure, we "stretch" it across T time steps. This gives us a chain of RNN cells $y_1, y_2, \ldots, y_T$, one for each time step, as shown in Algorithm 9.1.

The depicted structure exhibits a chain-like configuration that serves as a visual representation of the information propagation within the network. The significance lies in the manner in which the concealed state functions as a form of memory, facilitating the transmission of information from one temporal iteration to the subsequent one. The propagation of the hidden state throughout the unfolded RNN cells signifies the network's memory. The retention of information from previous inputs in the RNN enables the model to incorporate historical context while processing novel data.

The unfolding perspective additionally offers enhanced comprehension of the training procedure. During the process of adjusting the weights of the network, the error is not only propagated through the layers but also through the time steps. The process referred to as Backpropagation Through Time (BPTT) entails the computation of gradients for every time step within the unfolded sequence, followed by the subsequent adjustment of the weights based on these gradients. This is shown in Algorithm 9.2.

9.3.3 Challenges with Vanilla RNNs

Recurrent Neural Networks have been a cornerstone in the realm of sequential data processing. However, the basic or "vanilla" RNNs come with their own set of challenges that can hinder their performance, especially when dealing with long sequences or complex dependencies.

1. Vanishing and Exploding Gradient Problem

One of the most prominent challenges with vanilla RNNs is the vanishing and exploding gradient problem. This issue arises during the training phase, specifically during backpropagation through time (BPTT).

Vanishing Gradient: As the sequence length increases, the gradients of the loss with respect to the weights can become exceedingly small. This means that the weights of the network, especially those that connect the recurrent hidden states, receive very tiny updates. Over time, this leads to a scenario where the network stops learning or learns very slowly. The vanishing gradient problem is particularly troublesome when trying to capture long-term dependencies in sequences, as the influence of past inputs diminishes rapidly.

Exploding Gradient: Conversely, there are situations where the gradients can grow exponentially as they are propagated backward through time. This leads to very large weight updates, causing the model to diverge and making training unstable. In practice, when an exploding gradient is detected, techniques like gradient clipping are used to prevent the weights from updating too drastically.

The root cause of these problems lies in the recurrent nature of the RNNs. When the same weight matrix is applied repeatedly over time steps, small eigenvalues in the weight matrix can lead to vanishing gradients, while large eigenvalues can cause exploding gradients.

2. Short-term Memory Limitations

Vanilla RNNs inherently have a form of memory, given by their hidden states. However, this memory is often short-lived. This means that while RNNs can remember recent inputs or events in a sequence, they struggle to retain information from distant past inputs.

Why it's a Problem: Many tasks, such as language modeling or time-series prediction, require the model to understand and remember long-term dependencies. For instance, in a sentence, the meaning of a word might be influenced by another word that appeared several words before it. Vanilla RNNs find it challenging to capture such relationships due to their short-term memory limitations.

Cause: The primary reason for this limitation is the vanishing gradient problem. As gradients vanish during backpropagation, the influence of past inputs diminishes, making it hard for the RNN to retain long-term information.

Given these challenges, scholars have put forth more sophisticated recurrent neural network (RNN) structures, such as Long Short-Term Memory (LSTM) and Gated Recurrent Units (GRU). These architectural designs incorporate mechanisms to address the issue of the vanishing gradient problem and improve the memory capacity of the network, enabling it to capture and model long-term dependencies in the data.

9.4 Long Short-Term Memory (LSTM) Networks

The Long Short-Term Memory (LSTM) networks were first proposed by Hochreiter and Schmidhuber in 1997 as a variant of Recurrent Neural Networks. The primary objective of their design was to specifically target the limitations observed in vanilla RNNs, with a particular emphasis on mitigating the vanishing gradient problem. LSTM models have emerged as a fundamental component in the domain of deep learning due to their remarkable capability to effectively manage and capture long-term dependencies within sequential data.

9.4.1 Architecture and Components of LSTMs

LSTM networks, while being a type of RNN, have a distinct internal structure that sets them apart from vanilla RNNs. This structure is designed to help LSTMs remember patterns over long sequences, making them particularly effective for tasks that involve long-term dependencies.

9.4.1.1 Core Components of LSTMs

The LSTM cell, a specialized unit designed to capture long-term dependencies in sequences, comprises several intricately connected components. These components work in tandem to regulate the flow of information, ensuring that the LSTM can remember and forget details as needed.

Cell State (C):

The cell state, often denoted as C, is a vector that runs through the entire LSTM cell, from start to end. It can be visualized as a horizontal line at the top of the cell in most LSTM diagrams. The cell state carries information across time steps, ensuring that the LSTM retains relevant details and forgets the unnecessary ones. The primary role of the cell state is to store and transport information across time steps. However, its contents can be modified by the LSTM's gates, which can add, remove, or update information.

Gates in LSTMs:

Gates are fundamental to the LSTM's architecture, regulating the flow of information through the cell. Each gate uses a sigmoid activation function, ensuring its output values range between 0 (block all information) and 1 (allow all information). Here's a detailed breakdown:

1. Input Gate (i)

The input gate decides which values from the input should be updated or added to the cell state. It uses a sigmoid activation to determine the updated scale and a tanh activation to create a vector of new candidate values. The combination of these determines the final information that gets added to the cell state. Mathematically, Update scale:

$$i_t = \sigma(W_i \cdot [h_{t-1}, x_t] + b_i) \tag{9.12}$$

And new candidate values:

$$\tilde{C}_t = \tanh(W_C \cdot [h_{t-1}, x_t] + b_C) \tag{9.13}$$

where σ is the sigmoid function, W_i and W_C are weight matrices, b_i and b_C are bias vectors, h_{t-1} is the previous hidden state, and x_t is the current input.

2. Forget Gate (f)

The forget gate decides which parts of the cell state should be discarded or kept. It looks at the current input and the previous hidden state and outputs a number between 0 and 1 for each number in the cell state. A 1 represents "completely keep this" while a 0 represents "completely get rid of this." Mathematically,

$$f_t = \sigma(W_f \cdot [h_{t-1}, x_t] + b_f) \tag{9.14}$$

where W_f is the weight matrix for the forget gate and b_f is the bias vector for the forget gate.

3. Output Gate (o)

The output gate controls how much of the information in the cell state is used to determine the output of the LSTM cell. It decides what parts of the cell state make it to the hidden state (output) based on the current input and the previous hidden state. Mathematically,

$$o_t = \sigma(W_o \cdot [h_{t-1}, x_t] + b_o) \tag{9.15}$$

where W_o is the weight matrix for the output gate and b_o is the bias vector for the output gate.

9.4.2 Flow of Information in LSTMs

The information flow within LSTM networks is carefully crafted to enable the network to make well-informed choices regarding the retention of information, the discarding of irrelevant information, and the generation of appropriate outputs. The ability of LSTMs to effectively manage long-term dependencies in sequences is facilitated by the control exerted by the gates and the cell state. The flow of information is depicted as Algorithm 9.3.

When processing a sequence, the LSTM cell receives an input vector x_t, the previous hidden state h_{t-1}, and the previous cell state C_{t-1}. The LSTM cell then undergoes several stages to produce the current hidden state h_t and the updated cell state C_t.

Algorithm 9.3 Flow of Information in LSTM Cell

procedure LSTMFLOW($x_t, h_{t-1}, C_{t-1}, W_*, b_*$)
  Forget Gate's Decision:
  $f_t \leftarrow \sigma(W_f \cdot [h_{t-1}, x_t] + b_f)$
  $C_t' \leftarrow f_t \odot C_{t-1}$ ▷ Discard information in cell state based on f_t
  Input Gate's Decision:
  $i_t \leftarrow \sigma(W_i \cdot [h_{t-1}, x_t] + b_i)$
  $\tilde{C}_t \leftarrow \tanh(W_C \cdot [h_{t-1}, x_t] + b_C)$
  $C_t \leftarrow C_t' + i_t \odot \tilde{C}_t$ ▷ Update cell state with new information
  Output Gate's Decision:
  $o_t \leftarrow \sigma(W_o \cdot [h_{t-1}, x_t] + b_o)$
  $h_t \leftarrow o_t \odot \tanh(C_t)$ ▷ Decide and output the hidden state for this timestep
  return h_t, C_t
end procedure

1. Forget Gate's Decision

The first step involves the forget gate. This gate looks at the current input x_t and the previous hidden state h_{t-1} to decide which parts of the cell state C_{t-1} should be retained or discarded. The output of the forget gate, f_t, is a vector with values between 0 and 1. A value close to 1 means "retain this information," while a value close to 0 means "discard this information." The intermediate cell state C_t' is then computed by element-wise multiplying f_t with C_{t-1}.

2. Input Gate's Decision

Next, the LSTM cell determines what new information should be stored in the cell state. This decision is made in two parts. First, a sigmoid layer in the input gate decides which values in the cell state should be updated, producing a vector i_t. Concurrently, a tanh layer creates a vector of new candidate values, $\tilde{C}_t$, that could be added to the state. The cell state is then updated by adding the element-wise product of i_t and $\tilde{C}_t$ to C_t', resulting in the updated cell state C_t.

3. Output Gate's Decision

Finally, the LSTM cell determines its output, which is the hidden state h_t. The output gate looks at the current input x_t and the previous hidden state h_{t-1} to decide how much of the cell state C_t should be exposed as the output. The output of the output gate, o_t, is a vector with values between 0 and 1. The hidden state h_t is then computed by taking the element-wise product of o_t and the tanh activation of C_t.

9.4.3 Advantages of LSTMs over Vanilla RNNs in the Context of Neuro-Symbolic AI

Recurrent Neural Networks have significantly transformed the landscape of sequence modeling tasks, yet they encounter inherent difficulties, particularly when dealing with lengthy sequences. Long Short-Term Memory Networks were devised as a means to address these aforementioned challenges. LSTM models possess a distinct advantage over Vanilla RNNs, particularly when considering their application within the context of Neuro-Symbolic AI.

The utilization of LSTM models is highly effective in managing and leveraging information from inputs that occurred in the distant past. Although Vanilla RNNs encounter difficulties caused by the vanishing gradient problem, LSTMs, equipped with their specialized gating mechanisms, effectively retain important information for longer periods of time. The possession of this ability is crucial for tasks such as symbolic reasoning, wherein the significance of context and the sequence of operations or instructions holds importance.

The vanishing and exploding gradient problems pose challenges in the training of deep RNNs. LSTM networks, characterized by their specific architectural design, effectively regulate the propagation of gradients, thereby guaranteeing both stability and efficiency during the training process. The maintenance of stability is of utmost importance when integrating neural networks with symbolic systems, as it guarantees the optimal training of the neural component.

The utilization of LSTM models enables the incorporation of flexible memory management through the implementation of gates. These gates facilitate the dynamic determination of whether to store, retrieve, or delete information. The quality of adaptability is especially advantageous when interacting with symbolic systems, as these systems frequently necessitate the retention or elimination of specific information based on symbolic rules or logic.

LSTMs consistently exhibit superior performance compared to Vanilla RNNs when it comes to handling complex tasks. The capacity to effectively comprehend intricate patterns renders them highly suitable for tasks that necessitate a combination of neural learning and symbolic reasoning, such as program synthesis or logic puzzle solving.

The potential for integration between LSTM models and symbolic systems is considered a promising aspect within the field of Neuro-Symbolic Artificial Intelligence. The memory and gating mechanisms employed by LSTM networks can be regarded as a means of connecting the neural and symbolic domains. The neural component, specifically the LSTM model, is responsible for capturing patterns and acquiring representations. On the other hand, the symbolic component leverages this acquired information to engage in logical reasoning, rule-based deductions, and the formation of high-level abstractions. The integration of neural and symbolic methodologies enables a comprehensive problem-solving approach that leverages the strengths of both paradigms.

9.5 Gated Recurrent Units (GRUs)

The Gated Recurrent Units (GRUs) represent a specific form of RNN structure that was initially proposed by Cho et al. in 2014. The purpose of their design is to address comparable issues encountered by LSTMs, specifically the vanishing gradient problem observed in conventional RNNs, albeit with a more efficient architectural framework. GRUs have garnered significant attention in a range of sequence modeling applications owing to their notable computational efficiency and straightforward architecture.

The primary distinction between GRUs and LSTMs is the reduced complexity in the GRU's design. While LSTMs have three gates (input, forget, and output), GRUs simplify this with just two gates:

Update Gate: This gate determines how much of the previous hidden state should be passed to the current state. It's a blend between the forget and input gates of the LSTM.

Reset Gate: This gate decides how much of the previous hidden state should be forgotten. It allows the model to drop information that is no longer required.

Another significant difference is the absence of a cell state in GRUs. Instead, GRUs use a single mechanism to maintain their hidden state, making them more straightforward and often faster to compute. Algorithm 9.4 presents the GRU architecture.

Algorithm 9.4 Gated Recurrent Units (GRU) Architecture

procedure GRU(x_t, h_{t-1}, W_*, b_*)
  Update Gate:
  $z_t = \sigma(W_z \cdot [h_{t-1}, x_t] + b_z)$ ▷ Decides how much of the past information to keep.
  Reset Gate:
  $r_t = \sigma(W_r \cdot [h_{t-1}, x_t] + b_r)$ ▷ Decides how much past information to forget.
  Candidate Activation:
  $\tilde{h}_t = \tanh(W \cdot [r_t \odot h_{t-1}, x_t] + b)$ ▷ Computes the new memory content without considering the past information.
  Final Hidden State:
  $h_t = (1 - z_t) \odot h_{t-1} + z_t \odot \tilde{h}_t$ ▷ Combines the past information and the new memory content.
  return h_t
end procedure

Variables:

- x_t: Input vector at time t.
- h_{t-1}: Hidden state from the previous time step.
- W_*: Weight matrices associated with the respective gates.
- b_*: Bias terms associated with the respective gates.
- z_t: Update gate. It decides how much of the past hidden state h_{t-1} should be passed to the current state.
- r_t: Reset gate. It determines how much of the past hidden state is forgotten.
- $\tilde{h}_t$: Candidate activation. It represents the new memory content without considering the past information.
- h_t: Final hidden state for time t. It's a combination of the past information and the new memory content.

Comparing GRUs and LSTMs:

Complexity: LSTMs, with their three gates and a separate cell state, are inherently more complex than GRUs. This complexity can sometimes lead to better performance on certain tasks, but it also means increased computational cost and more parameters to train.

Performance: In many tasks, especially those with shorter sequences, GRUs have been found to perform on par with LSTMs while being more computationally efficient. However, for tasks with longer sequences or those requiring more intricate memory capabilities, LSTMs might have an edge.

Training Time: Due to their simplified architecture, GRUs often train faster than LSTMs. This can be particularly beneficial in projects with tight timelines or when computational resources are limited.

Memory Usage: GRUs generally have fewer parameters than LSTMs, leading to less memory consumption. This makes them more suitable for devices with limited memory or for applications where model size is a concern.

9.6 Advanced Topics in RNNs

9.6.1 Bidirectional RNNs

Bidirectional recurrent neural networks (BiRNNs) [5] represent an advancement over conventional RNNs and are specifically engineered to enhance performance on tasks that heavily rely on contextual information from both preceding and subsequent sequences. In contrast to conventional RNNs that transmit information unidirectionally, from the previous time steps to the subsequent ones, BiRNNs employ two distinct hidden layers to process data in both forward and backward directions. One layer progresses chronologically, acquiring historical context, while the other layer regresses chronologically, acquiring future context. The outputs of the two layers are subsequently combined at each time step, resulting in a more comprehensive representation that incorporates information from both preceding and succeeding time points.

The bidirectional context is highly valuable in tasks such as machine translation, as it enables the understanding of a word by considering the context provided by both the preceding and following words. In the given example sentence, "The cat, which was black, sat on the mat," the comprehension of the term "was" necessitates familiarity with both the preceding context, which pertains to the subject "cat," and the subsequent context, which relates to the attribute "black."

Variables:

- X: Input sequence of length T.
- W_{f*}, W_{b*}: Weight matrices for the forward and backward RNNs, respectively.
- b_{f*}, b_{b*}: Bias terms for the forward and backward RNNs, respectively.

Algorithm 9.5 Bidirectional RNNs

procedure BiRNN($X, W_{f*}, W_{b*}, b_{f*}, b_{b*}$)
  for $t = 1$ to T **do** ▷ Forward Pass
    $h_{f_t} = \phi(W_{fx} \cdot X[t] + W_{fh} \cdot h_{f_{t-1}} + b_f)$
  end for
  for $t = T$ to 1 **do** ▷ Backward Pass
    $h_{b_t} = \phi(W_{bx} \cdot X[t] + W_{bh} \cdot h_{b_{t+1}} + b_b)$
  end for
  for $t = 1$ to T **do** ▷ Combining Outputs
    $h_t = [h_{f_t}, h_{b_t}]$
  end for
  return h
end procedure

- h_{f_t}, h_{b_t}: Hidden states at time t for the forward and backward RNNs, respectively.
- h_t: Combined hidden state at time t, obtained by concatenating h_{f_t} and h_{b_t}.
- ϕ: Activation function, typically the hyperbolic tangent function.

Algorithm 9.5 provides a structured presentation of the BiRNN network architecture.

9.6.2 Deep RNNs

Deep recurrent neural networks (Deep RNNs) [6] expand upon the concept of deep learning by incorporating recurrent networks. Deep RNNs are characterized by the stacking of multiple recurrent layers, as opposed to the utilization of a single layer of recurrence. This facilitates the acquisition of hierarchical representations of the data, wherein each layer has the potential to capture distinct levels of abstraction.

In the context of a language modeling task, it is possible to observe that the initial layer of the model could potentially capture fundamental syntactic structures. Subsequently, the subsequent layer could exhibit an understanding of phrases, while a third layer could potentially comprehend more intricate sentence structures. By increasing the complexity of the network architecture, it becomes capable of capturing more intricate and nuanced patterns and relationships within the dataset, which frequently results in enhanced performance. Nevertheless, the process of training Deep RNNs can present significant challenges, primarily stemming from problems such as vanishing and exploding gradients [7]. In order to address these issues, additional techniques such as gradient clipping or specialized initializations may need to be employed.

Variables:

- X: Input sequence of length T.
- W_*: Weight matrices for each layer and connection (input and recurrent).
- b_*: Bias terms for each layer.

Algorithm 9.6 Deep RNNs

```
1: procedure DEEPRNN(X, W_*, b_*, L)
2:     h_t^(0) = X[t]                                          ▷ Initial input
3:     for l = 1 to L do                                       ▷ For each layer
4:         for t = 1 to T do                                   ▷ Time step iteration
5:             h_t^(l) = φ(W_x(l) · h_t^(l-1) + W_h(l) · h_{t-1}^(l) + b^(l))
6:         end for
7:     end for
8:     return h^(L)                                            ▷ Return output of the last layer
9: end procedure
```

- $h_t^{(l)}$: Hidden state at time t for the lth layer.
- L: Total number of layers in the deep RNN.
- ϕ: Activation function, typically the hyperbolic tangent function.

Algorithm 9.6 provides a structured presentation of the Deep RNN network architecture.

9.6.3 Attention Mechanisms

Attention mechanisms [8] have emerged as a fundamental component in the field of sequence-to-sequence models, particularly in applications such as machine translation. When encoding a sequence into a fixed-size vector, traditional Recurrent Neural Networks (RNNs) encounter difficulties in effectively capturing the intricate details of longer sequences. Attention mechanisms are employed to tackle this issue by enabling the model to selectively concentrate on various segments of the input sequence during the generation of an output, rather than solely relying on a predetermined context.

Fundamentally, attention serves the purpose of imparting a contextual significance to the model by assigning weights that determine the relative importance of different segments within the input sequence in generating a specific output. When translating the given sentence "She loves reading books" into a different language, the attention mechanism may assign greater importance to both "loves" and "books" while translating the term "reading." This is done in order to capture the contextual meaning accurately.

One notable advantage of attention mechanisms is their ability to enable the model to adaptively allocate its attention based on the information it is processing. This adaptability leads to notable enhancements in tasks that necessitate comprehension of long-term dependencies or intricate relationships within the data.

The attention mechanism of graph neural networks will be thoroughly examined in the forthcoming chapter.

9.7 Role of RNNs in Neuro-Symbolic AI

Neuro-Symbolic AI represents an intersection of two dominant paradigms in artificial intelligence: neural networks (the "neuro" part) and symbolic reasoning (the "symbolic" part). While neural networks, including RNNs, excel at handling raw data, learning patterns, and generalizing from examples, symbolic systems are adept at explicit reasoning, handling structured knowledge, and following predefined rules.

Recurrent Neural Networks play a pivotal role in the neuro-symbolic paradigm, particularly in bridging the gap between raw data processing and structured symbolic reasoning. Here's how.

Handling Sequential Data: One of the primary strengths of RNNs is their ability to process sequential data. In the realm of symbolic AI, sequences can represent chains of reasoning, ordered sets of instructions, or temporal patterns in data. RNNs can capture these sequences and transform them into a form suitable for symbolic processing.

Embedding Symbolic Knowledge: RNNs can be trained to embed symbolic knowledge within their weights and activations. For instance, an RNN trained on a language task can implicitly learn grammar rules, which are inherently symbolic. This embedded knowledge can then be extracted or interfaced with explicit symbolic systems for more advanced reasoning.

Temporal Reasoning: Neuro-symbolic systems often need to reason about events that occur over time. RNNs, with their inherent memory and sequential processing capabilities, can capture temporal dependencies, making them invaluable for tasks that require understanding the order or timing of events.

Bridging Low-level Perception and High-level Reasoning: In many AI tasks, there's a need to transition from low-level perception (like recognizing words in a sentence) to high-level reasoning (like understanding the meaning of the sentence). RNNs can act as this bridge, processing raw data and producing representations that symbolic systems can reason about.

Enhancing Symbolic Models with Context: Traditional symbolic models can be rigid, lacking the ability to adapt to context or nuances in data. RNNs, with their adaptive nature, can provide context to symbolic models, allowing them to make more nuanced decisions or reasonings.

Integration with Other Neural Architectures: RNNs can be seamlessly integrated with other neural architectures, like CNNs for image data or attention mechanisms for enhanced sequence processing. This integration allows neuro-symbolic systems to handle a diverse range of tasks, from image captioning (combining vision and language) to rule-based reasoning on visual data.

9.8 Conclusion

As we draw this chapter to a close, it's evident that Recurrent Neural Networks (RNNs) have carved a unique niche in the landscape of neural architectures. Their inherent ability to process sequences, remember past information, and predict future events has made them indispensable in a myriad of applications, from natural language processing to financial forecasting. The intricate dance of sequences and patterns that RNNs navigate is not just a testament to their design but also to the evolving nature of AI research that continually seeks to mimic and understand human cognition.

Throughout this chapter, we've journeyed from the foundational concepts of RNNs, delving into their architecture, challenges, and advanced variants. We've seen how they bridge the gap between raw data and symbolic reasoning, setting the stage for a deeper exploration into the world of Neuro-Symbolic AI. This integration of neural processing and symbolic logic is not just a fusion of two AI paradigms but a step towards building systems that can learn from data and reason like humans.

As we transition to the upcoming chapters, our focus will shift to Neuro-Symbolic Integration techniques. While RNNs have provided us with a robust tool for sequence modeling, the next frontier lies in seamlessly integrating neural learning with symbolic reasoning. This integration promises AI systems that are not only powerful pattern recognizers but also adept reasoners, capable of understanding, explaining, and generalizing in ways that are more aligned with human intelligence.

References

1. Pascanu, R., Gulcehre, C., Cho, K., Bengio, Y.: How to construct deep recurrent neural networks. arXiv preprint arXiv:1312.6026 (2013)
2. Salehinejad, H., Sankar, S., Barfett, J., Colak, E., Valaee, S.: Recent advances in recurrent neural networks. arXiv preprint arXiv:1801.01078 (2017)
3. Sherstinsky, A.: Fundamentals of recurrent neural network (rnn) and long short-term memory (lstm) network. Physica D: Nonlinear Phenomena **404**, 132306 (2020)
4. Chung, J., Gulcehre, C., Cho, K., Bengio, Y.: Empirical evaluation of gated recurrent neural networks on sequence modeling. arXiv preprint arXiv:1412.3555 (2014)
5. Schuster, M., Paliwal, K.K.: Bidirectional recurrent neural networks. IEEE Trans. Signal Process. **45**(11), 2673–2681 (1997)
6. Graves, A., Mohamed, A.r., Hinton, G.: Speech recognition with deep recurrent neural networks. In: 2013 IEEE International Conference on Acoustics, Speech and Signal Processing, pp. 6645–6649. IEEE (2013)
7. Mikhaeil, J., Monfared, Z., Durstewitz, D.: On the difficulty of learning chaotic dynamics with RNNS. Adv. Neural Inf. Process. Syst. **35**, 11297–11312 (2022)
8. Niu, Z., Zhong, G., Yu, H.: A review on the attention mechanism of deep learning. Neurocomputing **452**, 48–62 (2021)

Part IV
Neuro-Symbolic Integration Approaches

Chapter 10
Overview of Neuro-Symbolic Integration Frameworks

"Artificial Intelligence is about making machines think like humans. Neuro-Symbolic AI is about making them think logically like Spock and intuitively like Kirk."

10.1 Introduction

Imagine a self-driving car navigating through a busy city. It must not only recognize traffic signs and pedestrians through its sensors and cameras (a task well-suited for neural networks) but also make high-level decisions about routes and driving strategies based on traffic laws and ethical considerations (a task that requires symbolic reasoning). This real-world scenario underscores the need for AI systems that can both learn from data and reason symbolically. Welcome to the world of Neuro-Symbolic AI, a burgeoning field that aims to integrate the robust learning capabilities of neural networks with the logical reasoning power of symbolic systems.

The objective of this chapter is to provide a comprehensive overview of the frameworks and methodologies that are shaping the future of Neuro-Symbolic AI. We delve into Henry Kautz's taxonomy, which offers a structured way to understand the different types of neuro-symbolic systems, each with its own set of advantages and limitations. This taxonomy serves as a guide to the intricate landscape of neuro-symbolic integration, helping us understand how these systems can be designed, what they can achieve, and where they might fall short.

But understanding the types of neuro-symbolic systems is just the beginning. We also introduce a new perspective that focuses on the algorithmic and application-level considerations in Neuro-Symbolic AI. This perspective is crucial for evaluating the

B. P. Bhuyan et al., *Neuro-Symbolic Artificial Intelligence*, Studies in Computational Intelligence 1176, https://doi.org/10.1007/978-981-97-8171-3_10

effectiveness of these systems in real-world applications, from healthcare and finance to autonomous vehicles and robotics.

By the end of this chapter, you will have a well-rounded understanding of the current state of Neuro-Symbolic AI, the challenges it faces, and the promising directions in which it is headed. Whether you are a researcher, a practitioner, or simply someone interested in the future of AI, this chapter aims to equip you with the knowledge and insights needed to engage with this exciting and rapidly evolving field.

10.2 Henry Kautz's Taxonomy for Neuro-Symbolic AI

Henry Kautz, a prominent figure in the field of Artificial Intelligence, has proposed a taxonomy that categorizes various approaches to neuro-symbolic integration [1]. This taxonomy serves as a comprehensive framework for understanding the different methodologies that aim to bridge neural networks and symbolic reasoning systems.

The Taxonomy has six types.

Type 1: **Symbolic Neuro symbolic**: This is the standard approach for natural language processing. Symbolic inputs are converted to vectors by methods like word2vec [2] and then passed to a neural network. The output is then converted back to a symbolic category.

Type 2: **Symbolic[Neuro]**: This architecture employs a neural pattern recognition subroutine within a symbolic problem solver. AlphaGo [3] is a prime example, where the problem solver is the Monte-Carlo Tree Search [4] algorithm, and its heuristic evaluation function is a neural network.

Type 3: **Neuro|Symbolic**: In this system, a neural network converts non-symbolic input, like image pixels, into a symbolic data structure. This is then processed by a symbolic reasoning system. The Neuro-Symbolic Concept Learner is an example of this design.

Type 4: **Neuro: Symbolic → Neuro**: This approach uses a special training regime based on symbolic rules.

Type 5: $Neuro_{Symbolic}$: This architecture transforms symbolic rules into templates for structures within the neural network. Tensor product representations [5] and logic tensor networks [6] have been demonstrated for building abstraction and part of hierarchies into the network.

Type 6: **Neuro[Symbolic]**: This is the most promising approach according to Kautz. It aims to embed a symbolic reasoning engine inside a neural engine. The architecture is based on Daniel Kahneman's theory of "thinking fast and slow" [7], which states that the brain implements two distinct mechanisms for reasoning.

10.2.1 Type 1: Symbolic Neuro Symbolic

The "Symbolic Neuro symbolic" architecture is a fundamental approach within neuro-symbolic AI, primarily used in natural language processing (NLP) and other applications that require a seamless integration of symbolic reasoning and neural network-based learning. This architecture is designed to convert symbolic inputs into numerical vectors, process them through a neural network, and then convert the outputs back into symbolic forms. This section delves into the components, mechanisms, advantages, and limitations of this architecture.

The main components of this architecture (shown in Fig. 10.1) are as follows:

1. **Symbolic Input Layer**: The Symbolic Input Layer serves as the entry point for all symbolic data into the system. In the context of natural language processing, this could be a sequence of words or sentences that need to be processed. This layer is crucial for capturing the initial state of the data in its raw, symbolic form. It sets the stage for the subsequent transformation and processing steps, ensuring that the symbolic reasoning aspects of the task are adequately represented.
   In the context of large language models (LLMs) such as GPT-4 [8], this layer would take in lengthy documents and preprocess them into manageable symbolic chunks (e.g., sentences or paragraphs) that retain the essential information while discarding redundant or irrelevant details. In systems designed for question answering, such as those integrating with LLMs like BERT [9] or GPT-4, the Symbolic Input Layer would parse the question and related context, identifying

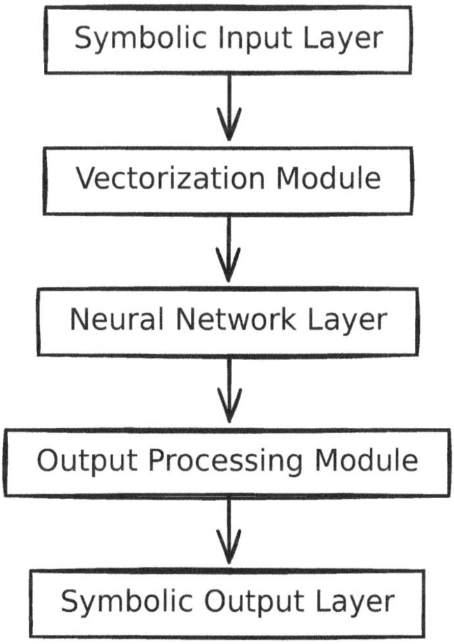

Fig. 10.1 Components of type 1: symbolic neuro symbolic

key symbolic elements (e.g., entities, actions) that need to be processed to generate a coherent answer. In applications like GitHub Copilot [10], the Symbolic Input Layer could process snippets of code and comments, translating them into a form that the neural network can understand and use to predict subsequent lines of code or suggest improvements.

2. **Vectorization Module**: Following the Symbolic Input Layer is the Vectorization Module, playing a pivotal role in transforming symbolic inputs into numerical vectors that can be processed by neural networks. This conversion is essential because neural networks require numerical data to function effectively. The quality of the vectorization process has a significant impact on the system's overall performance. Various techniques are employed to achieve this transformation, each with its strengths and specific applications.
   One of the foundational techniques is the Bag-of-Words (BoW) model [11], which represents text by converting it into a vector of word frequencies. Despite its simplicity and wide use in document classification and sentiment analysis, BoW does not capture the context or semantics of words. To address this limitation, the Term Frequency-Inverse Document Frequency (TF-IDF) method [12] is used. TF-IDF enhances the BoW approach by weighing words based on their frequency across the corpus, thus highlighting significant terms that distinguish documents more effectively.
   More sophisticated methods like Word2Vec and GloVe [9] are employed to capture semantic relationships between words. Word2Vec, developed by Google, generates dense word embeddings by predicting words within a context window. It effectively captures semantic similarities, allowing words used in similar contexts to have similar vectors. GloVe, on the other hand, creates embeddings by leveraging global word-word co-occurrence statistics from a corpus. This dual approach combines the strengths of local context methods and global statistical information, making it highly effective for various NLP tasks such as named entity recognition and machine translation.
   The advent of transformer models like BERT (Bidirectional Encoder Representations from Transformers) [9] has revolutionized text vectorization. BERT generates contextualized word embeddings that understand the context of words in both directions, providing deep and nuanced representations. This capability makes BERT particularly suitable for complex NLP tasks, including question answering, text summarization, and translation, where understanding context and semantics is crucial.
3. **Neural Network Layer**: The Neural Network Layer is the core computational unit of the architecture. Once the symbolic inputs are converted into numerical vectors by the Vectorization Module, they are fed into this layer for processing. This layer is where the "learning" happens, as the network adjusts its internal parameters during training to minimize the difference between its predictions and the actual outputs.
   Depending on the complexity of the task and the specific requirements, various types of neural networks can be employed. As seen in the previous chapters, Feed-Forward Neural Networks (FFNNs) are the simplest type, consisting of an

input layer, one or more hidden layers, and an output layer. For tasks involving spatial hierarchies, such as image and video recognition, Convolutional Neural Networks (CNNs) are particularly effective. CNNs apply filters to the input data to detect features, reduce dimensionality through pooling layers, and map the learned features to the output through fully connected layers. This architecture excels at capturing local dependencies in data, making CNNs the go-to choice for image classification, medical image analysis, and even certain NLP tasks.
When dealing with sequential data, Recurrent Neural Networks (RNNs) and their advanced variant, Long Short-Term Memory Networks (LSTMs), come into play. RNNs maintain a "memory" of previous inputs through internal states, which is crucial for tasks involving time series, text, and other sequential data. LSTMs, specifically designed to handle long-term dependencies, use memory cells to retain information over extended sequences, making them ideal for language modeling, speech recognition, and machine translation.
Transformers, particularly models like BERT, represent a significant advancement in neural network architecture, especially for NLP tasks. Unlike RNNs, transformers do not require sequential data processing, allowing for parallelization and greater efficiency. They use self-attention mechanisms to weigh the influence of different words in a sentence when producing a representation of a word. This ability to focus on relevant parts of the input sequence makes transformers highly effective at capturing long-range dependencies and context.

4. **Output Processing Module**: After the neural network layer comes the Output Processing Module. This component is tasked with converting the numerical outputs of the neural network back into a symbolic form. This is often achieved through a softmax operation, which converts the raw output of the neural network into a probability distribution over possible symbolic categories or sequences. This step is crucial for tasks that require a symbolic output, such as text classification or machine translation, as it bridges the gap between the neural and symbolic components of the system.
5. **Symbolic Output Layer**: The final component is the Symbolic Output Layer, which presents the processed data in a symbolic form. This layer completes the loop, taking the system from symbolic input to symbolic output. It serves as the endpoint of the architecture, delivering the final output that can be interpreted, analyzed, or used for further processing in a symbolic reasoning system.

Operational Mechanisms: The forward and backward passes constitute the operational mechanisms of the "Symbolic Neuro symbolic" architecture. They ensure not only the seamless integration of symbolic and neural elements but also the adaptability and learning capabilities of the system.

a. Forward Pass: The forward pass is the phase where the actual computation and transformation of data occur. It starts with the Symbolic Input Layer, where the raw symbolic data is fed into the system. This data is then passed to the Vectorization Module, which converts the symbolic elements into numerical vectors suitable for neural network processing. Once vectorized, the data moves to the Neural Network Layer, where it undergoes various transformations, activations, and computations based on the network's architecture and parameters. The output from the neural network is then passed to the Output Processing Module. Here, the numerical data is converted back into a symbolic form, often through a softmax operation that categorizes the output into different classes or sequences of symbols. Finally, the Symbolic Output Layer presents the processed data in its symbolic form, completing the forward pass.

During the forward pass, the primary goal is to transform the input data into a meaningful output. This is where the integration between symbolic and neural elements is most evident, as the data transitions from a symbolic form to a numerical form and back to a symbolic form.

b. Backward Pass: The backward pass is crucial for the training and optimization of the neural network. It begins at the Symbolic Output Layer, where the predicted output is compared with the actual or "ground truth" output to calculate the loss. This loss serves as a measure of how well the neural network is performing. The loss value is then propagated backward through the network, reaching all the way to the Neural Network Layer. During this phase, gradients are computed for each parameter in the neural network using techniques like backpropagation. These gradients indicate how much each parameter should be adjusted to minimize the loss.

Once the gradients are computed, the Neural Network Layer updates its parameters, typically using optimization algorithms like stochastic gradient descent (SGD) or its variants. The Vectorization Module and the Symbolic Input Layer usually remain unchanged during the backward pass, as they are not involved in the learning process. However, in some advanced setups, even the vectorization techniques can be fine-tuned during training.

The backward pass is essential for the "learning" aspect of the neural network. It adjusts the internal parameters of the Neural Network Layer to improve the model's performance iteratively. This is where the neural network "learns" from the data, optimizing its parameters to produce more accurate symbolic outputs in subsequent forward passes.

Advantages:

Interoperability: One of the most significant advantages of this architecture is its ability to seamlessly integrate symbolic reasoning with neural network-based learning. This allows the system to leverage the strengths of both paradigms, making it highly versatile and effective for a range of tasks.

Flexibility: The architecture is inherently flexible and can be adapted to various types of tasks and domains. While it is most commonly used in natural language processing, its modular design allows it to be easily extended to other fields like computer vision or robotics.

Scalability: Given that the architecture is built on neural networks, it can benefit from the scalability that neural networks offer. As more data becomes available, the neural network layer can be expanded or fine-tuned to improve performance.

End-to-End Learning: The architecture supports end-to-end learning, meaning that it can be trained to optimize for the final output directly. This is particularly useful for tasks where the intermediate steps are hard to quantify or optimize individually.

Limitations:

Computational Overhead: One of the main limitations is the computational cost associated with converting symbolic data to numerical vectors and back. This can be a bottleneck in real-time applications or systems with limited computational resources.

Information Loss: The process of converting symbolic data to numerical vectors and back can result in the loss of nuanced or complex information. This is especially problematic in tasks that require a high level of detail or precision.

Training Complexity: The presence of both symbolic and neural elements can make the training process more complex. Balancing the learning rates, layer sizes, and other hyperparameters for both types of elements can be challenging.

Interpretability: While the architecture does incorporate symbolic reasoning, the neural network layer often acts as a "black box," making it difficult to interpret or understand why specific decisions were made. This can be a significant drawback in applications that require high levels of transparency or explainability.

10.2.2 Type 2: Symbolic[Neuro]

The "Symbolic[Neuro]" architecture represents a unique approach to neuro-symbolic integration, where a neural pattern recognition subroutine is employed within a symbolic problem solver. This architecture is particularly relevant in applications like game playing, robotics, and autonomous vehicles. In this model, the primary system is symbolic, and it employs a neural network as a subroutine for specific tasks. For example, in the case of AlphaGo, the Monte-Carlo Tree Search algorithm, which is symbolic, uses a neural network for its heuristic evaluation function. The neural network operates as a specialized component within the broader symbolic framework, providing pattern recognition or heuristic evaluation capabilities. Its main components, as shown in Fig. 10.2, are as follows:

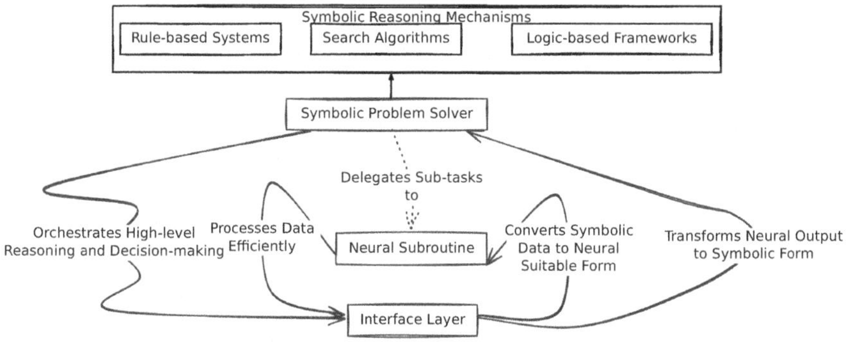

Fig. 10.2 Components of type 2: Symbolic[Neuro]

1. **Symbolic Problem Solver**: The Symbolic Problem Solver serves as the central hub of the architecture, responsible for orchestrating high-level reasoning and decision-making. This component can be implemented using various symbolic reasoning mechanisms, such as rule-based systems, search algorithms, or logic-based frameworks. It is designed to handle complex tasks that require structured thinking, planning, and decision-making. The Symbolic Problem Solver identifies specific sub-tasks or problems that would benefit from specialized pattern recognition or heuristic evaluation. These sub-tasks are then delegated to the Neural Subroutine for more efficient and specialized processing.
2. **Neural Subroutine**: The Neural Subroutine is a neural network trained to perform specific tasks that the Symbolic Problem Solver deems necessary for achieving its objectives. This could range from pattern recognition in images to heuristic evaluations in game-playing algorithms like AlphaGo. The Neural Subroutine acts as a specialized tool within the broader symbolic framework, providing capabilities that are computationally expensive or inefficient for symbolic systems to perform directly. It is this neural component that brings the power of machine learning into the architecture, allowing the system to adapt and improve over time based on data.
3. **Interface Layer**: The Interface Layer is perhaps the most crucial component when it comes to the seamless integration of the Symbolic Problem Solver and the Neural Subroutine. This layer is responsible for facilitating the data flow and operational coordination between the symbolic and neural components. It may involve converting symbolic data into a form suitable for neural network processing or transforming the neural network's output back into a symbolic form that the Symbolic Problem Solver can understand and utilize. The Interface Layer ensures that both components can operate in harmony, making the architecture truly neuro-symbolic.

Operational Mechanisms: Task Allocation allows the system to optimize its performance by leveraging the strengths of both symbolic and neural components. Data Transformation ensures data compatibility and integrity across the system, and Result Integration makes sure that the system's diverse components contribute to a unified objective.

a. Task Allocation: The first stage in the operational mechanism is Task Allocation, which is primarily handled by the Symbolic Problem Solver. During this stage, the Symbolic Problem Solver identifies specific sub-tasks or problems that require specialized attention, such as pattern recognition or heuristic evaluation. These are tasks that the symbolic system either cannot perform efficiently or lacks the capability to execute. Once identified, these tasks are delegated to the Neural Subroutine for processing. The Task Allocation stage is crucial for optimizing the system's performance, as it ensures that each component focuses on what it does best.

b. Data Transformation: The second stage is Data Transformation, facilitated by the Interface Layer. This stage is essential for ensuring seamless interoperability between the Symbolic Problem Solver and the Neural Subroutine. The Interface

Layer converts the symbolic data into a form suitable for neural network processing. This could involve encoding symbolic expressions into numerical vectors or normalizing input data to meet the neural network's requirements. Conversely, the Interface Layer may also convert the neural network's numerical output back into a symbolic form that the Symbolic Problem Solver can interpret and utilize. The Data Transformation stage is vital for maintaining data integrity and compatibility across the system.

c. Result Integration: The final stage is Result Integration, where the output from the Neural Subroutine is integrated back into the Symbolic Problem Solver. This is where the "loop" closes, and the system becomes a cohesive unit. The Symbolic Problem Solver takes the processed data or results from the Neural Subroutine and incorporates them into its ongoing reasoning and decision-making processes. This could involve updating its internal state, modifying its decision tree, or triggering specific actions based on the results. The Result Integration stage is critical for ensuring that the specialized processing done by the Neural Subroutine contributes meaningfully to the overall objectives of the Symbolic Problem Solver.

Advantages:

Optimized Performance: One of the most significant advantages of this architecture is its ability to optimize performance by allocating tasks to the component best suited for them. The Symbolic Problem Solver handles high-level reasoning, while the Neural Subroutine takes care of specialized tasks like pattern recognition or heuristic evaluation.

Modularity: The architecture is inherently modular, allowing for easy updates or replacements of individual components. This makes it highly adaptable to evolving requirements and technologies.

Data Efficiency: By leveraging the strengths of both symbolic and neural approaches, the architecture can often achieve high levels of performance with less data compared to purely neural or purely symbolic systems.

Interpretability: The presence of a Symbolic Problem Solver offers a level of transparency and interpretability that is often lacking in purely neural systems. This is particularly beneficial in applications that require explainability.

Limitations:

Implementation Complexity: One of the main challenges is the complexity involved in implementing and maintaining the architecture. Ensuring seamless integration between the Symbolic Problem Solver and the Neural Subroutine can be technically challenging.

Resource Intensive: The architecture can be computationally expensive, especially in scenarios where rapid data transformation and processing are required between the symbolic and neural components.

Limited Adaptability: While the Neural Subroutine can adapt and learn from data, the Symbolic Problem Solver is generally static unless manually updated. This could limit the system's ability to adapt to new or unforeseen challenges autonomously.

Data Compatibility: Ensuring that data is compatible and meaningful across both symbolic and neural components can be a significant challenge, requiring sophisticated data transformation and normalization techniques.

10.2.3 Type 3: Neuro | Symbolic

The "Neuro | Symbolic" architecture is a hybrid system where a neural network and a symbolic reasoning system work in tandem, each specializing in complementary tasks. Unlike architectures, where neural networks primarily serve as tools within a symbolic framework or vice versa, Type 3 systems facilitate a dynamic interaction between the two. This symbiosis allows the strengths of neural networks—such as handling unstructured data and learning from examples—to complement the structured, logical processing of symbolic reasoning systems. This type of integration is particularly advantageous in complex tasks that require both deep learning and logical inference.

In these systems, neural networks are often employed to generate or refine symbolic representations, which are then used by symbolic reasoning components to perform higher-level cognitive tasks. For instance, in program synthesis, deep learning models can produce symbolic programs and rule systems that fulfill high-level task specifications. The interaction between the neural and symbolic components helps ensure that the generated solutions are both accurate and efficient. This bidirectional cooperation enhances the overall performance and reliability of the AI system.

One prominent application of Type 3 neuro-symbolic AI is in reinforcement learning (RL). Here, symbolic planning is incorporated to improve decision-making processes. Neural networks handle the learning of task-specific patterns and actions, while symbolic planners ensure that the strategies adhere to logical constraints and long-term objectives. This integration allows for more sophisticated and reliable autonomous decision-making systems, which are crucial in fields such as robotics and autonomous vehicles.

Probabilistic logic programming is another area where Type 3 systems excel. Systems like NLProlog [13] and DeepProbLog [14] employ neural networks to calculate the probabilities of probabilistic facts and utilize the inference mechanism of ProbLog to compute the required loss gradient. This combination allows these systems to handle uncertainty and probabilistic reasoning more effectively, making them suitable for applications that require robust decision-making under uncertainty, such as natural language processing and knowledge graph reasoning.

The cooperative interaction between neural and symbolic layers in Type 3 systems offers several significant benefits. By leveraging the pattern recognition capabilities of neural networks and the precise, logical processing of symbolic reasoning, these systems can tackle complex problems more effectively. This synergy enhances problem-solving capabilities, making the AI more adaptable to various tasks. Moreover, the integration improves decision-making processes, ensuring that

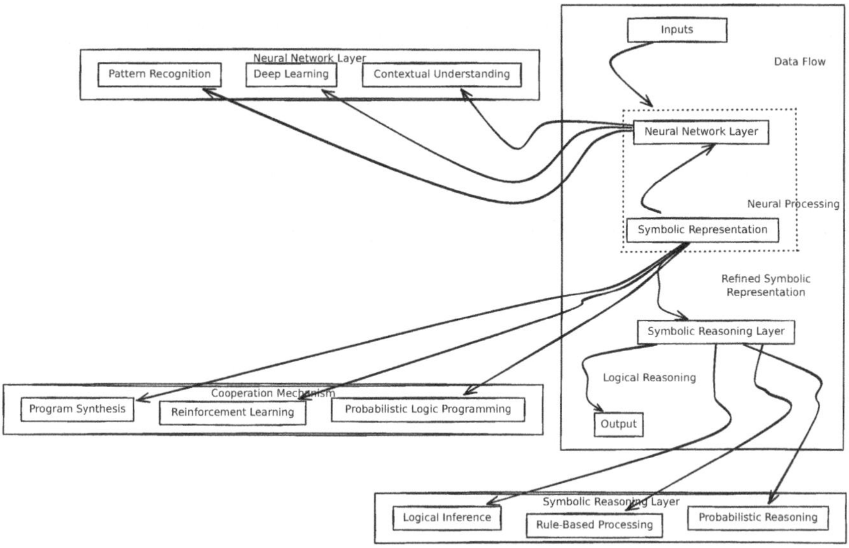

Fig. 10.3 Components of type 3: Neuro|Symbolic

decisions are both data-driven and logically sound, which is essential for reliability and trustworthiness in AI applications.

Its main components, as shown in Fig. 10.3, are as follows:

1. **Neural Network Layer**: The Neural Network Layer acts as the frontline for processing raw, non-symbolic data. Its primary role is to specialize in tasks such as object detection, feature extraction, and pattern recognition. Functionally, this layer is responsible for data preprocessing, where it cleanses and normalizes incoming raw data to make it suitable for neural processing. It employs specialized neural network architectures, such as convolutional or recurrent layers, to extract meaningful features from the data. Additionally, this layer may include a module that converts these features into a symbolic data structure, serving as an intermediate representation that can be further processed by the symbolic reasoning system. In terms of interconnections, the Neural Network Layer receives raw data from external sources or systems and sends either processed data or symbolic data structures to the Symbolic Reasoning System or directly to the output stage, depending on the task requirements.
2. **Symbolic Reasoning System** The Symbolic Reasoning System is the high-level engine of the architecture, specializing in tasks that require logical reasoning. This could range from query answering and decision-making to rule-based processing. Functionally, it operates by utilizing a set of predefined rules, axioms, or algorithms to perform logical reasoning on the symbolic data it receives. It is also capable of answering queries based on this symbolic data structure or its internal logic. For decision-making, the Symbolic Reasoning System can trigger

actions or make decisions based on the logical reasoning it has performed. In terms of interconnections, this component receives symbolic data structures from the Neural Network Layer or may also receive direct symbolic inputs, depending on the specific use case or task at hand.

Advantages:

Task Specialization: One of the most significant advantages of this architecture is the ability for each component to specialize in what it does best. The neural network excels at tasks like object detection and feature extraction, while the symbolic reasoning system is adept at high-level reasoning tasks like query answering. This division of labor allows for more efficient and effective task execution.

Adaptability: The architecture features a feedback mechanism that allows the system to adapt and improve over time. This makes the system more resilient and capable, as it can learn from past experiences and refine its performance.

Versatility: The architecture's dual-layered design makes it highly versatile. It can be applied to a wide range of tasks, from object detection to query answering, and can even be customized to suit specific application needs.

High-Level Reasoning: The inclusion of a symbolic reasoning system allows the architecture to perform high-level reasoning, which is often necessary for more complex tasks. This enables the system to handle tasks that purely neural or purely symbolic systems might struggle with.

Limitations:

Implementation Complexity: One of the major limitations is the complexity involved in implementing this architecture. Ensuring seamless integration and interaction between the neural and symbolic components can be a challenging task that requires specialized expertise.

Computational Cost: Due to the dual-layered nature of the architecture and the need for data transformation between the neural and symbolic components, the system can be computationally expensive. This might make it less suitable for real-time applications or systems with limited computational resources.

Data Compatibility: Ensuring that the symbolic data structure is compatible with the symbolic reasoning system can be a complex task. This often requires sophisticated data transformation techniques, adding another layer of complexity to the system.

Scalability Concerns: As the system grows in complexity or the volume of data it needs to process increases, scalability can become a concern. The architecture may require significant modifications or optimizations to handle larger-scale operations effectively.

10.2.4 Type 4: Neuro: Symbolic → Neuro

In a Type 4 neural-symbolic system, symbolic knowledge is compiled into the training set of a neural network. This type of architecture has been used in various applications like mathematical reasoning, natural language processing, and logical inference. For instance, an approach to learning and reasoning over mathematical constructions is proposed in [15], and a learning architecture that extrapolates to more complex symbolic math reasoning problems is introduced in [16].

Moreover, Type 4 also includes other tightly coupled but localist neural-symbolic systems. In these systems, various forms of symbolic knowledge, not restricted to if-then rules, are translated into the initial architecture and set of weights of a neural network. In some cases, this is done with guarantees of correctness.

A notable variant of this type is Logical Neural Networks. In this approach, the key concept is to create a one-to-one correspondence between neurons and the elements of logical formulas, as discussed in [16].

The primary components of Type 4 integration, as shown in 10.4, are as follows:

1. **Special Training Regime**: The "Neuro: Symbolic → Neuro" approach utilizes the Standard Operating Procedure (SOP) architecture but augments it with a specialized training regime based on symbolic rules. This specialized training regime is what sets it apart from other neural network architectures.
   A remarkable example of this approach is the work by Lample and Charton [16]. They trained a transformer sequence-to-sequence deep learning system using

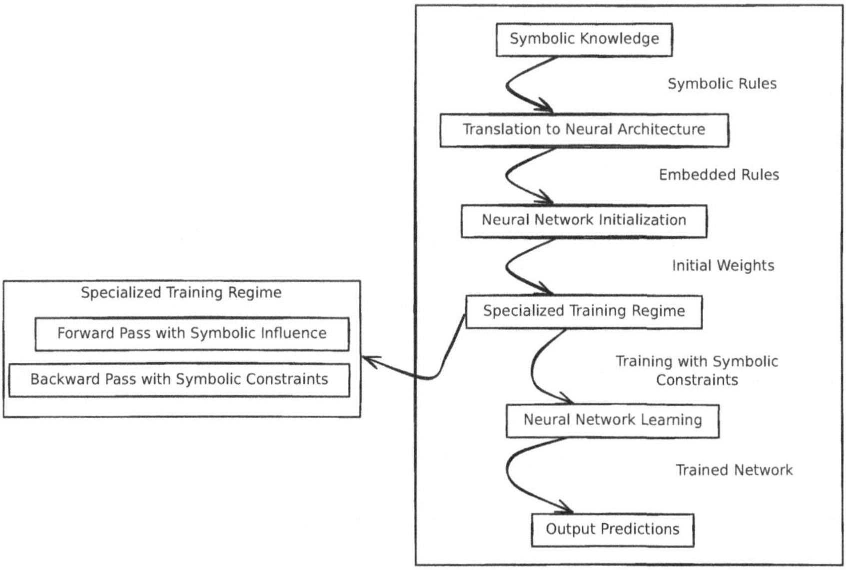

Fig. 10.4 Components of type 4: Neuro: Symbolic → Neuro

input-output pairs of the form ($\mathcal{A}$, $\mathcal{B}$), where $\mathcal{A}$ is a complex mathematical expression and $\mathcal{B}$ is its simplified form. The neural network was trained to simplify these mathematical expressions.
2. **Operational Mechanism**: After the training phase, the neural network could usually simplify a previously unseen mathematical expression correctly. Notably, the neural network did not generate a step-by-step derivation of the simplified expression. Instead, it had so thoroughly absorbed its training that it could "guess" the correct simplified form of the expression.

Advantages:

Efficient Symbolic Reasoning: One of the standout advantages is the ability to perform symbolic reasoning tasks with the efficiency of neural computation. This is particularly useful for tasks that require both speed and accuracy.

Specialized Training: The architecture allows for highly specialized training regimes, enabling the neural network to tackle tasks that are traditionally difficult for machine learning models, such as symbolic mathematics.

No Need for Step-by-Step Derivation: The architecture eliminates the need for step-by-step symbolic derivations, which can be time-consuming and computationally expensive. This makes the system faster and potentially more efficient.

Versatility: Given that various forms of symbolic knowledge can be incorporated into the training regime, the architecture is versatile and can be adapted for a wide range of applications.

Limitations:

Complex Training Regime: The specialized training regime can be complex and time-consuming, requiring a carefully curated set of input-output pairs and potentially specialized hardware for training.

Verification Challenges: Since the neural network is trained to "guess" the correct answer, additional verification steps may be necessary to ensure the output's accuracy, adding another layer of complexity.

Limited Generalizability: The neural network is highly specialized and trained for specific tasks, which may limit its ability to generalize to other, untrained tasks.

Resource Intensive: Given the specialized nature of the training, the architecture may require significant computational resources, making it less suitable for low-resource environments.

10.2.5 Type 5: $Neuro_{Symbolic}$

Type 5 architectures are tightly coupled but distributed neural-symbolic systems where a symbolic logic rule is mapped onto an embedding that acts as a soft constraint (a regularizer) on the network's loss function. This type of architecture aims to bridge the gap between symbolic reasoning and neural network-based learning by incorporating symbolic rules into the neural network's learning process.

Examples of Type 5 architectures include Logic Tensor Networks [6] and Tensor Product Representations [5], which are often referred to as tensorization methods. These methods allow for the incorporation of symbolic rules into the neural network's architecture, serving as a form of regularization during the learning process. The major components, as shown in Fig. 10.5, are as follows:

1. The first key component in this architecture is the **Symbolic Logic Rule**. These are predefined sets of rules or axioms that encapsulate some form of symbolic reasoning or domain-specific knowledge. The role of these rules is foundational, serving as the bedrock for the symbolic component of the architecture. They offer a structured way to incorporate domain-specific knowledge into the learning process.
2. The second component is the **Neural Network**. This is a machine learning model that is designed to learn from data and make predictions or decisions based on that data. The neural network takes on the role of the learner in this architecture, adapting its parameters to minimize a loss function during the training phase.
3. Next is the **Embedding Layer**, a specialized component for transforming the symbolic logic rules into a continuous vector space or an embedding. This layer serves as the bridge between the symbolic logic rules and the neural network, translating the discrete symbolic space into a form that can be integrated into the neural network.
4. The final component is the **Loss Function**, a mathematical function that the neural network aims to minimize during the training process. This function quantifies how well the model's predictions align with the actual data. However, this architecture is modified by embedding to include symbolic constraints, making it a hybrid loss function that the neural network aims to minimize.

The operational mechanisms start with the Embedding Transformation. In this step, the symbolic logic rules are transformed into an embedding via the specialized embedding layer. This involves a complex mapping process that translates the discrete symbolic space into a continuous vector space.

Once the embedding is generated, it acts as a Soft Constraint on the neural network's loss function. This means that the loss function is modified not just to minimize the error between the predicted and actual values but also to adhere to the symbolic rules encapsulated in the embedding.

During the Training Process, the neural network adjusts its parameters to minimize this modified loss function. The training data provides the primary source of learning, while the embedding provides a secondary source of information, acting as a regularizer. This dual-source learning makes the model robust and capable of handling complex reasoning tasks.

As the model trains, it undergoes Model Evaluation and Adaptation. It is evaluated on a validation set to monitor its performance. If the model is found to be overfitting to the training data or the symbolic rules, regularization techniques may be applied to improve its generalization capabilities.

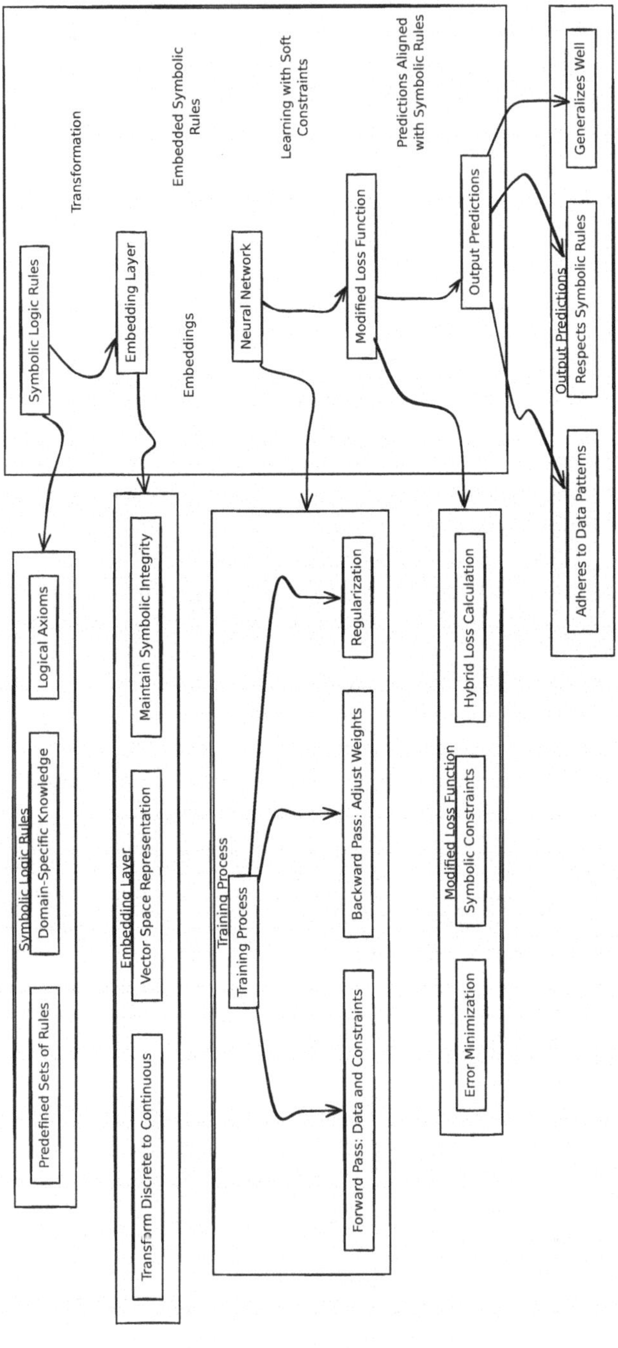

Fig. 10.5 Components of type 5: $Neuro_{Symbolic}$

The end result is a Final Model that is a tightly coupled neural-symbolic system. This model is capable of both data-driven prediction and symbolic reasoning, thanks to the integrated architecture that leverages both symbolic rules and neural network-based learning.

Advantages:

Robust Learning: One of the most significant advantages of the Type 5 architecture is its robust learning capabilities. By incorporating symbolic rules as soft constraints, the architecture gains an additional layer of information that supplements the training data. This dual-source learning approach enhances the model's robustness, making it more resilient to noise and anomalies in the data.

Improved Generalization: The soft constraints imposed by the symbolic rules serve as a form of regularization, helping the model generalize better to unseen data. This is particularly beneficial in applications where the training data may not fully capture the complexity of the problem space, allowing the model to make more accurate predictions on new data.

Complex Reasoning: The architecture's unique combination of symbolic reasoning and neural network-based learning enables it to tackle complex reasoning tasks. This hybrid approach allows the model to leverage the strengths of both symbolic and sub-symbolic methods, making it well-suited for tasks that require a blend of deductive reasoning and pattern recognition.

Limitations:

Computational Complexity: One of the main limitations is the increased computational complexity due to the integration of symbolic rules and their corresponding embeddings. This complexity can make the training process more resource-intensive, requiring specialized hardware or longer training times.

Risk of Overfitting: While the soft constraints can aid in regularization, there's also a risk of the model overfitting to these symbolic rules. If the model becomes too attuned to the rules at the expense of the training data, it may perform poorly on new, unseen data.

Interpretability Challenges: The tightly coupled nature of the architecture can make it difficult to dissect and understand how the model is making its decisions. This lack of interpretability can be a significant drawback in applications where understanding the model's reasoning process is crucial, such as in healthcare or finance.

For Type 5 systems, the focus of future research is likely to be on the development of rich embeddings. These embeddings could serve as soft constraints or regularizers on the network's loss function. Researchers are particularly interested in understanding the extent to which these embeddings can either correspond to predefined prior knowledge or be generated through learned attention mechanisms.

10.2.6 Type 6: Neuro[Symbolic]

The Type 6 Neuro[Symbolic] architecture aims to embed a symbolic reasoning engine directly within a neural engine. This architecture is designed to combine the strengths of both logic-based and neural-based AI. A Type 6 system should be capable of true combinatorial reasoning, possibly facilitated by an attention schema. According to Henry Kautz, a fully fledged Type 6 system for combinatorial reasoning does not yet exist, but significant progress has been made towards this goal.

The major components, as shown in Fig. 10.6, are as follows:

1. **Neural Engine**: The Neural Engine is the cornerstone of the Type 6 architecture, designed to excel in tasks that are inherently suited for neural networks. This includes a range of activities from pattern recognition to natural language understanding. The engine is usually constructed using advanced deep learning techniques, incorporating specialized neural network architectures like Convolutional Neural Networks (CNNs) for image-related tasks or Recurrent Neural Networks (RNNs) for tasks involving sequences. Within this engine, the input layer serves as the gateway for raw data, preparing it for the computational journey ahead. The hidden layers, which form the bulk of the neural network, are where the magic happens—these layers perform complex computations that transform the input data. Finally, the output layer takes these computations and converts them into a format that can either be used for decision-making or fed into the symbolic reasoning engine for more complex tasks.
2. **Symbolic Reasoning Engine**: Nested within the neural engine is the Symbolic Reasoning Engine, a specialized unit designed for logical reasoning and decision-making. This engine can be invoked by the neural engine when the task at hand requires more than just data-driven insights, such as complex problem-solving or logical deductions. The Symbolic Reasoning Engine is composed of several sub-components. The Logic Processor is the brain of this engine, responsible for executing logical operations and deductive reasoning. It works in conjunction with a Knowledge Base, which stores a set of predefined rules, facts, and relationships that the logic processor can draw upon. The Decision-Making Unit then takes the outcomes of these logical operations and uses them to make informed decisions or predictions.
3. **Attention Schema**: The Attention Schema is perhaps the most innovative component of the Type 6 architecture. It acts as a cognitive mechanism that bridges the gap between the neural and symbolic engines. The schema is hypothesized to generate a symbolic representation of the task at hand, serving as an interface that enables effective combinatorial reasoning. Within the Attention Schema, the Task Encoder plays a pivotal role by converting the problem or task into a symbolic format. This encoded task is then passed to the Interface Unit, which ensures seamless communication between the neural and symbolic engines. Lastly, the Feedback Decoder translates the output from the symbolic reasoning engine back into a neural-friendly format, allowing the neural engine to understand and act upon the results.

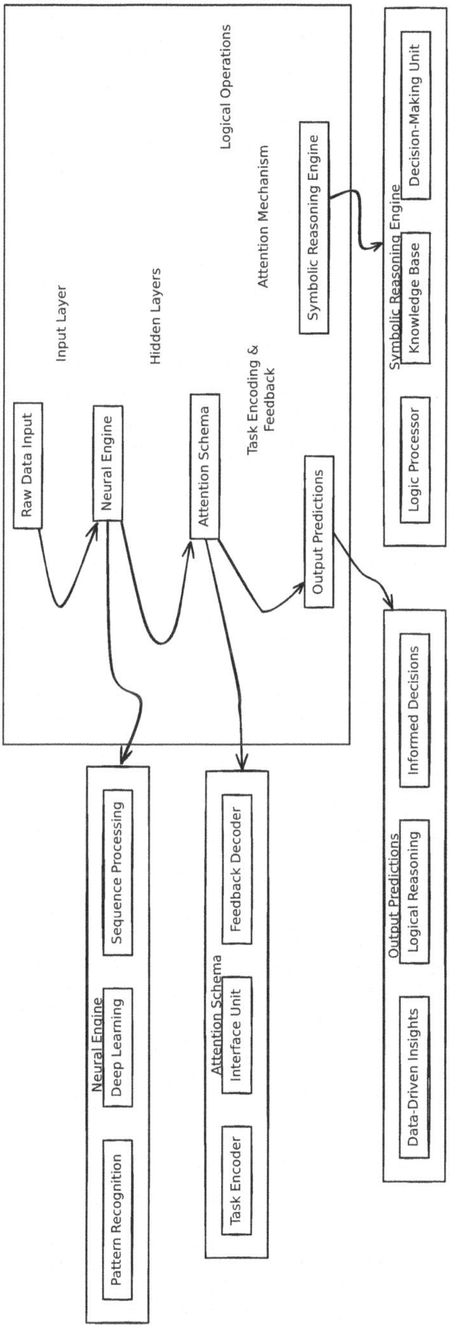

Fig. 10.6 Components of type 6: Neuro[Symbolic]

Type 6 systems hold significant relevance for the theory of neural-symbolic computing. Future research in this area is expected to delve into the tension between effective learning and sound reasoning. This may lead to the exploration of more hybrid approaches, possibly involving elements from Types 3–5. For instance, the use of attention mechanisms in conjunction with tensorization methods could be a promising avenue.

Yoshi Bengio and collaborators have proposed an architecture called "recurrent independent mechanisms" [17] which includes an Attention Schema-like module. However, this module contains a reduced vector representation of the Artificial Neural Network's (ANN) state rather than a fully symbolic representation. The practical challenge here is understanding how even these reduced vector representations could support fast combinatorial search.

Criticisms and Limitations: One of the criticisms of the Neuro[Symbolic] architecture is that the symbolic solver is generally not differentiable, making it incompatible with gradient-descent-based learning for the neural part of the system. However, this neural part, often referred to as System 1, could still be trained on input/output pairs from the symbolic solver. Over time, System 1 could become increasingly proficient at predicting solutions, reducing the need to invoke the symbolic reasoning engine, or System 2.

Another point of critique is that the Neuro[Symbolic] approach is generally limited to logical reasoning or problems that can be reduced to logical reasoning. However, there's no fundamental reason why it couldn't be extended to include probabilistic reasoning. For example, the Attention Schema could be instantiated with a Bayesian network where the System 2 engine is a probabilistic reasoning engine.

Lastly, while the Neuro[Symbolic] approach is inspired by cognitive models, it doesn't necessarily model the brain at what David Marr called the "implementation level" [18]. It may even differ at the algorithmic level. For instance, it seems unlikely that when solving a logic problem, we perform a DPLL search with clause learning, as some computational models suggest [19]. However, this isn't necessarily a limitation if the ultimate goal is to solve complex problems in fields like science, engineering, and commerce rather than to model human intelligence.

10.3 A New Perspective on Neuro-Symbolic AI: Algorithmic and Application-Level Considerations

The evolving landscape of Neuro-symbolic AI has been enriched by new perspectives that emphasize not just the integration but also the functionalities and capabilities that such integration can offer. One such perspective is presented in the paper [20] by Amit Sheth, Kaushik Roy, and Manas Gaur. This perspective is particularly insightful for understanding the need to combine neural and symbolic approaches, focusing on two key dimensions: algorithmic-level and application-level considerations.

10.3.1 Algorithmic-Level Considerations in Neuro-Symbolic AI

Abstraction: Neural networks are unparalleled in their ability to recognize patterns in data. However, they often struggle to abstract these patterns into higher-level concepts that can be generalized across different contexts. Symbolic systems, on the other hand, excel at this. They can take the patterns recognized by neural networks and abstract them into higher-level concepts, rules, or principles. This ability for abstraction is crucial for tasks that require understanding the underlying structure of data, such as natural language understanding and complex problem-solving.

Analogy: Another area where neural networks fall short is analogical reasoning. While they can identify similarities between different data points, they lack the structural framework to reason by analogy. Symbolic systems can fill this gap effectively. They can map the similarities recognized by neural networks into a structured form, enabling the system to make analogical inferences. This is particularly important for tasks that involve transferring knowledge from one domain to another, a common requirement in advanced AI applications.

Long-term Planning: Neural networks are designed to make immediate predictions or decisions based on the data they have been trained on. They are not inherently equipped for long-term planning or reasoning over extended periods. Symbolic systems can provide this capability. By using logical rules and constraints, symbolic systems can plan actions and make decisions that consider long-term outcomes. This is vital for applications like robotics, supply chain management, and healthcare, where decisions have long-term implications.

By focusing on these algorithmic-level considerations, the integration of neural and symbolic systems in Neuro-symbolic AI can be more purposeful and effective. It allows for the development of AI systems that are not only powerful but also capable of higher-level reasoning and decision-making. This makes the field of Neuro-symbolic AI a promising avenue for future research and development.

10.3.2 Application-Level Considerations in Neuro-Symbolic AI

While the algorithmic-level considerations provide a theoretical foundation for the integration of neural and symbolic systems, it is the application-level considerations that dictate the practical utility of Neuro-symbolic AI. The paper [20] emphasizes three key aspects at the application level: explainability, interpretability, and safety. This section delves into these considerations, outlining their significance in the deployment of Neuro-symbolic AI systems.

Explainability: One of the most pressing challenges in the field of AI is the "black box" nature of neural networks, which makes it difficult to understand how they arrive at specific decisions or predictions. This lack of transparency is a significant hurdle in sensitive applications like healthcare, finance, and legal systems, where stakeholders require clear explanations for AI-generated outcomes. The symbolic component in Neuro-symbolic AI systems can provide this much-needed explainability. By translating the neural network's complex computations into human-understandable rules or guidelines, these systems make it easier to understand and trust AI decisions.

Interpretability: Closely related to explainability is the concept of interpretability, which refers to the ease with which a human can understand the inner workings of an AI system. While neural networks excel at pattern recognition and prediction, their complex architectures often make it challenging to interpret how they function. Symbolic systems, with their rule-based logic, offer a structured framework that can be easily interpreted. This interpretability is crucial for debugging, improving, and fine-tuning AI systems, especially in critical applications where even a small error can have significant consequences.

Safety: The third key consideration at the application level is safety. As AI systems become increasingly autonomous, ensuring that they operate safely within predefined boundaries becomes paramount. Neural networks, trained solely on past data, can sometimes produce unpredictable or unsafe outcomes. Symbolic systems can mitigate this risk by incorporating hard constraints and rules that guide the neural network's behavior. This ensures that the AI system operates within safe and acceptable parameters, making it more reliable for real-world applications.

10.3.3 Categories of Neuro-Symbolic AI Methods

The paper [20] provides a comprehensive framework for categorizing Neuro-symbolic AI methods. This categorization is crucial for understanding how these methods can address both algorithmic-level and application-level considerations. Below, we explore these categories in detail.

10.3.3.1 Methods for Compressing Structured Symbolic Knowledge in Neuro-Symbolic AI

The integration of structured symbolic knowledge with neural networks is a cornerstone of Neuro-symbolic AI. This integration aims to leverage the best of both worlds: the pattern recognition capabilities of neural networks and the rule-based reasoning of symbolic systems. The category of methods that focus on compressing structured symbolic knowledge into neural patterns is pivotal in achieving this integration. Below, we delve into the two main sub-categories of these methods: Compressed Knowledge Graph Representations and Compressed Formal Logic-Based Representations.

Compressed Knowledge Graph Representations: Knowledge graphs serve as a structured repository of relationships and entities, capturing the semantics of a domain in a way that is both rich and interpretable. In the context of Neuro-symbolic AI, compressed knowledge graph representations are used to encapsulate this structured information into a form that can be readily processed by neural networks.

How it Works: The compression techniques reduce the complexity of the knowledge graph while retaining its essential features. These compressed representations are then integrated into the neural network's architecture, serving as an additional layer or as part of the input data.

Applications: This approach is particularly useful in natural language processing, recommendation systems, and any application that requires a deep understanding of relationships between various entities.

Challenges: The main challenge lies in the compression process itself, which must be done in a way that retains the essential semantic features of the knowledge graph. Additionally, the integration with neural networks must be seamless to ensure effective reasoning.

Compressed Formal Logic-Based Representations: Formal logic provides a rigorous framework for rule-based reasoning. In Neuro-symbolic AI, formal logic-based representations are compressed and integrated into neural networks to enable more structured and reliable reasoning.

How it Works: Logical rules and axioms are transformed into a compressed format, often through techniques like quantization or encoding. These compressed logic representations are then incorporated into the neural network, either as a separate layer or directly into the network's weights and biases.

Applications: This method is highly effective in domains that require strict rule adherence, such as medical diagnosis, legal reasoning, and safety-critical systems.

Challenges: The primary challenge is ensuring that the compressed logic retains its original meaning and efficacy. Additionally, the neural network must be capable of interpreting these compressed logic forms to perform effective reasoning.

10.3.3.2 Methods for Extracting Information from Neural Patterns

The other side of the Neuro-symbolic AI coin involves methods that focus on extracting structured symbolic knowledge from neural patterns. These methods aim to translate the learned features and representations within neural networks into a form that can be understood and manipulated symbolically. This extraction process is crucial for enabling more transparent, interpretable, and modular AI systems. Below, we explore the two main sub-categories of these methods: Decoupled Integration and Intertwined Integration.

Decoupled Integration: In decoupled integration, the neural and symbolic components operate somewhat independently, with the neural component serving as a feature extractor that feeds into the symbolic reasoning system.

How it Works: The neural network processes raw data to extract features or patterns, which are then translated into symbolic representations. These symbolic forms are then used for further reasoning or decision-making in a separate symbolic system.

Applications: This approach is commonly used in computer vision tasks where neural networks identify objects or features, and symbolic systems handle higher-level reasoning, such as scene understanding or activity recognition.

Challenges: The primary challenge is ensuring a seamless translation from neural patterns to symbolic representations. Any loss in this translation could lead to incorrect or suboptimal reasoning.

Intertwined Integration: Unlike decoupled integration, intertwined integration aims for a more seamless and continuous interaction between the neural and symbolic components.

How it Works: In this approach, the neural network not only extracts features but also dynamically interacts with the symbolic system. The symbolic system can influence the neural network's learning process and vice versa, creating a more holistic system.

Applications: Intertwined integration is particularly useful in complex tasks that require both pattern recognition and high-level reasoning, such as autonomous driving or advanced natural language understanding.

Challenges: The main challenge here is the complexity of the integration. Both the neural and symbolic components need to be designed to work in harmony, which can be computationally intensive and require sophisticated engineering.

10.3.3.3 Effectiveness Across Scales

The concept of "Effectiveness Across Scales" in the paper refers to the varying degrees of effectiveness that neuro-symbolic AI methods can achieve at both algorithmic and application levels. These effectiveness levels are categorized into low (L), medium (M), and high (H) scales. The paper discusses how different methods in neuro-symbolic AI can be effective in different ways, depending on their focus and implementation.

For instance, methods that focus on compressing knowledge graphs for integration with neural patterns are highly effective in large-scale perception (H) but are only moderately effective (M) in aspects like abstraction, analogy, and planning. On the application level, these methods score low (L) in user-explainability but medium (M) in domain constraints.

Similarly, methods that focus on lifting neural patterns for integration with symbolic structured knowledge also vary in their effectiveness. These methods are highly effective in large-scale perception (H) and moderately effective in abstraction and analogy (M). On the application level, they score medium (M) in user-explainability and domain constraints.

Algorithmic Level:

Large-Scale Perception: Methods that excel in large-scale perception are usually those that can process and analyze vast amounts of data efficiently. These methods are often rated high (H) in effectiveness for tasks like image recognition, natural language processing, and other data-intensive tasks.

Abstraction, Analogy, and Planning: The effectiveness in these areas is often moderate (M) for most neuro-symbolic methods. These tasks require a blend of neural networks for pattern recognition and symbolic systems for logical reasoning, making them more challenging to optimize.

Application Level:

User-Explainability: Neuro-symbolic methods that focus on lifting neural patterns to symbolic knowledge often score medium (M) in user-explainability. These methods can provide some level of interpretability but may not be as transparent as purely symbolic methods, which would score high (H).

Domain Constraints: Methods that compress structured symbolic knowledge into neural patterns usually have a medium (M) effectiveness in adhering to domain-specific constraints. They can incorporate some rules and constraints but may not cover all the nuances of a particular domain.

Interplay Between Scales:

The effectiveness across scales also indicates that there's an interplay between the algorithmic and application levels. For example, a method that is highly effective in large-scale perception may not necessarily be as effective in user-explainability. This is because the neural networks, while excellent at pattern recognition, may not provide the level of transparency or reasoning that a symbolic system can offer.

The concept of "Effectiveness Across Scales" suggests that future research in neuro-symbolic AI should aim for a balanced approach. Researchers should strive to develop methods that are not only technically robust but also practically effective across different scales and domains.

10.4 Conclusion

The landscape of Neuro-Symbolic AI is both complex and promising, offering a multitude of frameworks and approaches that aim to combine the best of both symbolic reasoning and neural network-based learning. This chapter has provided an in-depth exploration of Henry Kautz's taxonomy, which serves as a foundational guide to understanding the various types of neuro-symbolic systems. Each type offers unique advantages and limitations, emphasizing the need for a multi-faceted approach to AI development.

We also look into a new perspective that focuses on algorithmic and application-level considerations. This perspective serves as a lens through which we can evaluate the effectiveness of neuro-symbolic methods across different scales—low, medium, and high. It emphasizes that the future of neuro-symbolic AI is not just about developing technically robust methods but also about ensuring that these methods are practically effective and ethically sound.

The categories of neuro-symbolic AI methods further enrich our understanding by highlighting the different ways structured symbolic knowledge can be compressed into neural patterns and vice versa. These methods are not mutually exclusive but can be combined in innovative ways to address both algorithmic and application-level challenges.

References

1. Kautz, H.: The third AI summer: Aaai robert s. engelmore memorial lecture. AI Mag. **43**(1), 105–125 (2022)
2. Church, K.W.: Word2vec. Nat. Language Eng. **23**(1), 155–162 (2017)
3. Wang, F.Y., Zhang, J.J., Zheng, X., Wang, X., Yuan, Y., Dai, X., Zhang, J., Yang, L.: Where does alphago go: from church-turing thesis to alphago thesis and beyond. IEEE/CAA J. Automatica Sinica **3**(2), 113–120 (2016)
4. Browne, C.B., Powley, E., Whitehouse, D., Lucas, S.M., Cowling, P.I., Rohlfshagen, P., Tavener, S., Perez, D., Samothrakis, S., Colton, S.: A survey of Monte Carlo tree search methods. IEEE Trans. Comput. Intell. AI Games **4**(1), 1–43 (2012)
5. Smolensky, P.: Tensor product variable binding and the representation of symbolic structures in connectionist systems. Artif. Intell. **46**(1–2), 159–216 (1990)
6. Badreddine, S., Garcez, A.d., Serafini, L., Spranger, M.: Logic tensor networks. Artif. Intell. **303**, 103649 (2022)
7. Kahneman, D.: Thinking, Fast and Slow. Macmillan (2011)
8. Achiam, J., Adler, S., Agarwal, S., Ahmad, L., Akkaya, I., Aleman, F.L., Almeida, D., Altenschmidt, J., Altman, S., Anadkat, S., et al.: Gpt-4 technical report. arXiv preprint arXiv:2303.08774 (2023)
9. Sanh, V., Debut, L., Chaumond, J., Wolf, T.: Distilbert, a distilled version of bert: smaller, faster, cheaper and lighter. arXiv preprint arXiv:1910.01108 (2019)
10. Wermelinger, M.: Using github copilot to solve simple programming problems. In: Proceedings of the 54th ACM Technical Symposium on Computer Science Education, vol. 1, pp. 172–178 (2023)
11. Qader, W.A., Ameen, M.M., Ahmed, B.I.: An overview of bag of words; importance, implementation, applications, and challenges. In: 2019 International Engineering Conference (IEC), pp. 200–204. IEEE (2019)
12. Christian, H., Agus, M.P., Suhartono, D.: Single document automatic text summarization using term frequency-inverse document frequency (tf-idf). ComTech: Comput., Math. Eng. Appl. **7**(4), 285–294 (2016)
13. Weber, L., Minervini, P., Münchmeyer, J., Leser, U., Rocktäschel, T.: Nlprolog: reasoning with weak unification for question answering in natural language. arXiv preprint arXiv:1906.06187 (2019)
14. Manhaeve, R., Dumancic, S., Kimmig, A., Demeester, T., De Raedt, L.: Deepproblog: neural probabilistic logic programming. In: Advances in Neural Information Processing Systems, vol. 31 (2018)
15. Lample, G., Charton, F.: Deep learning for symbolic mathematics. arXiv preprint arXiv:1912.01412 (2019)
16. Riegel, R., Gray, A., Luus, F., Khan, N., Makondo, N., Akhalwaya, I.Y., Qian, H., Fagin, R., Barahona, F., Sharma, U., et al.: Logical neural networks. arXiv preprint arXiv:2006.13155 (2020)
17. Goyal, A., Lamb, A., Hoffmann, J., Sodhani, S., Levine, S., Bengio, Y., Schölkopf, B.: Recurrent independent mechanisms. arXiv preprint arXiv:1909.10893 (2019)

18. Hardcastle, V.G., Hardcastle, K.: Marr's levels revisited: understanding how brains break. Topics Cognit. Sci. **7**(2), 259–273 (2015)
19. Beame, P., Kautz, H., Sabharwal, A.: Understanding the power of clause learning. In: IJCAI, pp. 1194–1201. Citeseer (2003)
20. Sheth, A., Roy, K., Gaur, M.: Neurosymbolic artificial intelligence (why, what, and how). IEEE Intell. Syst. **38**(3), 56–62 (2023)

Chapter 11
Learning from Symbolic Knowledge for Neural Networks

> *"Logic is the beginning of wisdom, not the end."*
> Spock, *Star Wars*

11.1 Introduction

In the realm of advanced recommendation systems for specialized scientific journals, a unique challenge arises. While deep learning models adeptly analyze researcher behavior, citation patterns, and content embeddings to predict reading preferences, there exist inherent domain-specific rules. For instance, certain articles are foundational and must be recommended when specific topics are queried, or some papers have interdependencies due to shared research or rebuttals. These symbolic constraints, deeply rooted in the academic structure, cannot be inferred solely from reading patterns or citation networks. They necessitate an explicit integration to ensure that recommendations are not only personalized but also academically coherent.

This chapter delves into the intricate methodologies that facilitate the integration of such symbolic knowledge into neural architectures. We explore the technical nuances of approaches that treat symbolic rules as guiding constraints during the learning process and those that embed this knowledge directly into the neural structure.

From regularization techniques that mold the learning trajectory based on symbolic constraints to model-based strategies that inherently weave symbolic knowledge into the neural fabric, we aim to provide a deep dive into the mechanisms, challenges, and innovations of each approach. Through this exploration, we seek to offer insights into crafting AI systems that are both data-adaptive and rule-compliant.

B. P. Bhuyan et al., *Neuro-Symbolic Artificial Intelligence*, Studies in Computational Intelligence 1176, https://doi.org/10.1007/978-981-97-8171-3_11

11.2 The Dual Approaches to Learning with Prior Knowledge

In the realm of artificial intelligence, the integration of symbolic knowledge with neural networks has emerged as a pivotal area of research. This integration seeks to harness the structured and explicit nature of symbolic knowledge to enhance the learning capabilities of neural networks. Predominantly, there are two primary methodologies employed to achieve this integration, each with its unique advantages and challenges.

1. Regularization Approaches: **Treating Logical Rules as Constraints on Neural Network Predictions** This approach is grounded in the idea of using symbolic knowledge as a guiding force for neural networks. By treating logical rules derived from symbolic knowledge as constraints, the neural network is directed to produce predictions that adhere to these rules.

Advantages:

Consistency with Prior Knowledge: Ensures that the neural network's predictions are consistent with known symbolic rules, enhancing the reliability of the model.

Flexibility: Allows for the integration of varying degrees of symbolic knowledge, from simple constraints to complex logical formulations.

Challenges:

Balancing Act: Striking the right balance between adhering to constraints and allowing the neural network to learn from data can be intricate. Overemphasis on constraints might hinder the network's ability to generalize from data.

2. Model-Based Approaches: **Modifying the Neural Network by Infusing Symbolic Knowledge into its Structure** Rather than externally guiding the neural network using constraints, this approach seeks to modify the very architecture of the network by embedding symbolic knowledge into it. This can be achieved by designing specific layers or modules that are informed by symbolic rules.

Advantages:

Inherent Knowledge: The neural network inherently understands and respects the symbolic rules, leading to more consistent and reliable predictions. Efficiency: By integrating knowledge into the structure, the need for external checks or post-processing steps is reduced.

Challenges:

Complexity: Designing and training such specialized neural network architectures can be complex and computationally intensive.

Flexibility Concerns: Making subsequent modifications or updates to the embedded knowledge might require significant changes to the network structure.

11.3 Regularization Approaches: Maximizing Constraint Satisfiability

Regularization approaches [1] in the context of integrating symbolic knowledge with neural networks revolve around the principle of constraining the network's predictions to align with predefined logical rules. This is achieved by introducing a regularization term in the loss function, which penalizes deviations from these rules.

Mathematical Formulation:

Given a neural network with a standard loss function $\mathcal{L}_{\text{data}}$, which measures the difference between the network's predictions and the actual data, we introduce a regularization term $\mathcal{L}_{\text{reg}}$ to account for the symbolic constraints. The combined loss function is then

$$\mathcal{L}_{\text{total}} = \mathcal{L}_{\text{data}} + \lambda \mathcal{L}_{\text{reg}} \tag{11.1}$$

where

- $\mathcal{L}_{\text{data}}$ is the original loss without symbolic knowledge. For instance, in a regression task, this could be the mean squared error.
- $\mathcal{L}_{\text{reg}}$ represents the penalty for not adhering to the symbolic constraints.
- λ is a hyperparameter that determines the weight or importance of the regularization term. A higher value of λ means that the symbolic constraints are deemed more important relative to the original loss.

For example, imagine we're training a neural network to predict house prices based on various features. Our original loss function, $\mathcal{L}_{\text{data}}$, might be the mean squared error between the predicted and actual house prices.

Now, let's say we have some symbolic knowledge: "Houses with swimming pools are always more expensive than those without." We want our neural network to respect this rule.

To integrate this rule, we introduce a regularization term $\mathcal{L}_{\text{reg}}$. For every pair of data points where one house has a swimming pool and the other doesn't, if the predicted price of the house with the pool is not higher, $\mathcal{L}_{\text{reg}}$ will penalize the prediction.

Mathematically, for a pair of data points $(x_{\text{pool}}, x_{\text{no-pool}})$ where x_{pool} is a house with a pool and $x_{\text{no-pool}}$ is a house without a pool, the regularization term for this pair is

$$L_{\text{reg\_pair}} = \max(0, f(x_{\text{no-pool}}) - f(x_{\text{pool}}) + \epsilon)$$

where

- f is the neural network's prediction function.
- ϵ is a small margin to ensure that houses with pools are not just slightly but significantly more expensive in the predictions.

The overall $\mathcal{L}_{\text{reg}}$ is the average of $L_{\text{reg\_pair}}$ over all such pairs in the dataset. By adding this $\mathcal{L}_{\text{reg}}$ to our original loss $\mathcal{L}_{\text{data}}$, we ensure that our neural network respects the symbolic rule about swimming pools while still learning from the data.

11.3.1 Logic Tensor Network (LTN) and Semantic-Based Regularization (SBR)

LTN [2] and SBR [3] are both innovative approaches that seek to bridge the gap between symbolic logic and neural networks. By embedding First-Order Logic (FOL) within the framework of tensor-based computations, LTNs provide a mechanism to represent and reason with symbolic knowledge in a differentiable manner, making it compatible with gradient-based learning methods like backpropagation. In the context of regularization, LTN introduces a penalty term to the loss function based on the degree to which the network's predictions violate the given logical formulas.

Core Principles of LTN:

Differentiable Logic Semantics: Traditional symbolic logic operates with crisp truth values (true or false). LTN, however, employs a fuzzy logic semantic, where truth values are continuous and lie in the range [0, 1]. This continuous nature allows for differentiability, which is essential for training neural networks.

Tensor Representations: In LTN, logical constants, variables, and functions are represented as tensors. This tensor-based representation facilitates the integration of logical formulas into neural network architectures.

Grounding: Grounding is the process of mapping logical symbols to their corresponding tensor representations. For instance, a logical predicate might be grounded in a neural network, and logical constants might be grounded in specific tensor values.

The LTN regularization term is derived from the truth values of the logical formulas. For a set of FOL formulas Φ, the regularization term $\mathcal{L}_{\text{reg}}$ in the context of LTN is

$$\mathcal{L}_{\text{reg}} = 1 - \frac{1}{|\Phi|} \sum_{\phi \in \Phi} \mu(\phi) \tag{11.2}$$

where

- $\mu(\phi)$ represents the degree of truth of formula ϕ under the fuzzy logic semantic.
- $|\Phi|$ is the total number of formulas.

Suppose, we're building a neural network to predict the performance of students based on features like study hours, participation in extracurricular activities, and attendance. We also have some symbolic knowledge about student performance:

a. Students who study more than 5 h a day and have an attendance above 90% always perform better than those who study less than 2 h and have an attendance below 60%.
b. Students who participate in extracurricular activities and study more than 3 h a day perform better than those who neither participate in extracurricular activities nor study more than 1 hour a day.

Regularization with LTN:
Grounding:

- Let $s(x)$ be the neural network's prediction of a student's performance based on input features x.
- Let study(x) represent the number of study hours of a student.
- Let attend(x) represent the attendance percentage of a student.
- Let extra(x) be a binary value indicating participation in extracurricular activities.

Logical Formulas:

- $\phi_1 : \forall x(\text{study}(x) > 5 \wedge \text{attend}(x) > 90) \rightarrow s(x) > s(y)$ for all y with $\text{study}(y) < 2 \wedge \text{attend}(y) < 60$.
- $\phi_2 : \forall x(\text{extra}(x) = 1 \wedge \text{study}(x) > 3) \rightarrow s(x) > s(z)$ for all z with $\text{extra}(z) = 0 \wedge \text{study}(z) < 1$.

Regularization Term:

- For each pair of students x and y (or z), compute the truth value of ϕ_1 and ϕ_2 using the fuzzy logic semantic.
- The regularization term $\mathcal{L}_{\text{reg}}$ is

$$\mathcal{L}_{\text{reg}} = 1 - \frac{1}{2}(\mu(\phi_1) + \mu(\phi_2))$$

where $\mu(\phi)$ represents the degree of truth of formula ϕ under the fuzzy logic semantic.
Total Loss Function:

- If $\mathcal{L}_{\text{data}}$ is the mean squared error between the predicted and actual student performance, the total loss function becomes

$$\mathcal{L}_{\text{total}} = \mathcal{L}_{\text{data}} + \lambda \mathcal{L}_{\text{reg}}$$

where λ is a hyperparameter determining the weight of the regularization term.

Semantic-Based Regularization and Logic Tensor Networks both aim to integrate symbolic knowledge, represented as First-Order Logic (FOL) formulas, into neural networks. However, there are key differences in their methodologies and underlying principles.

Foundational Approach:

LTN: LTNs are designed as a comprehensive framework that embeds the entire logic system within the tensor-based computations of neural networks. It's not just about adding a regularization term but about representing and processing logical entities (like predicates) directly as tensor operations within the network.

SBR: SBR is primarily a regularization technique. It focuses on adding a penalty to the loss function based on the degree to which the network's predictions violate the given logical formulas.

Grounding Mechanism:

LTN: In LTN, grounding is a central concept. Logical constants, variables, and functions are represented as tensors, and logical predicates might be grounded in specific neural network structures.

SBR: Grounding in SBR is more about mapping the symbolic knowledge to the data space and then evaluating the satisfaction of the formulas based on the network's predictions.

Flexibility and Scope:

LTN: LTNs provide a broader framework that can be extended to various tasks and can handle a wide range of logical formulas. It's more about creating a hybrid system where logic and learning coexist seamlessly.

SBR: SBR is more specialized in its approach, focusing on ensuring that the neural network's predictions align with the provided symbolic knowledge through the regularization term.

Implementation Complexity:

LTN: Implementing LTNs can be more complex due to the need to represent logical entities as tensors and integrate them into the neural network's architecture.

SBR: SBR is relatively straightforward to implement, as it primarily involves adding a regularization term to the existing loss function.

Interpretability:

LTN: LTNs offer a level of interpretability by grounding logical predicates to specific neural network structures, allowing for an understanding of the network's decisions in terms of logical reasoning.

SBR: While SBR ensures that the network's predictions align with symbolic knowledge, it might not offer the same level of detailed interpretability as LTN.

11.3.2 DL2: Deep Learning with Logic

DL2 [4] is a method that integrates domain-specific knowledge, represented as logical constraints, into deep learning models. It's particularly tailored for regression tasks, where the logical predicates correspond to comparison constraints.

DL2 operates on the premise that certain domain-specific knowledge can be represented as logical constraints on the predictions of a neural network. These constraints are then incorporated into the training process to guide the model towards solutions that not only fit the data but also respect the given symbolic knowledge.

Mathematical Formulation:

1. **Logical Constraints**:
   Let's consider a set of logical constraints Φ. Each constraint ϕ_i in Φ can be represented as a comparison operation, such as

   - $f(x) \leq c$
   - $f(x) \geq c$
   - $f(x) = c$
   - $f(x) \neq c$

   where $f(x)$ is the prediction of the neural network for input x, and c is a constant derived from the domain-specific knowledge.
2. **Constraint Violation**:
   For each constraint ϕ_i, we can define a violation function $V_i(x)$ that measures the degree to which the constraint is violated. For instance, for the constraint $f(x) \leq c$, the violation can be defined as follows:

   $$V_i(x) = \max(0, f(x) - c)$$

   This function will output a positive value if the constraint is violated and zero otherwise.
3. **Regularization Term**:
   The regularization term $\mathcal{L}_{\text{constraints}}$ is the aggregated violation of all constraints for a given input:

   $$\mathcal{L}_{\text{constraints}}(x) = \sum_i V_i(x)$$

   This term measures the total degree to which the network's prediction for input x violates the set of constraints Φ.
4. **Total Loss Function**:
   The combined loss function, which the neural network is trained on, becomes

   $$\mathcal{L}_{\text{total}}(x, y) = \mathcal{L}_{\text{data}}(x, y) + \lambda \mathcal{L}_{\text{constraints}}(x)$$

   where

- $\mathcal{L}_{\text{data}}(x, y)$ is the original loss function, measuring the discrepancy between the prediction $f(x)$ and the true output y.
- λ is a hyperparameter that determines the weight or importance of the regularization term.

Example Scenario: We have data for three houses. We want to predict their prices based on their square footage. We also have domain-specific knowledge that houses with more than 1000 sq. ft. should be priced higher than houses with less than 500 sq. ft (Tables 11.1 and 11.2).

Table 11.1 Given data

House	Square footage (sq. ft.)	True price ($)
A	1100	150,000
B	450	50,000
C	600	80,000

Table 11.2 Neural network predictions (initial, untrained)

House	Predicted price ($)
A	100,000
B	70,000
C	85,000

Logical Constraint:

$$\phi : \text{If sq. ft.} > 1000 \rightarrow \text{Price} > 100,000$$

$$\psi : \text{If sq. ft.} < 500 \rightarrow \text{Price} < 60,000$$

Mathematical Formulation using DL2

1. **Constraint Violation for House A**: Given the constraint ϕ, House A's price should be more than $100,000. The violation is

$$V_A = \max(0, 100,000 - f(A)) = \max(0, 100,000 - 100,000) = 0$$

2. **Constraint Violation for House B**: Given the constraint ψ, House B's price should be less than $60,000. The violation is

$$V_B = \max(0, f(B) - 60,000) = \max(0, 70,000 - 60,000) = 10,000$$

3. **Regularization Term**: The regularization term $\mathcal{L}_{\text{constraints}}$ is

$$\mathcal{L}_{\text{constraints}} = V_A + V_B = 0 + 10,000 = 10,000$$

4. **Total Loss Function**: Assuming a simple mean squared error for the data loss, and given the true prices:

$$\mathcal{L}_{\text{data}} = \frac{1}{3}[(150,000 - 100,000)^2 + (50,000 - 70,000)^2 + (80,000 - 85,000)^2]$$

$$\mathcal{L}_{\text{data}} = \frac{1}{3}[2,500,000 + 400,000 + 25,000] = 975,000$$

Using $\lambda = 0.01$ as the weight for the regularization term:

$$\mathcal{L}_{\text{total}} = \mathcal{L}_{data} + \lambda\mathcal{L}_{constraints}$$

$$\mathcal{L}_{\text{total}} = 975,000 + 0.01(10,000) = 975,100$$

Interpretation

The total loss $\mathcal{L}_{\text{total}}$ reflects both the error in the neural network's predictions relative to the true prices and the degree to which the predictions violate the domain-specific constraints. By minimizing this combined loss during training, the neural network will adjust its predictions to align with both the data and the given logical constraints. In this example, the network would be guided to increase its predicted price for House A and decrease its predicted price for House B to reduce the overall loss.

11.3.3 MultiplexNet

MultiplexNet [5] is a neural network architecture designed to incorporate domain-specific knowledge into the learning process. It's particularly tailored for safety-critical domains where complete satisfaction of constraints is desirable. The knowledge in MultiplexNet is expressed in disjunctive normal form (DNF).

The primary motivation behind MultiplexNet is to ensure that the neural network's predictions adhere to a set of predefined logical rules or constraints. These constraints, expressed in DNF, play a crucial role in guiding the network towards solutions that are not only consistent with the data but also with domain-specific knowledge.
Mathematical Formulation:

1. **Disjunctive Normal Form (DNF)**:
   A logical formula is said to be in DNF if it is a disjunction of one or more conjunctions of literals. For instance, the formula:

   $$(A \wedge B) \vee (C \wedge \neg D)$$

   is in DNF.
2. **Constraints in MultiplexNet**:
   Let's consider a set of constraints Φ expressed in DNF. Each constraint ϕ_i in Φ can be a disjunction of conjunctions of literals.
3. **Incorporating Constraints**:
   MultiplexNet modifies the neural network's structure to ensure that the output adheres to the constraints in Φ. This is achieved by introducing auxiliary nodes and connections in the network that correspond to the literals and conjunctions in the DNF constraints.

4. **Loss Function**:
   The loss function in MultiplexNet is designed to penalize predictions that violate the constraints. This is achieved by introducing a regularization term $\mathcal{L}_{\text{constraints}}$ that measures the degree of violation of the constraints.
   The combined loss function becomes

$$\mathcal{L}_{\text{total}} = \mathcal{L}_{\text{data}} + \lambda \mathcal{L}_{\text{constraints}}$$

   where

- $\mathcal{L}_{\text{data}}$ is the original loss function, measuring the discrepancy between the network's predictions and the true outputs.
- λ is a hyperparameter that determines the weight or importance of the regularization term.

Example Scenario: We want to predict whether a person should be approved for a credit card based on two features: their income and credit score. We have domain-specific knowledge that people with high income and high credit scores should always be approved (Tables 11.3 and 11.4).

Logical Constraint (expressed in DNF):

$$\phi : (\text{Income} > 60 \wedge \text{Credit Score} > 700) \rightarrow \text{Approval} = \text{Yes}$$

Mathematical Formulation using MultiplexNet

1. **Constraint Violation for Person P1**: Given the constraint ϕ, P1 should be approved. The violation is

Table 11.3 Given data

Person	Income (in thousands)	Credit score	True approval
P1	80	750	Yes
P2	30	600	No
P3	50	700	Yes

Table 11.4 Neural network predictions (initial, untrained)

Person	Predicted approval probability
P1	0.4
P2	0.6
P3	0.7

$$V_{P1} = \begin{cases} 1 - \text{pred}(P1) & \text{if Approval(P1) = No} \\ 0 & \text{otherwise} \end{cases}$$

$$V_{P1} = 1 - 0.4 = 0.6$$

2. **Regularization Term**: The regularization term $\mathcal{L}_{\text{constraints}}$ is

$$\mathcal{L}_{\text{constraints}} = V_{P1} = 0.6$$

3. **Data Loss**: Assuming a simple binary cross-entropy for the data loss:

$$\mathcal{L}_{\text{data}} = -\frac{1}{3}[1 \log(0.4) + 0 \log(1 - 0.6) + 1 \log(0.7)]$$

$$\mathcal{L}_{\text{data}} \approx 0.801$$

Using $\lambda = 0.1$ as the weight for the regularization term:

$$\mathcal{L}_{\text{total}} = \mathcal{L}_{\text{data}} + \lambda \mathcal{L}_{\text{constraints}}$$

$$\mathcal{L}_{\text{total}} \approx 0.801 + 0.1(0.6) = 0.861$$

Training

Given the combined loss $\mathcal{L}_{\text{total}}$, we would use an optimization algorithm (like gradient descent) to adjust the weights of the neural network. The goal is to minimize this loss.

In the first iteration, the network would recognize that its prediction for P1 is causing a high constraint violation. As training progresses, the network will adjust its weights to increase the predicted approval probability for P1, moving it closer to 1 (Yes). This adjustment will continue until the constraint violation is minimized or eliminated.

After several iterations, the network's predictions might look something like this (Table 11.5).

The network has now learned to respect the domain-specific constraint for P1 while also trying to fit the data for all persons.

Table 11.5 Neural network predictions (after training)

Person	Predicted approval probability
P1	0.9
P2	0.3
P3	0.8

Advantages of MultiplexNet:

Safety in Critical Domains: By ensuring that the network's predictions adhere to domain-specific constraints, MultiplexNet is particularly suited for safety-critical applications where deviations from known behavior can have severe consequences.

Interpretability: The constraints, being expressed in logical form, provide a level of interpretability to the network's predictions. Stakeholders can understand why certain predictions are made based on the satisfaction of the constraints.

Robustness: By adhering to domain-specific knowledge, MultiplexNet can potentially make more robust predictions, especially in scenarios where the training data might be sparse or noisy.

11.3.4 Other Notable Approaches

11.3.4.1 Semantic Loss Function

The Semantic Loss Function [6] aims to optimize the probability of the rules being true. Given a set of rules R where each rule r_i has an associated true probability $p(r_i)$, the loss function can be formulated as follows:

$$\mathcal{L}_{\text{semantic}} = -\sum_{i=1}^{n} p(r_i)\log(q(r_i)) \tag{11.3}$$

where

- $p(r_i)$ is the true probability of rule r_i being satisfied.
- $q(r_i)$ is the predicted probability of rule r_i by the neural network.

11.3.4.2 Adversarial Regularization

In adversarial regularization [7], the model is trained to be robust against adversarial perturbations that aim to violate symbolic knowledge. The loss function can be expressed as follows:

$$\mathcal{L}_{\text{adv}} = \mathcal{L}(\mathbf{x}, y) + \lambda\mathcal{L}(\mathbf{x} + \delta, y) \tag{11.4}$$

where

- $\mathbf{x}$ is the input data.
- y is the true label.
- δ is the adversarial perturbation.
- λ is a hyperparameter controlling the importance of the adversarial term.

11.3.4.3 Distillation Mechanism

In the teacher-student paradigm [8], the student model is trained to mimic the teacher model (which encodes the rules). The loss function can be described as follows:

$$\mathcal{L}_{\text{distill}} = (1 - \alpha)\mathcal{L}_{\text{data}} + \alpha\mathcal{L}_{T/S} \tag{11.5}$$

where

- $\mathcal{L}_{\text{data}}$ is the original loss function.
- $\mathcal{L}_{T/S}$ is the Kullback-Leibler divergence between the teacher's output and the student's output.
- α is a hyperparameter controlling the balance between the two terms.

11.4 Model-Based Approaches: Knowledge as an Integral Classifier Component

Model-based approaches to integrating symbolic knowledge into neural networks stand in contrast to regularization techniques. Instead of treating symbolic knowledge as constraints that guide the training process, model-based methods incorporate this knowledge directly into the model's architecture. This makes the knowledge an integral part of the classifier, ensuring that it plays a role both during training and inference.

Regularization techniques, as previously discussed, add a penalty term to the loss function to ensure that the model's predictions adhere to the provided symbolic knowledge. This approach is indirect, as the symbolic knowledge influences the model's training but is not a part of the model itself.

On the other hand, model-based methods embed the symbolic knowledge directly into the model's structure. This ensures that the knowledge is always considered, not just during training but also during inference. As a result, model-based methods can offer more robust guarantees about the model's behavior, especially in scenarios where symbolic knowledge is crucial.

11.4.1 KENN: Knowledge Enhanced Neural Networks

KENN [9] is a neural-symbolic integration framework designed to incorporate prior logical knowledge into neural networks. This is achieved by adding an additional layer to the neural network, termed the logical layer. This layer is responsible for adjusting the neural network's predictions based on the provided logical knowledge. The logical knowledge is represented as First-Order Logic (FOL) clauses. Each clause is associated with a weight, which signifies its importance or reliability. These clause weights are learnable parameters, allowing the model to determine the impact of each rule on the final predictions during the training process.

One of the strengths of KENN is its ability to handle scenarios where the provided knowledge might contain errors or when rules are not strictly adhered to in real-world situations. If training data contradicts a constraint, KENN can learn to downweight or ignore it. This makes the model robust against potentially incorrect or noisy knowledge.

Mathematical Formulation:

1. Initial Predictions: Given an input x, the neural network produces an initial prediction vector $\mathbf{p}$ where each element p_i represents the prediction for the ith label.

$$\mathbf{p} = NN(x) \tag{11.6}$$

2. Logical Adjustments: The logical layer in KENN adjusts these predictions based on the provided logical knowledge. For each clause k in the knowledge set $\mathcal{K}$, there's an associated weight w_k which is learned during training.

The adjustment for each prediction p_i based on a clause k is given by

$$\Delta p_i^k = w_k \times \mathcal{F}(p_i, k) \tag{11.7}$$

where $\mathcal{F}$ is a function that computes the adjustment for the prediction p_i based on the clause k.

The total adjustment for p_i is the sum of adjustments from all clauses:

$$\Delta p_i = \sum_{k \in \mathcal{K}} \Delta p_i^k \tag{11.8}$$

3. Modified Predictions: The modified predictions $\mathbf{p}'$ are obtained by adding the logical adjustments to the initial predictions:

$$p_i' = p_i + \alpha \times \Delta p_i \tag{11.9}$$

where α is a learnable scaling factor that determines the overall impact of the logical adjustments.

4. Loss Function: The loss function for KENN is a combination of the base neural network's loss and a regularization term based on the clause weights:

$$\mathcal{L} = \mathcal{L}_{NN}(\mathbf{p}') + \lambda \sum_{k \in \mathcal{K}} w_k^2 \tag{11.10}$$

where $\mathcal{L}_{NN}$ is the loss of the neural network given the modified predictions $\mathbf{p}'$, and λ is a regularization coefficient.

Example Scenario: Imagine we have a neural network trained to predict if a fruit is edible based on its color and size. The network produces an initial prediction p where $p = 0.7$ indicates that the fruit is 70% likely to be edible.

Logical Knowledge: We have the following logical knowledge about fruits:

1. If a fruit is red and small, it's likely to be a cherry and is edible.
2. If a fruit is green and large, it's likely to be a watermelon and is edible.

These rules can be represented as follows:

1. Red $\wedge$ Small $\rightarrow$ Edible
2. Green $\wedge$ Large $\rightarrow$ Edible

Applying KENN:

1. **Initial Prediction**:
$$p = 0.7$$
2. **Logical Adjustments**: Suppose our fruit is red and small. Based on our logical knowledge, it should be edible. The adjustment for the prediction based on the first rule is
$$\Delta p_1 = w_1 \times \mathcal{F}(p, \text{rule } 1)$$
Assuming $w_1 = 0.2$ (indicating the strength of our belief in this rule) and $\mathcal{F}$ increases the prediction by 0.1, we get
$$\Delta p_1 = 0.2 \times 0.1 = 0.02$$
3. **Modified Predictions**: The modified prediction is
$$p' = p + \alpha \times \Delta p_1$$
Assuming $\alpha = 1$ (indicating full trust in the logical adjustment), we get
$$p' = 0.7 + 1 \times 0.02 = 0.72$$

Thus, after applying KENN, our modified prediction for the fruit being edible is 72%.

The updated version of KENN is designed to handle relational data, addressing challenges like binary predicates. This is achieved by introducing multiple logical layers, allowing for a scalable and flexible treatment of rule dependencies.

11.4.2 Li and Srikumar's Method: Direct Codification of Logical Constraints

Li and Srikumar [10] proposed a unique approach to integrate logical knowledge directly into the structure of neural networks. Unlike other methods that adjust the predictions post-hoc or add regularization terms to the loss function, their method encodes the logical constraints as part of the network's architecture.

In their methodology, logical rules are transformed into a format that can be directly encoded into the neural network's architecture. This ensures that the network's predictions always adhere to the provided logical constraints, without the need for additional layers or post-processing steps. One limitation of their approach is that they restrict the logical rules to implications with exactly one consequence. This means that rules of the form A ∧ B → C can be encoded, but more complex rules with multiple consequents or disjunctions may not be directly supported. In their system, the importance or weight of each logical rule is not learned during training. Instead, these weights are set as hyperparameters. This contrasts with methods like KENN, where clause weights are learnable parameters.

Mathematical Formulation:

1. Neural Network Predictions:

Given an input x, the neural network produces a prediction vector $\mathbf{p}$ where each element p_i represents the prediction for the ith label.

$$\mathbf{p} = NN(x) \tag{11.11}$$

2. Logical Constraints:

Let's consider a set of logical rules $\mathcal{R}$ where each rule r is of the form:

$$A_1 \wedge A_2 \wedge \cdots \wedge A_n \rightarrow C \tag{11.12}$$

This can be represented as a function f_r such that

$$f_r(\mathbf{p}) = \begin{cases} 1 & \text{if } A_1, A_2, \ldots, A_n \text{ are satisfied by } \mathbf{p} \\ 0 & \text{otherwise} \end{cases} \tag{11.13}$$

3. Encoding Logical Constraints:

The logical constraints are encoded into the network's predictions using a function $\mathcal{G}$:

$$\mathcal{G}(\mathbf{p}) = \sum_{r \in \mathcal{R}} w_r \times f_r(\mathbf{p}) \tag{11.14}$$

where w_r is the weight or importance of rule r, set as a hyperparameter.

4. Modified Predictions:

The modified predictions $\mathbf{p}'$ are obtained by adding the logical constraints to the initial predictions:

$$\mathbf{p}' = \mathbf{p} + \mathcal{G}(\mathbf{p}) \tag{11.15}$$

5. Loss Function:

The loss function $\mathcal{L}$ is computed based on the modified predictions $\mathbf{p}'$:

$$\mathcal{L} = \mathcal{L}_{\text{base}}(\mathbf{p}') + \lambda \sum_{r \in \mathcal{R}} w_r^2 \tag{11.16}$$

where $\mathcal{L}_{\text{base}}$ is the base loss of the neural network given the modified predictions $\mathbf{p}'$, and λ is a regularization coefficient.

Example Scenario: Imagine we have a neural network trained to predict if an animal is a mammal based on its features like having fur and giving live birth. The network produces an initial prediction p for a given animal.

Logical Knowledge: We have the following logical knowledge about animals:

1. If an animal has fur and gives live birth, it's likely to be a mammal.
2. If an animal lays eggs and has scales, it's unlikely to be a mammal.

These rules can be represented as follows:

1. Fur $\wedge$ LiveBirth $\rightarrow$ Mammal
2. Eggs $\wedge$ Scales $\rightarrow$ $\neg$Mammal

Applying Li and Srikumar's Method:

1. **Initial Prediction**: For a given animal with fur and that gives live birth:

   $$p = 0.6$$

   (60% confidence that it's a mammal)
2. **Logical Adjustments**: Based on our logical knowledge, the animal should be a mammal. The adjustment for the prediction based on the first rule is

   $$f_1(p) = \begin{cases} 1 & \text{if Fur and LiveBirth are true} \\ 0 & \text{otherwise} \end{cases}$$

   Given that both conditions are true, $f_1(p) = 1$.
3. **Encoding Logical Constraints**: Using the function $\mathcal{G}$:

   $$\mathcal{G}(p) = w_1 \times f_1(p)$$

   Assuming $w_1 = 0.3$ (indicating the strength of our belief in this rule), we get

   $$\mathcal{G}(p) = 0.3 \times 1 = 0.3$$

4. **Modified Predictions**: The modified prediction is

$$p' = p + \mathcal{G}(p)$$

$$p' = 0.6 + 0.3 = 0.9$$

Thus, after applying Li and Srikumar's method, our modified prediction for the animal being a mammal is 90%.

11.4.3 Relational Neural Machines (RNM)

RNM [11] integrates a neural network model with a First-Order Logic (FOL) reasoner. This dual-stage approach allows for the seamless combination of deep learning's expressive power with the structured reasoning capabilities of FOL. In the first stage, the neural network is used to calculate initial predictions for the atomic formulas. These predictions serve as the base upon which the logical reasoning is built. In the second stage, a graphical model represents a probability distribution over the set of atomic formulas. This model captures the relationships and constraints specified by the logical knowledge. To obtain the final predictions, a Maximum a Posteriori (MAP) estimation is performed. This finds the most probable assignment to the grounded atoms, given the output of the neural network and the set of constraints. The MAP estimation ensures that the final predictions respect the logical constraints while also being consistent with the neural network's initial predictions.

One of the distinguishing features of RNM is its ability to learn the weights associated with logical clauses. This adaptability makes it suitable for scenarios where the exact importance or reliability of each logical rule is not known a priori. The learnable clause weights allow the model to adjust the influence of each rule based on the data, providing a balance between data-driven learning and symbolic reasoning.

Mathematical Formulation:
Given:

- $\mathcal{X}$: The set of input features.
- $\mathcal{Y}$: The set of atomic formulas or predicates.
- $\mathbf{p}$: A vector representing the initial predictions from the neural network for each atomic formula in $\mathcal{Y}$.

1. **Neural Network Predictions**: The neural network f_{NN} maps input features to initial predictions:

$$\mathbf{p} = f_{NN}(\mathcal{X})$$

2. **Logical Constraints**: For each rule r in the knowledge base, there exists a function f_r that captures the logical constraint imposed by the rule. This function takes the initial predictions $\mathbf{p}$ and returns an adjustment based on the rule.
3. **Graphical Model**: A graphical model G is constructed where nodes represent atomic formulas in $\mathcal{Y}$ and edges capture the relationships specified by the logical rules. The probability distribution over $\mathcal{Y}$ is given by

$$P(\mathcal{Y}|\mathcal{X}) = \frac{1}{Z} \exp\left(\sum_r w_r \times f_r(\mathbf{p})\right)$$

where Z is a normalization constant and w_r represents the weight of rule r.
4. **Maximum a Posteriori (MAP) Estimation**: The goal is to find the most probable assignment to the grounded atoms in $\mathcal{Y}$ given the neural network's predictions and the logical constraints:

$$\mathbf{p}' = \arg\max_{\mathcal{Y}} P(\mathcal{Y}|\mathcal{X})$$

5. **Learning Clause Weights**: The weights w_r associated with each rule r can be learned during training by optimizing the likelihood of the observed data given the predictions:

$$\mathcal{L}(\mathcal{X}, \mathcal{Y}; w) = \sum_{i=1}^{N} \log P(y_i|x_i; w)$$

where N is the number of training examples, and x_i, y_i are the input features and corresponding atomic formula for the ith example, respectively.

Example Scenario: Suppose we want to predict whether a student will pass or fail based on their study hours and participation in extracurricular activities. We have prior knowledge in the form of logical rules that students who study more than a certain number of hours are likely to pass and those who participate in too many extracurricular activities are likely to fail due to lack of study time.
Data:

Student	Study hours	Extracurricular hours	Pass/fail
A	5	1	?
B	2	4	?
C	4	2	?

Logical Rules:

1. If Study Hours > 4, then likely to Pass.
2. If Extracurricular Hours > 3, then likely to Fail.

Neural Network Predictions: Suppose our neural network, based on its training, predicts the following probabilities of passing for each student:

- $P(A) = 0.7$
- $P(B) = 0.3$
- $P(C) = 0.6$

Graphical Model Adjustments: Based on our logical rules:

- Student A's prediction is increased by a weight w_1 because they studied for more than 4 h.
- Student B's prediction is decreased by a weight w_2 because they participated in extracurricular activities for more than 3 h.
- Student C's prediction is increased by w_1 but also slightly decreased by w_2 as they fall into both rule categories.

Final Predictions: Using the MAP estimation, we adjust the initial predictions:

- $P'(A) = 0.7 + w_1$
- $P'(B) = 0.3 - w_2$
- $P'(C) = 0.6 + w_1 - w_2$

Assuming $w_1 = 0.1$ and $w_2 = 0.2$:

- $P'(A) = 0.8$
- $P'(B) = 0.1$
- $P'(C) = 0.5$

Thus, based on the RNM methodology, Students A and C are predicted to pass, while Student B is predicted to fail.

11.5 Comparative Analysis of Symbolic Knowledge Integration Methods

In the realm of integrating symbolic knowledge with neural networks, various methodologies have been proposed, each with its unique strengths, weaknesses, and applicability. This section provides a comparative analysis of the methods discussed, aiming to offer insights into their relative advantages and potential use cases.

Performance Metrics

- **Logic Tensor Network (LTN) and Semantic-Based Regularization (SBR)**: These methods, relying on fuzzy logic semantics, have shown robust performance in scenarios where the knowledge is imprecise. However, they might be computationally intensive for large-scale problems.
- **DL2**: While DL2 offers a unique perspective by focusing on regression tasks, its specificity might limit its applicability in broader domains.
- **MultiplexNet**: Its strength lies in safety-critical domains, ensuring complete satisfaction of constraints. However, the complexity might increase with the number of constraints.

- **KENN**: KENN stands out in scenarios where the given knowledge contains errors or when rules are softly satisfied. Its adaptability to real-world imprecisions is commendable.
- **Li and Srikumar's Method**: The direct codification of logical constraints makes it efficient, but the restriction to implications with one consequent might be a limitation for complex knowledge bases.
- **Relational Neural Networks (RNM)**: RNM's two-stage approach, integrating a neural network model with a FOL reasoner, makes it versatile. However, the optimization problem at inference time can be computationally demanding.

Suitability for Specific Tasks

- **LTN and SBR**: Best suited for problems where knowledge can be expressed in First-Order Logic and where imprecision is expected.
- **DL2**: Specifically designed for regression tasks with comparison constraints.
- **MultiplexNet**: Ideal for safety-critical domains where constraint satisfaction is paramount.
- **KENN**: Highly adaptable, suitable for scenarios with imprecise rules or potential knowledge errors.
- **Li and Srikumar's Method**: Suitable for tasks where knowledge can be expressed as implications with a single consequent.
- **RNM**: Versatile and can be applied to a wide range of tasks, especially where a combination of initial predictions and logical constraints is needed.

Computational Efficiency

- **LTN and SBR**: Might be computationally intensive for large datasets or complex knowledge bases.
- **DL2**: Efficient for regression tasks but might not scale well for broader applications.
- **MultiplexNet**: Efficiency decreases as the number of constraints increases.
- **KENN**: Offers a good balance between computational efficiency and adaptability.
- **Li and Srikumar's Method**: Efficient due to direct codification but might face challenges with complex knowledge bases.
- **RNM**: The two-stage approach can be computationally demanding, especially during the optimization phase.

Ease of Implementation and Scalability

- **LTN and SBR**: Require expertise in fuzzy logic and might face challenges in scalability.
- **DL2**: Straightforward for regression tasks but might not scale well for broader applications.
- **MultiplexNet**: Implementation complexity increases with the number of constraints.
- **KENN**: Relatively easy to implement and scales well for real-world scenarios.
- **Li and Srikumar's Method**: Implementation is straightforward, but scalability might be a concern for complex knowledge bases.

- **RNM**: Implementation can be challenging due to the two-stage approach, and scalability might be an issue for large-scale problems.

Interpretability

- **LTN and SBR**: The use of fuzzy logic semantics provides a degree of interpretability, allowing for an understanding of how constraints influence the model.
- **DL2**: Its focus on regression tasks with comparison constraints offers clear interpretability in terms of how the model respects these constraints.
- **MultiplexNet**: The emphasis on safety-critical domains ensures that the model's decisions are interpretable within the context of the constraints.
- **KENN**: The ability to adapt to imprecise rules provides insights into how the model adjusts to real-world uncertainties.
- **Li and Srikumar's Method**: The direct codification approach offers clarity, but the restriction to specific implications might limit broader interpretability.
- **RNM**: The two-stage approach, especially the integration with a FOL reasoner, enhances the model's interpretability.

Flexibility and Adaptability

- **LTN and SBR**: Their reliance on fuzzy logic semantics allows for flexibility in handling imprecise knowledge.
- **DL2**: While efficient for its specific domain, its adaptability to other tasks might be limited.
- **MultiplexNet**: Its design ensures adaptability within safety-critical domains, but might be less flexible outside this context.
- **KENN**: Stands out in terms of flexibility, especially in scenarios with uncertain or imprecise rules.
- **Li and Srikumar's Method**: The method is flexible within its domain of implications with a single consequent but might face challenges outside this scope.
- **RNM**: Its design ensures adaptability across a range of tasks, but the two-stage approach might limit flexibility in some scenarios.

In conclusion, the choice of methodology largely depends on the specific requirements of the task at hand, the nature of the symbolic knowledge available, and the computational resources. While some methods excel in precision and adaptability, others stand out in terms of computational efficiency and scalability. It's crucial to align the method's strengths with the problem's demands to achieve optimal results.

11.6 Conclusion

The integration of symbolic knowledge with neural networks has emerged as a pivotal area of research in the domain of artificial intelligence. This synthesis aims to combine the interpretability and structured reasoning of symbolic systems with the adaptability and learning capabilities of neural networks. Throughout this chapter, we delved deep into various methodologies that facilitate this integration, each offering unique perspectives and solutions to the challenges posed by the fusion of symbolic and sub-symbolic paradigms.

We began our exploration with an understanding of the dual approaches to learning with prior knowledge. The distinction between regularization approaches, which treat logical rules as constraints on neural network predictions, and model-based approaches, where knowledge becomes an integral part of the classifier, set the stage for our subsequent discussions.

In our journey through regularization approaches, we encountered methodologies like Logic Tensor Network (LTN), Semantic-Based Regularization (SBR), DL2, and MultiplexNet. Each of these methods, while sharing the common goal of maximizing constraint satisfiability, brought unique strategies and nuances to the table. From leveraging fuzzy logic semantics in LTN and SBR to focusing on regression tasks in DL2 and ensuring complete constraint satisfaction in safety-critical domains with MultiplexNet, the landscape of regularization approaches is rich and varied.

On the other hand, model-based approaches, including KENN, Li and Srikumar's method, and Relational Neural Networks (RNM), showcased the power of embedding knowledge directly into the model structure. These methods not only ensure knowledge enforcement during inference but also offer the potential to derive rule weights directly from data, a feature not commonly found in regularization-based methods.

As we conclude our exploration of integrating symbolic knowledge into neural networks, it's evident that the harmonization of symbolic reasoning with neural adaptability holds immense potential. However, while we've delved into methods that utilize pre-existing symbolic rules, an intriguing question arises: Can neural networks learn or extract symbolic knowledge directly from data? This leads us to our next chapter, where we will embark on an exploration of cutting-edge methods that aim to derive symbolic knowledge from data-driven contexts.

References

1. Tian, Y., Zhang, Y.: A comprehensive survey on regularization strategies in machine learning. Inf. Fusion **80**, 146–166 (2022)
2. Badreddine, S., Garcez, A.d., Serafini, L., Spranger, M.: Logic tensor networks. Artif. Intell. **303**, 103649 (2022)
3. Diligenti, M., Gori, M., Sacca, C.: Semantic-based regularization for learning and inference. Artif. Intell. **244**, 143–165 (2017)
4. Fischer, M., Balunovic, M., Drachsler-Cohen, D., Gehr, T., Zhang, C., Vechev, M.: Dl2: training and querying neural networks with logic. In: International Conference on Machine Learning, pp. 1931–1941. PMLR (2019)
5. Hoernle, N., Karampatsis, R.M., Belle, V., Gal, K.: Multiplexnet: Towards fully satisfied logical constraints in neural networks. In: Proceedings of the AAAI Conference on Artificial Intelligence, vol. 36, pp. 5700–5709 (2022)

6. Xu, J., Zhang, Z., Friedman, T., Liang, Y., Broeck, G.: A semantic loss function for deep learning with symbolic knowledge. In: International Conference on Machine Learning, pp. 5502–5511. PMLR (2018)
7. Sato, M., Suzuki, J., Kiyono, S.: Effective adversarial regularization for neural machine translation. In: Proceedings of the 57th Annual Meeting of the Association for Computational Linguistics, pp. 204–210 (2019)
8. Qin, J., Liang, X., Hong, Y., Tang, J., Lin, L.: Neural-symbolic solver for math word problems with auxiliary tasks. arXiv preprint arXiv:2107.01431 (2021)
9. Daniele, A., Serafini, L.: Knowledge enhanced neural networks. In: PRICAI 2019: Trends in Artificial Intelligence: 16th Pacific Rim International Conference on Artificial Intelligence, Cuvu, Yanuca Island, Fiji, August 26–30, 2019, Proceedings, Part I 16, pp. 542–554. Springer (2019)
10. Li, T., Srikumar, V.: Augmenting neural networks with first-order logic. arXiv preprint arXiv:1906.06298 (2019)
11. Marra, G., Diligenti, M., Giannini, F., Gori, M., Maggini, M.: Relational neural machines. In: ECAI 2020, pp. 1340–1347. IOS Press (2020)

Chapter 12
Neural Extraction of Symbolic Knowledge

"Much to learn, you still have."

– Yoda, *Star Wars*

12.1 Introduction

As we venture into the intricate world of neural networks, one of the most compelling challenges we face is deciphering the complex, often inscrutable layers of learning and decision-making processes these networks undergo. This chapter aims to demystify this aspect by delving into the realm of symbolic knowledge extraction from neural networks.

Symbolic knowledge extraction is akin to translating a cryptic, ancient manuscript into a modern-day language. Consider a neural network trained to diagnose diseases from medical images. While it may perform with high accuracy, the rationale behind its diagnoses often remains hidden within its multiple layers. The goal of symbolic knowledge extraction is to unravel these layers and translate them into a form that is comprehensible, much like extracting specific diagnostic criteria from the network's intricate processing patterns.

In this chapter, we will explore various approaches to this translation process. We start with Direct Rule Extraction Approaches, where we look into methods like TREPAN [1], DeepRED [2], and REANN [3], each offering a unique lens to view and understand the decision-making fabric of neural networks. Then, we delve into Relational Knowledge Extraction Approaches, focusing on how relationships and patterns within data are deciphered by these advanced systems.

B. P. Bhuyan et al., *Neuro-Symbolic Artificial Intelligence*, Studies in Computational Intelligence 1176, https://doi.org/10.1007/978-981-97-8171-3_12

Throughout, we aim not only to provide a comprehensive understanding of these methodologies but also to illuminate the significance of such extraction in making AI more transparent and trustworthy in critical applications, from healthcare to finance.

By the end of this chapter, the reader will gain a deeper appreciation of the efforts and ingenuity involved in translating the complex language of neural networks into the clear dialect of symbolic knowledge, making AI's decision-making process not just effective but also interpretable and accountable.

12.2 Approaches for Symbolic Knowledge Extraction

We explore the various methodologies used to extract symbolic knowledge from neural networks, offering insights into how these techniques convert the complex, often hidden layers of neural processing into interpretable and actionable insights. The basic approaches are as follows:

Direct Rule Extraction Approaches: Direct Rule Extraction Approaches in neural networks focus on translating the complex and often opaque decision-making processes of these networks into explicit, understandable rules. This method involves dissecting the network's structure and operations to distill its learned knowledge into a set of logical rules or decision trees that are human-readable. Such approaches are particularly valuable in domains where interpretability and justification of AI decisions are crucial, such as in healthcare diagnostics, financial decision-making, or legal assessments. By transforming the intricate patterns recognized by a neural network into clear, articulated rules, Direct Rule Extraction makes the AI's reasoning process more transparent and accessible to users.

Feature Visualization and Analysis: Feature Visualization and Analysis in the context of neural networks is a technique focused on understanding and interpreting the features learned by these networks. By visualizing the activations and representations within the network layers, this approach aims to demystify what the network is "seeing" or "learning" from the input data. Techniques like activation maximization, where the input that maximally activates certain neurons is generated, help uncover what features the network focuses on when making decisions. Additionally, layer-wise visualization can reveal how different layers of the network contribute to the final decision, shedding light on the hierarchical process of feature extraction and combination in deep learning models. This method is particularly beneficial for enhancing the interpretability of complex models in fields like computer vision and natural language processing, where understanding the internal workings of the model is key for trust and further improvement.

Knowledge Distillation: Knowledge distillation [4] is a technique used in machine learning where knowledge from a complex, often large, and opaque model (known as the "teacher" model) is transferred to a simpler, more interpretable model (the "student" model). The goal is to make the insights gained by the complex model accessible and understandable without significantly compromising performance.

In this process, the student model is trained not just to predict the same outputs as the teacher model but also to mimic how the teacher model processes data. This is achieved by using the outputs of the teacher model as a form of "soft labels" or guidance for training the student model. As a result, the student model learns to approximate the complex decision boundaries and the nuances captured by the teacher model but in a simpler framework.

Knowledge distillation is particularly valuable in scenarios where the interpretability and efficiency of a model are as important as its accuracy. By distilling the knowledge of a sophisticated neural network into a more straightforward model, such as a shallow neural network or a decision tree, it becomes easier to understand, explain, and deploy the model in resource-constrained environments. This method bridges the gap between achieving state-of-the-art performance with complex models and the need for practical, interpretable AI solutions.

Relational Knowledge Extraction: Relational Knowledge Extraction in the realm of neural networks refers to the process of identifying and understanding the relationships and patterns within data as learned by these networks. Unlike methods that focus on individual features or direct rule extraction, this approach emphasizes the extraction of knowledge in the form of relational structures or patterns.

This method often involves leveraging techniques from relational databases, knowledge graphs, or graph neural networks to analyze and interpret the relationships between different entities in the data. For instance, by constructing a knowledge graph from the outputs of a neural network, one can visualize and understand how different concepts or entities are interrelated as per the model's learning.

The primary advantage of Relational Knowledge Extraction is its ability to capture and represent complex dependencies and relationships within data, which is crucial for tasks requiring an in-depth understanding of the data's inherent structure. This approach is particularly useful in fields such as natural language processing, where understanding the relationships between words and phrases is essential, or in social network analysis, where the interconnections between individuals or groups play a significant role.

By focusing on relational patterns, Relational Knowledge Extraction provides a deeper and more holistic view of the learned knowledge, enabling a more nuanced interpretation of neural network models. This enhances the ability to derive meaningful insights, especially in complex domains where the relationships within the data are as important as the individual data points themselves.

12.3 Direct Rule Extraction Approaches

In the realm of neural networks, interpretability remains a significant challenge. While these models are known for their high accuracy and adaptability, understanding their decision-making process is not straightforward. Direct Rule Extraction Approaches aim to bridge this gap by converting the learned knowledge of neural networks into human-readable symbolic rules.

Given a neural network f that maps an input x to an output y, i.e., $y = f(x)$, the goal of direct rule extraction is to find a symbolic representation g such that $y \approx g(x)$.

1. **Network Simplification**: Before extracting rules, it's often beneficial to simplify the network. This can involve pruning less important neurons or connections. Mathematically, given a weight matrix W, we aim to find a simplified matrix W' such that the performance difference $|f(x; W) - f(x; W')|$ is minimized.
2. **Rule Extraction**: Once the network is simplified, the next step is to extract rules. For a neuron n with activation function σ, weights w, and bias b, a basic rule can be represented as follows:

$$\sigma(w \cdot x + b) > \theta \implies \text{Rule is activated}$$

   where θ is a threshold.
3. **Rule Aggregation**: After extracting rules from individual neurons, they can be aggregated based on the network's architecture. For instance, in a feedforward network, rules from earlier layers can influence those from later layers.

We will delve deeper into specific methods and their intricacies in the subsequent subsections.

12.3.1 TREPAN: Decision Trees from Trained Neural Networks

TREPAN [1] is a method designed to extract decision trees from trained neural networks. It was introduced as a way to interpret the complex decision boundaries learned by neural networks in a more human-understandable format. Let us dive into TREPAN's methodology.

Initialization in TREPAN:

The initialization phase sets the foundation for the tree extraction process. Begin by defining the dataset on which the neural network was trained.

$$D = \{(x_1, y_1), (x_2, y_2), \ldots, (x_n, y_n)\}$$

where x_i represents the input features and y_i is the corresponding output or label. The dataset D contains n samples.

The decision tree starts with a single node, the root.

$$\text{root} = \text{Node}(D)$$

Here, the function Node() creates a new tree node containing the dataset D.

Set the current node to the root.

$$\text{node}_{\text{current}} = \text{root}$$

Depending on the specific implementation or variant of TREPAN, initialize other parameters such as the following:

- Maximum depth of the tree, d_{max}.
- Minimum samples required to split a node, s_{min}.
- A threshold for node purity.

Node splitting is the next crucial step in the TREPAN algorithm, as it determines how the decision tree grows and refines its structure.

Node Splitting in TREPAN:

1. **Selection of the Current Node**:
   Start by selecting the current node, $\text{node}_{\text{current}}$, which was initialized as the root in the beginning. As the algorithm progresses, this node will change based on the tree's growth.
2. **Determine Splitting Necessity**:
   Check various criteria to decide if the node should be split:

   - Node purity. There's no need to split if the node's dataset is pure (all samples belong to a single class).
   - Maximum depth, d_{max}. If the node's depth reaches the maximum allowed depth, stop splitting.
   - Minimum samples, s_{min}. If the number of samples in the node is less than the minimum threshold, avoid splitting.

3. **Feature Selection for Splitting**:
   The crux of the node-splitting process is determining the optimal feature to split on. The algorithm gauges the significance of each feature by querying the neural network with slightly altered samples and observing the resultant output variations. The feature, say $\mathcal{F}$, that induces the most pronounced change in the network's output is earmarked for splitting. This can be represented as follows:

   $$\mathcal{F}^* = \arg\max_{\mathcal{F}} \Delta O(\mathcal{F})$$

   where $\Delta O(\mathcal{F})$ denotes the change in output due to perturbations in feature $\mathcal{F}$.
4. **Determine Split Points**:
   With the feature in hand, the next step is pinpointing the exact values or ranges at which splits should occur. Methods like binary splits or histogram-based multi-way splits can be employed for continuous features. For categorical features, each unique value can serve as a split point. An M-of-N split can be represented as follows:

$$S = \bigvee_{i=1}^{M} \bigwedge_{j=1}^{N} c_{ij}$$

where c_{ij} are conditions on the input features, and S is the split condition.

5. **Query Neural Network**:
For each potential split S, query the neural network with samples x that satisfy the conditions of the split. Let $NN(x)$ be the response of the neural network for input x.
6. **Quality of Split**:
The quality of a split, Q, can be determined using a metric such as fidelity to the neural network's decisions. For a split S with subsets D_1 and D_2 (resulting from the split), the quality is

$$Q(S) = \frac{|D_1|}{|D|} \times I(D_1) + \frac{|D_2|}{|D|} \times I(D_2)$$

where $I(D)$ is the information content (or purity) of dataset D.
7. **Create Child Nodes**:
The dataset of the current node is then partitioned based on the chosen split points, leading to the formation of child nodes.
8. **Recursion**:
Recursively apply the node-splitting process to each child node. The recursion continues until a stopping criterion, such as a maximum tree depth $d_{\max}$ or a minimum node purity $I_{\min}$, is met:

$$\text{Stop if depth}(\text{node}_{\text{current}}) \geq d_{\max} \text{ or } I(\text{node}_{\text{current}}) \geq I_{\min}$$

9. **Tree Pruning (Optional)**:
Optionally prune the tree to avoid overfitting.

For a given dataset D,

Legs	Can fly	Has gills	Label
2	1	0	Bird
4	0	0	Mammal
0	0	1	Fish
2	0	0	Mammal
4	0	1	Fish
2	1	0	Bird
4	0	0	Mammal

Entropy for a dataset D with classes $C_1, C_2, ...C_k$ is given by

$$E(D) = -\sum_{i=1}^{k} p(C_i) \log_2 p(C_i)$$

where $p(C_i)$ is the proportion of samples in class C_i in D.

Entropy of the Entire Dataset D

$$E(D) = -\left(\frac{2}{7}\log_2\frac{2}{7} + \frac{3}{7}\log_2\frac{3}{7} + \frac{2}{7}\log_2\frac{2}{7}\right)$$

$$E(D) \approx 1.5567$$

Node Splitting at Root based on "has_gills":

- D_1: Samples with "has_gills" = 1
- D_2: Samples with "has_gills" = 0

Entropy of D_1 and D_2:

$$E(D_1) = 0$$

$$E(D_2) \approx 0.9709$$

Quality of Split based on "has_gills":

$$Q(S) \approx 0.6935$$

Node Splitting of D_2 based on "can_fly":

- $D_{2.1}$: Samples in D_2 with "can_fly" = 1
- $D_{2.2}$: Samples in D_2 with "can_fly" = 0

Entropy of $D_{2.1}$ and $D_{2.2}$:

$$E(D_{2.1}) = 0$$

$$E(D_{2.2}) = 0$$

Quality of Split based on "can_fly" for D_2:

$$Q(S) = 0$$

(since both subsets are pure)

The decision tree is then constructed based on these splits and their qualities.

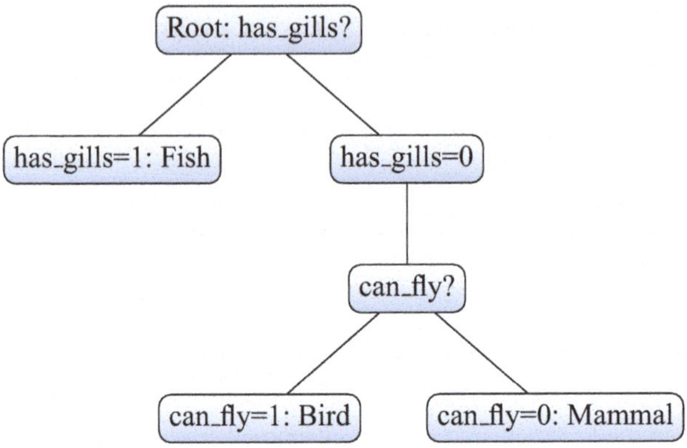

12.3.2 DeepRED (Deep Rule Extraction from Deep Neural Networks)

DeepRED [2] is another method to extract symbolic rules from trained deep neural networks. Let's delve deeper into the methodology of DeepRED, starting with the Network Transformation.

Network Transformation

The primary goal of the network transformation step is to simplify the deep neural network structure to facilitate rule extraction. This is achieved by reducing the network to a single hidden layer. DeepRED ensures that each neuron in this layer can be directly mapped to a symbolic rule.

Given two consecutive layers with weight matrices W_i and W_{i+1} and bias vectors b_i and b_{i+1}, the transformation for the weights and biases is

$$W_i' = W_{i+1} \times \sigma(W_i)$$

$$b_i' = W_{i+1} \times \sigma(b_i) + b_{i+1}$$

Here, σ denotes the activation function. This transformation ensures that the behavior of the network remains unchanged, even though its structure is simplified.

The type of activation function can influence the transformation process. For instance, ReLU (Rectified Linear Unit) activations introduce non-linearity without bounding the output. In contrast, sigmoid or tanh activations bound the output between specific ranges. DeepRED handles these differences by adjusting the transformation equations accordingly.

After the weight transformation, the biases are recalculated. These biases play a crucial role in the subsequent rule extraction process as they help determine the thresholds for the input features in the rules.

Rule Extraction

Each neuron in the single hidden layer can be directly mapped to a rule. For a given neuron j, the rule is determined based on the weights connecting the input features to the neuron and the neuron's bias. Positive weights indicate a positive correlation with the neuron's activation, while negative weights indicate a negative correlation. The rule for neuron j can be represented as follows:

$$R_j : \bigwedge_{i:W'_{ij}>0} (x_i > \theta_{ij}) \wedge \bigwedge_{i:W'_{ij}<0} (x_i < \theta_{ij}) \rightarrow y = c_j$$

Here,

- $\bigwedge_{i:W'_{ij}>0}(x_i > \theta_{ij})$ is the part of the rule that checks for all input features i with positive weights W'_{ij}. The condition $(x_i > \theta_{ij})$ means that for the rule to be satisfied, the value of feature i should be greater than a threshold θ_{ij}.
- $\bigwedge_{i:W'_{ij}<0}(x_i < \theta_{ij})$ is the part of the rule that checks for all input features i with negative weights W'_{ij}. The condition $(x_i < \theta_{ij})$ means that for the rule to be satisfied, the value of feature i should be less than a threshold θ_{ij}.
- $y = c_j$ considers that if all the conditions in the rule are satisfied, the output y is predicted to be a certain class c_j.

The rule says: "If these particular conditions on the input features are met, then the output is this specific class."

Rule Pruning

Post-rule extraction, DeepRED identifies and removes rules that might be redundant or not significantly contributing to the network's decision-making. This pruning is based on the coverage and accuracy of each rule on the training data.

12.3.3 REANN (Rule Extraction from ANNs)

REANN [3] aims to extract comprehensible rules from trained neural networks, especially from feedforward networks. The primary objective is to represent the decision boundaries learned by the neural network in a more interpretable form, typically as IF-THEN rules.

Before extracting rules, the continuous-valued activation functions of the neurons are discretized. This is done to simplify the rule extraction process and to ensure that the rules are interpretable. Given an activation function $f(x)$, it is transformed into a binary function $g(x)$ such that

$$g(x) = \begin{cases} 1 & \text{if } f(x) \geq \theta \\ 0 & \text{otherwise} \end{cases} \tag{12.1}$$

where θ is a threshold value.

Rule Extraction in REANN

In REANN, the rule extraction process is a crucial step that translates the discretized neural network's behavior into symbolic rules. This process is designed to capture the essence of the network's decision-making in a human-understandable format.

Given the discretized neural network, each neuron's output can be seen as a binary decision based on its input. Specifically, for a neuron i with input x_i and weight w_i, the decision boundary is defined by the following equation:

$$w_i \cdot x_i + b_i = 0 \tag{12.2}$$

where b_i is the bias of the neuron.

The output of the neuron, after discretization, can be represented as follows:

$$y_i = \begin{cases} 1 & \text{if } w_i \cdot x_i + b_i \geq 0 \\ 0 & \text{otherwise} \end{cases} \tag{12.3}$$

This binary output, y_i, serves as the input to the subsequent layer. By traversing the network layer by layer, we can construct compound rules that describe the behavior of the network. Each neuron's decision boundary contributes a clause to the rule, and the conjunction of these clauses forms the complete rule for a particular output class.

For instance, consider a simple neural network with two input neurons x_1 and x_2, and one output neuron y. If the decision boundaries for the input neurons are $w_1 \cdot x_1 + b_1 = 0$ and $w_2 \cdot x_2 + b_2 = 0$, respectively, then the rule for the output neuron y can be constructed as follows:

$$y = 1 \iff (w_1 \cdot x_1 + b_1 \geq 0) \land (w_2 \cdot x_2 + b_2 \geq 0) \tag{12.4}$$

This rule states that the output neuron y will activate (i.e., have an output of 1) if and only if both input neurons' conditions are satisfied.

In more complex networks with multiple layers, the rule extraction process follows a similar logic but involves a recursive construction of rules, starting from the input layer and moving towards the output layer. Each layer's rules are constructed based on the previous layer's outputs and the current layer's weights and biases.

Rule Pruning in REANN

After extracting rules from a neural network, it's common to end up with a large set of rules, some of which might be redundant or unnecessary. Rule pruning in REANN aims to simplify the extracted rule set without compromising the accuracy of the representation.

The primary objective of rule pruning is to maintain the fidelity of the original neural network's decisions while reducing the complexity of the rule set. This is achieved through a two-step process:

1. **Redundancy Removal**: This step identifies and removes rules that are redundant. A rule is considered redundant if its removal does not affect the overall decision-making of the rule set. Mathematically, given a rule set R and a rule $r \in R$, r is redundant if
$$R - \{r\} \equiv R \tag{12.5}$$
where $\equiv$ denotes equivalence in decision-making.
2. **Simplification**: After removing redundant rules, the next step is to simplify the remaining rules. This involves reducing the number of clauses in each rule without altering its meaning. Given a rule r with clauses $C_1, C_2, \ldots, C_n$, a clause C_i can be removed if
$$r - C_i \equiv r \tag{12.6}$$
This step ensures that each rule is as concise as possible while still accurately representing the neural network's decisions.

By applying these two steps iteratively, REANN produces a pruned and simplified rule set that captures the essence of the neural network's behavior in a more human-understandable format.

Suppose we have a trained neural network that classifies whether a given fruit is an apple or an orange based on two features: `color` and `size`. The neural network has been trained with the following data:

Color (Red = 1, Orange = 0)	Size (Big = 1, Small = 0)	Fruit (Apple = 1, Orange = 0)
1	1	1
1	0	1
0	1	0
0	0	0

We discretize the continuous outputs of the neural network into binary decisions. For simplicity, let's assume the neural network's output is already binary.

From the trained neural network, we extract the following rules:

- If `color` is Red AND `size` is Big, then Fruit is Apple.
- If `color` is Red AND `size` is Small, then Fruit is Apple.
- If `color` is Orange AND `size` is Big, then Fruit is Orange.
- If `color` is Orange AND `size` is Small, then Fruit is Orange.

Mathematically, these rules can be represented as follows:

$$
\begin{aligned}
R_1 &: (\text{color} = 1 \wedge \text{size} = 1) \rightarrow \text{Fruit} = 1 \\
R_2 &: (\text{color} = 1 \wedge \text{size} = 0) \rightarrow \text{Fruit} = 1 \\
R_3 &: (\text{color} = 0 \wedge \text{size} = 1) \rightarrow \text{Fruit} = 0 \\
R_4 &: (\text{color} = 0 \wedge \text{size} = 0) \rightarrow \text{Fruit} = 0
\end{aligned}
$$

Upon analyzing the rules, we notice that the `size` feature doesn't affect the classification when the `color` is Red. Thus, we can prune the rule set by removing the redundancy:

- If `color` is Red, then Fruit is Apple.
- If `color` is Orange AND `size` is Big, then Fruit is Orange.
- If `color` is Orange AND `size` is Small, then Fruit is Orange.

Mathematically,

$$
\begin{aligned}
R_1' &: (\text{color} = 1) \rightarrow \text{Fruit} = 1 \\
R_3 &: (\text{color} = 0 \wedge \text{size} = 1) \rightarrow \text{Fruit} = 0 \\
R_4 &: (\text{color} = 0 \wedge \text{size} = 0) \rightarrow \text{Fruit} = 0.
\end{aligned}
$$

12.4 Relational Knowledge Extraction Approaches

While direct rule extraction methods focus on deriving explicit rules from neural networks, relational knowledge extraction approaches emphasize the extraction of knowledge in the form of relational patterns or structures. These methods often leverage the power of relational databases, knowledge graphs, and other structured data representations to extract, represent, and utilize symbolic knowledge. The primary advantage of these approaches is their ability to capture complex relationships and dependencies between entities, which can be crucial for tasks that require a deeper understanding of the data's underlying structure.

Relational knowledge extraction methods typically involve the following:

- Representation Learning: Transforming the data into a suitable relational representation, often leveraging embeddings or other feature extraction techniques.
- Relational Pattern Mining: Identifying frequent patterns, associations, or other relational structures from the data.
- Knowledge Graph Construction: Building a knowledge graph or other structured knowledge representation based on the identified patterns and relationships.
- Neural Network Integration: Integrating the extracted relational knowledge into neural network architectures, either as auxiliary information, constraints, or hybrid models that combine symbolic and sub-symbolic reasoning.

In summary, relational knowledge extraction methods provide powerful tools for uncovering and utilizing complex relational patterns and structures in data. By transforming data into relational formats, mining for significant patterns, constructing knowledge graphs, and integrating these insights into neural networks, we can achieve a deeper and more nuanced understanding of the underlying relationships in various datasets.

As we conclude this discussion on relational knowledge extraction, it's important to recognize the evolving landscape of neural network architectures that further enhance our ability to interpret and utilize complex data relationships. This sets the stage for our next chapter, where we delve into the world of Graph Neural Networks (GNNs). GNNs represent a pivotal advancement in this realm, offering sophisticated frameworks that are specifically designed to handle graph-structured data. We will explore how these networks operate, their applications, and the unique advantages they bring to the field of symbolic knowledge extraction and representation, building upon the foundations laid in our exploration of relational knowledge extraction in the next chapter.

12.5 Conclusion

As we conclude our exploration of Neural Extraction of Symbolic Knowledge, we have traversed the landscape of translating neural networks' complex decision-making into a more interpretable and human-understandable form. From the direct rule extraction methods like TREPAN, DeepRED, and REANN to the relational knowledge extraction approaches, we have seen how these methodologies not only enhance the transparency of AI but also contribute significantly to its applicability in real-world scenarios.

The journey through this chapter has illuminated the intricate process of making sense of the abstract patterns learned by neural networks, akin to translating a sophisticated language into a simpler, more accessible one. This endeavor is not just about achieving clarity but also about building trust and reliability in AI applications, particularly in sectors where understanding the "why" behind an AI's decision is as crucial as the decision itself.

As we move forward, the next chapter, "Frameworks based on Graph Neural Network," promises to open a new frontier. Graph Neural Networks (GNNs) stand at the cutting edge of AI research, representing a significant leap in how we model and interpret complex data structures. By focusing on data that is inherently graph-structured, GNNs offer a powerful framework for extracting relational knowledge and understanding interconnected systems, from social networks to molecular structures.

References

1. Confalonieri, R., Weyde, T., Besold, T.R., Moscoso del Prado Martín, F.: Trepan reloaded: a knowledge-driven approach to explaining black-box models. In: ECAI 2020, pp. 2457–2464. IOS Press (2020)
2. Zilke, J.R., Loza Mencía, E., Janssen, F.: Deepred–rule extraction from deep neural networks. In: Discovery Science: 19th International Conference, DS 2016, Bari, Italy, October 19–21, 2016, Proceedings 19, pp. 457–473. Springer (2016)
3. Zhang, Y., Xia, J., Jiang, B.: Reann: A pytorch-based end-to-end multi-functional deep neural network package for molecular, reactive, and periodic systems. The J. Chem. Phys. **156**(11) (2022)
4. Gou, J., Yu, B., Maybank, S.J., Tao, D.: Knowledge distillation: a survey. Int. J. Comput. Vis. **129**(6), 1789–1819 (2021)

Chapter 13
Graph Neural Networks in Neural-Symbolic Computing

"Great Scott! It's all connected, just like the flux capacitor!"
– Doc Brown, *Back to the Future*

13.1 Introduction

In the dynamic field of Artificial Intelligence, Neural-Symbolic Computing (NSC) stands as a beacon of innovation, effectively bridging the gap between the robustness of machine learning and the precision of symbolic reasoning. At the heart of this convergence lies the role of Graph Neural Networks (GNNs), which have gained prominence for their unparalleled capability to process and interpret graph-structured data. The amalgamation of GNNs within the NSC framework is not merely a fusion of technologies but a strategic integration that leverages the intrinsic strengths of both domains [1–3].

This chapter embarks on a comprehensive exploration of how GNNs are harmonized within the NSC paradigm, shedding light on their fundamental concepts and evolutionary journey. We delve into the core mechanics of GNNs, underscoring their unique proficiency in encapsulating relational information inherent in symbolic data. The discourse then navigates through a spectrum of GNN models, encompassing Logic Tensor Networks, Pointer Networks, Graph Convolutional Networks, and Graph Attention Networks. Each model is dissected to reveal its distinct contribution to the neural-symbolic integration, highlighting how they collectively enhance the interpretability and efficiency of learning from graph-structured data.

B. P. Bhuyan et al., *Neuro-Symbolic Artificial Intelligence*, Studies in Computational Intelligence 1176, https://doi.org/10.1007/978-981-97-8171-3_13

As we traverse the landscape of these models, we accentuate the synergistic relationship between GNNs and NSC. This symbiosis not only propels the frontiers of AI research but also catalyzes innovative applications across diverse domains. From dissecting complex social networks to unraveling the intricacies of knowledge graphs, the union of GNNs and NSC equips us with advanced tools for understanding and interpreting multifaceted relationships and patterns embedded within data.

Concluding the chapter, readers will emerge with an enriched perspective on the transformative role of GNNs in NSC. This narrative not only highlights their theoretical significance but also underscores their practical impact in shaping the future trajectory of artificial intelligence.

13.2 Graph Neural Networks in Neural-Symbolic Computing Frameworks

GNNs are integrated into NSC frameworks to leverage their strength in handling relational data. This involves using GNNs to encode symbolic information in a manner that is amenable to neural network processing. The integration often takes the form of embedding symbolic elements (like nodes and edges in a graph) into a continuous space, where neural networks can efficiently operate.

Specific models such as Logic Tensor Networks (LTNs) [4] and Graph Convolutional Networks (GCNs) [5] demonstrate the potential of GNNs in NSC. LTNs, for example, use GNNs to handle logical predicates defined on the graph, blending symbolic logic with neural learning. GCNs, on the other hand, effectively capture the local graph structure and feature information, making them suitable for tasks that require reasoning over complex relational data.

At the core of this integration are the mathematical operations that GNNs perform on graphs. For instance, consider a graph $G = (V, E)$ where V and E represent the sets of vertices and edges, respectively. GNNs apply a series of transformations to the feature vectors of these vertices. A typical layer in a GNN might compute new features h'_v for each vertex v based on its own features h_v and the features of its neighbors $N(v)$, often using a function of the form:

$$h'_v = \sigma\left(W \cdot \text{AGGREGATE}\left(\{h_u : u \in N(v)\}\right) + B \cdot h_v\right)$$

where σ is a nonlinear activation function, W and B are learnable parameters, and AGGREGATE is an aggregation function like sum, mean, or max.

Some advanced models expand the repertoire of GNNs in NSC, offering sophisticated tools for tackling complex, graph-based symbolic reasoning tasks.

- *Pointer Networks* [6]: These networks, within the NSC framework, extend the capability to learn the conditional probability of an output sequence with elements that are discrete tokens corresponding to positions in an input sequence. They are particularly effective in sequence-to-sequence models where the output sequence is a permutation of the input.

– *Message-Passing Neural Networks (MPNNs)* [7]: MPNNs generalize various GNN approaches by framing the neural network layers as a message-passing scheme. This allows for a flexible and powerful mechanism to handle graph-structured data, making them highly suitable for NSC tasks that require dynamic information propagation.
– *Graph Attention Networks (GATs)* [8]: GATs introduce an attention mechanism into GNNs, enabling the model to assign importance to nodes in a graph dynamically. This feature makes GATs particularly useful in NSC for tasks where the relevance of different parts of the graph may change depending on the context.

A key aspect of GNNs in NSC is their ability to bridge symbolic and sub-symbolic learning. They can learn to represent and process symbolic data (like relationships and entities) in a way that's compatible with sub-symbolic AI models, facilitating complex reasoning tasks.

By focusing on these aspects, we can understand how GNNs contribute significantly to the progress of NSC, enhancing its ability to model and reason over complex, graph-structured data.

13.2.1 Logic Tensor Networks

Logic Tensor Networks (LTNs) [4] are a significant model in the realm of Graph Neural Networks (GNNs) within Neural-Symbolic Computing (NSC). LTNs effectively blend deep learning with symbolic logic, providing a framework for representing logical predicates and functions using the machinery of neural networks.

LTNs ground logical predicates and functions in the vector space. For a predicate P, an LTN learns a function $P : \mathbb{R}^n \to [0, 1]$, mapping entities represented as vectors in $\mathbb{R}^n$ to truth values. The truth value of a ground predicate $P(a)$ for an entity a is computed by a neural network:

$$P(a) = \sigma(W \cdot \phi(a) + b)$$

Here, σ is a sigmoid function, $\phi(a)$ is a feature vector representing a, and W and b are learnable weights and biases of the neural network.

Consider an example where we use a Logic Tensor Network (LTN) to model a simple relational logic problem. Suppose we have a domain of animals, and we want to model the predicate "IsMammal." In our domain, let's say we have two entities: "Cat" and "Lizard."

1. **Representation**: In an LTN, each entity is represented as a vector in $\mathbb{R}^n$. For simplicity, assume: – $\phi(\text{Cat}) = [1.0, 0.0]$ – $\phi(\text{Lizard}) = [0.0, 1.0]$

2. **Predicate Modeling**: The predicate "IsMammal" is modeled as a function:

$$\text{IsMammal}(x) = \sigma(W \cdot x + b)$$

Here, W and b are parameters learned during training.

3. **Learning**: Through training, the LTN learns to approximate the truth values for the predicate. Ideally, it learns weights such that: – IsMammal(Cat) outputs a value close to 1 (true). – IsMammal(Lizard) outputs a value close to 0 (false).

LTNs allow for the incorporation of logical reasoning within the learning process. They can handle complex relational data and are capable of learning from both symbolic and sub-symbolic data, making them highly versatile for tasks in NSC.

13.2.2 Pointer Networks

Pointer Networks (Ptr-Nets) are a novel architecture in neural networks designed to address problems where the output is a discrete sequence of elements from the input sequence. They are particularly useful in problems where the length of the output sequence is variable and depends on the input sequence.

Ptr-Nets use a combination of Recurrent Neural Networks (RNNs) like LSTM for encoding input sequences and an attention mechanism for generating the output sequence. The attention mechanism allows the network to learn a conditional probability distribution over the input sequence's positions.

The core of Ptr-Nets is the attention mechanism, which is used to select members of the input sequence as elements of the output sequence. The mechanism can be described as follows:

$$u_j^i = v^T \tanh(W_1 e_j + W_2 d_i)$$

$$\alpha_j^i = \text{softmax}(u_j^i)$$

Here, e_j is the encoder hidden state, d_i is the decoder state at step i, and u_j^i is a scalar that scores how well the inputs around position j and the output at position i match. The softmax function then turns these scores u_j^i into a probability distribution α_j^i, which is used to form a weighted sum over all the encoder hidden states to produce the next output.

Ptr-Nets are particularly effective in combinatorial optimization problems like the Traveling Salesman Problem (TSP) or in tasks like sorting variable-sized sequences, where the output is a permutation of the input.

Consider an example where Pointer Networks are used to solve a simplified version of the Traveling Salesman Problem (TSP):

1. **Problem Setup**: In TSP, we want to find the shortest possible route that visits a set of cities and returns to the origin city. Suppose we have four cities, represented as points in a 2D space.

2. **Input Representation**: The input sequence to the Ptr-Net is the coordinates of these cities. For example, $\text{City}_1 = (x_1, y_1), \text{City}_2 = (x_2, y_2), \ldots$.

3. **Pointer Network Operation**: The Ptr-Net, after encoding these inputs, uses its attention mechanism to create a sequence of pointers. Each pointer indicates the next city to visit. For instance, if the first output points to City_2, the salesman should travel there next.

4. **Output Sequence**: The output of the Ptr-Net is a sequence of indices corresponding to the cities in the order they should be visited to minimize travel distance.

While powerful, Ptr-Nets have limitations in processing very long sequences due to the complexities associated with RNNs and attention mechanisms.

13.2.3 Graph Convolutional Networks

In the context of Neural-Symbolic Computing (NSC), understanding convolutions as self-attention within Graph Neural Networks (GNNs) is crucial. This perspective allows for the effective integration of symbolic reasoning with sub-symbolic learning processes. Traditional convolutional layers apply filters to local neighborhoods in data (like pixels in images). In graph data, a similar operation can be seen as focusing (or attending) to a node's neighbors.

In NSC, this interpretation of convolution as self-attention in GNNs allows the model to focus on different parts of the graph (symbolic structures) based on the context of the problem. It aligns with symbolic reasoning, where the importance of relations and entities varies depending on the logical context.

The self-attention mechanism in GNNs can be mathematically represented as follows:

$$h_i' = \text{ReLU}\left(\sum_{j \in N(i)} \frac{1}{c_{ij}} W h_j\right)$$

Here, h_i' is the updated feature vector of node i, $N(i)$ denotes the neighbors of node i, c_{ij} is a normalization constant (like the degree of the node in simple cases), W is a learnable weight matrix, and h_j are the features of the neighboring nodes.

Each node in the graph attends to its neighbors, and the convolution operation is akin to weighting these neighbors' features, followed by aggregation and transformation. This is analogous to the attention mechanism, where the attention weights are implicitly defined by the graph's structure and the convolution operation. In NSC, this equation represents how each node (symbolic entity) updates its state by aggregating and transforming the features (symbolic attributes) of its neighboring nodes, considering the graph structure (symbolic relations).

This convolution mechanism enables the model to dynamically adjust its focus on different parts of the graph, mimicking the process of symbolic reasoning where certain relations or entities become more relevant based on the reasoning task.

Let's consider an example of using a Graph Convolutional Network (GCN) for classifying scientific papers into categories based on a citation network.

1. **Graph Setup**: – Nodes: Scientific papers. – Edges: Citations between papers. – Node Features: Text embeddings of paper abstracts. – Labels: Categories like "Biology," "Physics," etc.

2. **GCN Operation**: – For paper i, the GCN updates its feature h_i' based on its own abstract and the abstracts of papers it cites:

$$h_i' = \text{ReLU}\left(\sum_{j \in N(i)} \frac{1}{c_{ij}} W h_j\right)$$

– $N(i)$ includes papers cited by i, c_{ij} is the normalization factor, W is a learnable weight matrix, and h_j are the features of the cited papers.

3. **Classification**: – The GCN learns to classify papers into categories based on their updated feature vectors h_i'.

13.2.4 Message-Passing Neural Networks

Message-Passing Neural Networks (MPNNs) are a sophisticated class of Graph Neural Networks designed to generalize the concept of neural networks for graph-structured data. The core idea of MPNNs is to simulate the process of message passing in a network, where each node exchanges information with its neighbors and updates its state based on these interactions. This method allows for effective and versatile learning on graphs.

The mechanism of MPNNs typically involves two phases: a message-passing phase and a readout phase. During the message-passing phase, each node v in the graph receives messages from its neighboring nodes. These messages are formulated as functions of the node features and the features of its neighbors. The message function at a given time step t can be mathematically expressed as follows:

$$m_v^{(t+1)} = \sum_{w \in N(v)} M_t(h_v^{(t)}, h_w^{(t)}, e_{vw})$$

where M_t represents the message function, $h_v^{(t)}$ and $h_w^{(t)}$ are the features of nodes v and w, respectively, at time step t, and e_{vw} denotes the edge feature between nodes v and w.

After the message-passing phase, each node updates its state. This update phase is defined by an update function, represented as follows:

$$h_v^{(t+1)} = U_t(h_v^{(t)}, m_v^{(t+1)})$$

Here, U_t is the update function that modifies the feature of the node v at the next time step $t + 1$, based on its current state and the aggregated messages received.

Finally, the readout phase aggregates the features of nodes to produce a graph-level output, which is useful for tasks like graph classification or regression. This phase is crucial for extracting meaningful and global information from the graph after several iterations of message passing and state updates.

MPNNs provide a flexible and powerful framework for handling graph-structured data, making them suitable for diverse applications across various domains, including bioinformatics, social network analysis, and chemistry.

13.2.5 Graph Attention Networks

Graph Attention Networks (GATs) are a novel type of Graph Neural Network that integrate attention mechanisms, enabling nodes to focus on different parts of their local neighborhood when performing computations.

1. **Attention Mechanism**: – GATs assign different importance (attention) to nodes in a neighborhood. – The attention coefficients, determining how much focus a node gives to its neighbors, are computed as follows:

$$\alpha_{ij} = \frac{\exp\left(\text{LeakyReLU}\left(\mathbf{a}^T[\mathbf{W}h_i \mid\mid \mathbf{W}h_j]\right)\right)}{\sum_{k\in N(i)} \exp\left(\text{LeakyReLU}\left(\mathbf{a}^T[\mathbf{W}h_i \mid\mid \mathbf{W}h_k]\right)\right)}$$

– Here, α_{ij} is the attention coefficient between nodes i and j, $\mathbf{a}$ is a learnable weight vector, $\mathbf{W}$ is a shared linear transformation applied to every node, and $||$ denotes concatenation.

2. **Feature Update**: – Each node's features are updated by aggregating its neighbors' features (weighted by attention coefficients). – The updated features for node i are given by

$$h_i' = \sigma\left(\sum_{j\in N(i)} \alpha_{ij}\mathbf{W}h_j\right)$$

GATs, through this mechanism, enable more efficient learning on graph data by focusing on the most relevant parts of the input. In the context of Neural-Symbolic Computing (NSC), Graph Attention Networks (GATs) play a crucial role in enhancing the integration of symbolic and sub-symbolic learning. By allowing nodes to dynamically assign importance to their neighbors, GATs enable NSC frameworks to handle complex reasoning tasks on graph-structured data efficiently. This capability is especially valuable when dealing with symbolic data, such as knowledge graphs, and facilitates the seamless interaction between symbolic and neural representations within NSC, contributing to its ability to reason over intricate, graph-based information.

13.3 Relational Learning and Reasoning

In the domain of Neural-Symbolic Computing (NSC), relational learning and reasoning are fundamental components. NSC frameworks leverage Graph Neural Networks (GNNs), including Graph Attention Networks (GATs), to model and reason over complex, relational data. These networks enable the integration of symbolic and sub-symbolic information, making NSC systems adept at understanding and processing structured knowledge. Relational learning techniques facilitate the extraction of meaningful patterns and dependencies from knowledge graphs, while reasoning mechanisms empower NSC to perform intricate, context-aware decision-making tasks. Together, they form the backbone of NSC's ability to bridge symbolic and neural representations, enhancing its capacity for intelligent, knowledge-driven computations. This section is divided into several categories, each addressing different methods and strategies. Each of these categories and subcategories represents a unique approach to integrating symbolic logic with neural network methodologies, aiming to enhance the reasoning capabilities of graph structures. The variety of these approaches, as shown in Fig. 13.1, shows how flexible and creative neuro-symbolic AI is in graph reasoning, providing different ways to combine the ease of understanding of symbolic AI with the powerful pattern recognition abilities of neural networks.

13.3.1 Logically Informed Embedding Approaches

This category includes approaches that combine symbolic methods (like logic rules) with neural networks. The idea is to use symbolic reasoning to augment or control the scope of a graph, which is then fed into a neural network for further processing. These approaches are divided into the following:

13.3.1.1 Modular Two-Step Approaches

Modular two-step approaches in logically informed embedding methodologies involve two distinct phases: a symbolic logic phase and a neural network-based embedding phase.

1. **Symbolic Logic Phase**:
    - This phase uses symbolic reasoning, often based on expert-defined rules or ontologies, to augment the knowledge graph.
    - Logical rules or ontological reasoning are employed to infer new information or relationships within the graph.
    - The output is an augmented knowledge graph with new edges or nodes added based on logical inferences.

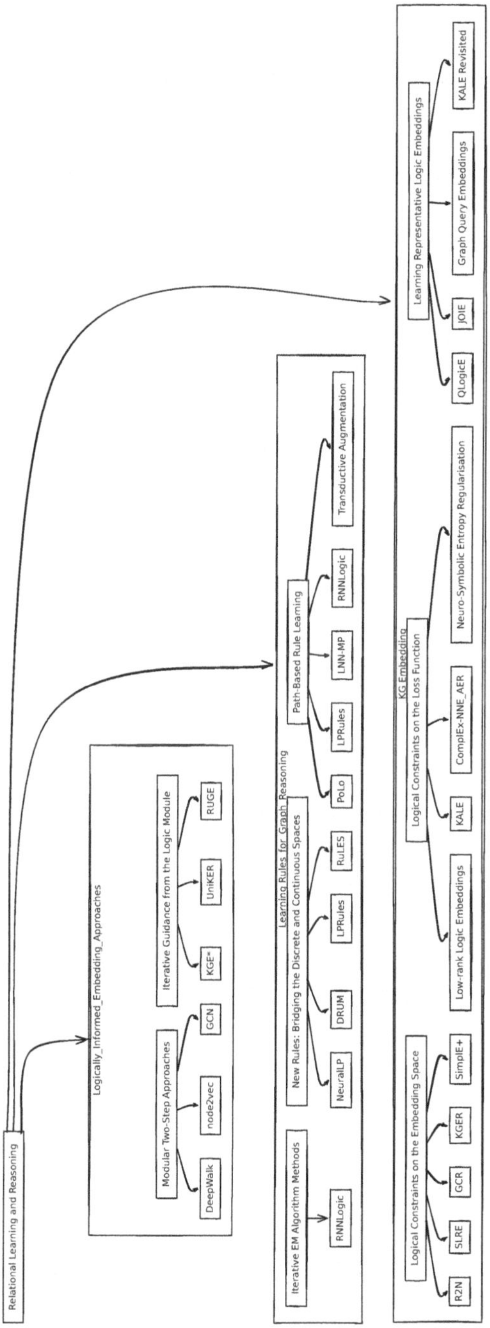

Fig. 13.1 Approaches for learning and reasoning in graph

2. **Neural Network-Based Embedding Phase**:
   - This phase involves generating embeddings, which are low-dimensional vector representations of the graph's nodes and edges.
   - Algorithms like DeepWalk, node2vec, or Graph Convolutional Networks (GCNs) are used.
   - The embeddings are utilized for tasks like link prediction or node classification.

Equations and Models:

- DeepWalk algorithm can be represented as follows:

$$\text{Embedding}(v) = \text{Optimize}\left(\sum_{u \in \text{Walks}(v)} \log \Pr(u|\text{Embedding}(v))\right)$$

  where Embedding(v) is the embedding of node v, and Walks(v) are the random walks starting at v.
- For GCNs, an example equation is

$$\mathbf{H}^{(l+1)} = \sigma\left(\mathbf{D}^{-\frac{1}{2}}\mathbf{A}\mathbf{D}^{-\frac{1}{2}}\mathbf{H}^{(l)}\mathbf{W}^{(l)}\right)$$

  where $\mathbf{H}^{(l)}$ is the matrix of activations in the l-th layer, $\mathbf{A}$ is the adjacency matrix, $\mathbf{D}$ is the degree matrix, $\mathbf{W}^{(l)}$ is the weight matrix for the l-th layer, and σ is a non-linear activation function.

Modular two-step approaches require balancing the influence of symbolic logic with the flexibility of neural embeddings. The quality and relevance of symbolic knowledge in the first phase significantly impact their success.

13.3.1.2 Iterative Guidance from the Logic Module

This subsubsection discusses approaches that extend the modular two-step pattern to enhance interaction between the symbolic logic and neural network modules, resulting in more dynamic and iterative guidance from the logic module.

- **Conceptual Framework**:
  - Unlike standard two-step approaches, these methods allow for a bi-directional flow of information between the logic and neural modules.
  - The output of the neural network module is used to refine and adjust the logic module, creating an iterative process of mutual guidance and improvement.

- **Key Approaches**:
  - **KGE* by Kaoudi et al.** [9]: Utilizes a two-way interaction between the logic and neural modules. The neural module's output informs the logic module in subsequent iterations, refining the symbolic reasoning process.
  - **UniKER by Cheng et al.** [10]: A unified framework that iteratively improves the knowledge graph embeddings using feedback from the logic module.
  - **RUGE by Guo et al.** [11]: Rule-Guided Embedding (RUGE) leverages soft rules with corresponding confidences learned from the knowledge graph to iteratively guide the embedding process.
- **Advantages and Challenges**:
  - **Advantages**: These approaches can lead to more accurate and robust graph reasoning models by continuously integrating and refining symbolic logic with neural network insights.
  - **Challenges**: Requires careful tuning to balance the iterative interaction between logic and neural modules. The complexity of the models can increase, demanding more computational resources and fine-tuning.

These iterative guidance approaches represent an advanced strategy in neuro-symbolic AI, demonstrating the potential for more integrated and synergistic use of symbolic and neural methodologies in graph reasoning tasks.

13.3.2 KG Embedding and Deep Learning with Logical Constraints

This subsection focuses on approaches in neuro-symbolic AI that apply logical constraints to knowledge graph embeddings and deep learning processes, enhancing the interpretability and effectiveness of the models.

13.3.2.1 Logical Constraints on the Embedding Space

This subsubsection delves into methods that impose logical constraints directly on the embedding space of knowledge graphs, thereby ensuring that the learned embeddings adhere to certain logical rules or properties.

- **Approach Overview**:
  - These methods integrate logical constraints into the process of generating embeddings for knowledge graphs.
  - The aim is to ensure that the resulting embeddings not only capture the structural and relational aspects of the graph but also conform to predefined logical rules or constraints.

- **Key Techniques and Models**:
  - **R2N** [12]: Incorporates relational logic directly into the embedding process, ensuring that relationships within the graph follow specified logical patterns.
  - **SLRE** [13]: Utilizes soft logic rules to guide the embedding process, allowing for the inclusion of probabilistic or uncertain information.
  - **GCR** [14]: Graph Constraint Reasoning, a method that enforces constraints on the graph structure during the embedding generation.
  - **KGER** [15]: Embedding technique that integrates knowledge graph information with external logical constraints to produce more comprehensive embeddings.
  - **SimplE+** [16]: An extension of the SimplE embedding method, adding the capability to embed logical constraints alongside graph data.
- **Advantages and Limitations**:
  - **Advantages**: Enhances the reliability and interpretability of embeddings by ensuring they are consistent with logical rules, improving their applicability in sensitive domains like healthcare or finance.
  - **Challenges**: The incorporation of logical constraints can increase the complexity of the embedding process and may require additional computational resources. Balancing the rigidity of logical rules with the flexibility of neural network-based embeddings can be challenging.

By imposing logical constraints on the embedding space, these approaches strive to create knowledge graph embeddings that are not only structurally and relationally accurate but also logically coherent, significantly enhancing the utility of neuro-symbolic methods in complex reasoning tasks.

13.3.2.2 Logical Constraints on the Loss Function

This subsubsection addresses methods that incorporate logical constraints into the loss function of neural network models during the learning process of knowledge graph embeddings.

- **Approach Overview**:
  - In these methods, logical constraints are embedded into the loss function of the neural network model. This integration guides the learning process to ensure that the model's predictions adhere to predefined logical rules.
  - The key idea is to penalize the model during training if it generates predictions that violate the logical constraints, thereby aligning the model's learning objective with logical consistency.
- **Key Techniques and Examples**:
  - **Low-rank Logic Embeddings** [17]: Implements a loss function that encourages the learned embeddings to respect logical constraints, emphasizing lower-dimensional representations for efficient learning.

  - **KALE (Knowledge-Aware Latent Embeddings)** [18]: Integrates logical knowledge into the embedding process by modifying the loss function to account for both the data-driven learning and the logical rules.
  - **ComplEx-NNE_AER** [19]: Enhances the standard ComplEx embedding method with a loss function that incorporates non-negative embedding (NNE) and approximate entailment rules (AER), ensuring logical validity in the embeddings.
  - **Neuro-Symbolic Entropy Regularization** [20]: Applies entropy-based regularization to the loss function, encouraging the model to learn representations that are not only accurate but also logically coherent.

- **Advantages and Limitations**:
  - **Advantages**: By integrating logical constraints into the loss function, these methods ensure that the learning process inherently respects logical rules, leading to more interpretable and reliable models.
  - **Challenges**: The design of an effective loss function that balances data-driven learning and logical constraints can be complex. These methods might require sophisticated tuning and can be computationally intensive, especially for large-scale graphs or intricate logical rules.

Embedding logical constraints into the loss function represents a novel approach in neuro-symbolic AI, ensuring that the learning process of neural networks is guided not only by data but also by logical consistency, significantly enhancing the applicability of these models in reasoning tasks.

13.3.2.3 Learning Representative Logic Embeddings

This subsubsection explores methods that focus on learning embeddings that are representative of logical rules or facts, effectively bridging the gap between symbolic logic and neural network embeddings in knowledge graphs.

- **Conceptual Framework**:
  - These methods aim to learn embeddings that not only capture the structural and relational aspects of a knowledge graph but also inherently represent logical rules or facts.
  - The goal is to encode logical information directly into the embeddings, allowing the model to leverage both graph structure and logical reasoning in its predictions.

- **Key Methods and Implementations**:
  - **QLogicE (Quantum Logic Embeddings)** [21]: Utilizes principles from quantum logic to create embeddings that encapsulate both relational data and logical structures.

  - **JOIE (Jointly Learning Knowledge Graph Embeddings)** [22]: An approach that simultaneously learns entity, relation, and hierarchical embeddings, aligning them with logical knowledge.
  - **Graph Query Embeddings** [23]: Focuses on representing complex graph queries as embeddings, integrating logical query constraints into the embedding learning process.
  - **KALE (Knowledge-Aware Latent Embeddings) Revisited** [24]: An extension of KALE, further emphasizing the integration of knowledge graph information with logical reasoning in the embeddings.

- **Advantages and Challenges**:
  - **Advantages**: By embedding logical rules directly, these methods can perform more sophisticated reasoning, improving both the interpretability and accuracy of the models.
  - **Challenges**: The complexity of these models can be high, requiring careful design and tuning to effectively learn embeddings that represent both graph structure and logical rules. Scalability to large graphs with complex logical structures can also be a concern.

The approach of learning representative logic embeddings represents a significant stride in neuro-symbolic AI, enabling models to inherently understand and utilize logical rules alongside traditional graph embeddings, thus enhancing their reasoning capabilities.

13.3.3 Learning Rules for Graph Reasoning

This subsection delves into neuro-symbolic approaches that focus on learning logical rules for graph reasoning. These methods enhance the reasoning capabilities of neural networks by integrating rule-based learning, offering a sophisticated blend of symbolic and neural methodologies in graph analysis.

13.3.3.1 Iterative EM Algorithm Methods

Iterative EM (Expectation-Maximization) Algorithm Methods in neuro-symbolic AI involve using iterative processes to learn logical rules that can be applied for reasoning on graph structures.

- **Conceptual Approach**:
  - These methods employ the EM algorithm, which iteratively improves the learning of logical rules by alternating between the expectation (E) and maximization (M) steps.

  - In the E step, the algorithm estimates the probabilities or weights of the rules given the current model parameters. In the M step, it optimizes the model parameters to maximize the likelihood of the observed data under these estimated probabilities.

- **Example Implementations**:
  - **RNNLogic** [25]: An approach that uses recurrent neural networks in conjunction with the EM algorithm to iteratively learn and refine logical rules from graph data.

- **Advantages and Limitations**:
  - **Advantages**: These methods can dynamically adapt and refine rule learning, leading to more accurate and contextually relevant logical rules for graph reasoning.
  - **Challenges**: The iterative nature of these methods can be computationally intensive, especially for large graphs or complex rule sets. Balancing the convergence and computational efficiency of the EM algorithm is crucial.

Iterative EM Algorithm Methods represent a powerful approach in neuro-symbolic AI, enabling the dynamic learning and refinement of logical rules for effective reasoning on graph structures.

13.3.3.2 New Rules: Bridging the Discrete and Continuous Spaces

This subsubsection focuses on neuro-symbolic methods that aim to bridge the gap between discrete logical spaces and continuous embedding spaces, thereby creating new rules that enhance graph reasoning.

- **Methodological Overview**:
  - These methods are centered on developing new rules that operate effectively in both discrete (logical rules) and continuous (neural embeddings) spaces.
  - The core idea is to translate or adapt symbolic logic into a format that can be seamlessly integrated with continuous representations used in neural networks, facilitating a more cohesive learning process.

- **Key Techniques and Examples**:
  - **NeuralLP** [26]: Utilizes neural networks to learn logic programming-like rules, enabling the representation of traditional symbolic rules in a continuous vector space.
  - **DRUM** [27]: A method that dynamically refines and updates rules by bridging logical reasoning with neural network outputs, maintaining consistency between discrete rules and continuous embeddings.

- **LPRules** [28]: Focuses on learning logical rules in the form of continuous vectors, enabling the application of these rules in graph neural network architectures.
- **RuLES** [29]: An approach that combines rule learning with embedding learning, ensuring that learned embeddings are consistent with logical rules derived from the graph data.

- **Advantages and Challenges**:
  - **Advantages**: These methods enhance the expressiveness and reasoning capabilities of graph neural networks by incorporating logical rules in a form that is congruent with neural embeddings.
  - **Challenges**: Designing algorithms that effectively translate logical rules into continuous representations can be complex. Ensuring that these new rules are both logically sound and effective in the continuous space poses a significant challenge.

The approach of creating new rules that bridge discrete and continuous spaces is a significant advancement in neuro-symbolic AI, offering a sophisticated mechanism to incorporate symbolic logic into neural network-based graph reasoning.

13.3.3.3 Path-Based Rule Learning

This subsubsection examines path-based rule learning methods in neuro-symbolic AI, focusing on how these approaches utilize paths in graphs to infer logical rules for graph reasoning.

- **Approach Overview**:
  - Path-based rule learning involves analyzing the paths within a graph structure to infer logical rules. These methods leverage the connectivity patterns in the graph to deduce relationships and rules that govern the graph's structure.
  - The key concept is to use sequences of edges and nodes (paths) in a graph to learn logical rules, which can then be applied for various reasoning tasks, such as link prediction or node classification.
- **Key Techniques and Implementations**:
  - **PoLo (Path-based Logic Learning)** [30]: A method that extracts and learns logical rules based on the paths observed in the graph, integrating these rules into a neural network framework.
  - **LPRules (Learning Path-based Rules)** [28]: Focuses on identifying and learning important paths in the graph, converting these into logical rules for reasoning.
  - **LNN-MP (Logic Neural Network—Message Passing)** [31]: Combines logic-based reasoning with message passing in neural networks, using paths to inform the message-passing process.

  - **RNNLogic** [32]: Extends its approach to incorporate path-based rule learning, enhancing its rule inference capabilities with path information from the graph.
  - **Transductive Augmentation** [33]: Uses path-based rules to augment transductive learning processes in graph neural networks, improving reasoning over unseen nodes or edges.

- **Advantages and Challenges**:
  - **Advantages**: Path-based rule learning can unveil intricate relational patterns within graphs, leading to more accurate and context-aware reasoning. It is particularly effective in capturing complex relational structures that are not immediately apparent.
  - **Challenges**: Identifying relevant and meaningful paths for rule learning can be computationally demanding. Balancing the complexity of paths and the generalizability of the learned rules is critical for effective learning.

Path-based rule learning represents a nuanced approach in neuro-symbolic AI, enabling the discovery of complex relational patterns and logical rules through the analysis of paths within graph structures, thereby enhancing the reasoning capabilities of neural network models.

13.3.4 Prospective Directions

This section outlines the prospective directions for future research in neuro-symbolic AI, particularly in the context of graph reasoning. The authors highlight several promising areas that offer opportunities for significant advancements and innovations.

13.3.4.1 Underexplored Application Areas

The realm of neuro-symbolic AI, particularly in graph reasoning, presents vast potential when applied to a spectrum of underexplored application domains. These areas offer fertile ground for innovative research and development, extending beyond the traditional boundaries. Among these domains, healthcare emerges as a prominent field. In healthcare, neuro-symbolic AI can revolutionize the way medical data is analyzed, leading to more accurate diagnosis, personalized treatment plans, and predictive models for disease progression. The complex nature of medical data, comprising patient records, genomic information, and clinical studies, makes it an ideal candidate for the sophisticated analytical capabilities of neuro-symbolic AI.

Financial modeling is another domain ripe for exploration. Here, neuro-symbolic AI can be instrumental in analyzing market trends, assessing risk, and making predictive models for investment strategies. The financial sector, with its intricate networks of transactions and economic indicators, stands to benefit greatly from the enhanced pattern recognition and reasoning abilities of neuro-symbolic systems.

Social network analysis, with its vast and complex datasets, is yet another area where neuro-symbolic AI can make a significant impact. By applying graph reasoning to social media data, researchers can gain insights into human behavior, social dynamics, and information dissemination. This can have far-reaching implications for marketing, sociology, and even political science.

Lastly, environmental monitoring presents a unique opportunity for applying neuro-symbolic AI. In this field, AI can play a crucial role in analyzing environmental data, predicting climate change patterns, and managing natural resources more effectively. The integration of various data types, such as satellite imagery, sensor data, and historical climate records, can be adeptly handled by neuro-symbolic models, offering more nuanced and accurate predictions about environmental phenomena.

Exploring these underexplored areas with neuro-symbolic AI not only challenges and expands the capabilities of current models but also opens up new avenues for practical applications. The unique characteristics and challenges inherent in each of these domains provide a platform for developing novel neuro-symbolic approaches and techniques, driving forward both the field of AI and the respective domains themselves.

13.3.4.2 EM-Based End-to-End Methods

The advancement of end-to-end methods founded on the Expectation-Maximization (EM) algorithm presents a promising frontier in the field of neuro-symbolic AI, particularly in graph reasoning. The EM algorithm, renowned for its efficacy in handling incomplete data, offers a robust framework for developing integrated models that can learn and apply logical rules within graph structures more efficiently. By employing the EM algorithm, these methods can iteratively improve the model's understanding of the graph, refining both the logical rules and the model parameters in a cohesive manner.

End-to-end EM-based methods aim to create a more streamlined architecture, where the process of learning logical rules and applying them to graph reasoning tasks is unified. This integrated approach reduces the complexity typically associated with separate learning and application phases, thereby simplifying the overall model architecture. The seamless nature of these methods enables a continuous feedback loop where the learning of rules and their application inform and enhance each other iteratively.

Furthermore, these EM-based methods hold the potential to significantly boost the performance of neuro-symbolic systems in complex reasoning tasks. By optimizing the learning process and ensuring that the logical rules are deeply embedded in the model's reasoning capabilities, these methods can achieve a higher level of accuracy and reliability. The ability to efficiently process and reason over graph structures while maintaining a coherent and simplified model architecture makes these EM-based end-to-end methods particularly suited for complex applications, such as in bioinformatics, network security, or intricate systems modeling.

The exploration and development of EM-based end-to-end methods could mark a substantial leap in the evolution of neuro-symbolic AI. By enhancing the integration, efficiency, and performance of these systems, such methods have the potential to address some of the most challenging problems in graph reasoning and beyond.

13.3.4.3 Multimodal Data Integration

The integration of multimodal data stands as a pivotal and promising direction in advancing neuro-symbolic AI, particularly in the realm of graph reasoning. This approach involves the amalgamation of diverse data types, encompassing text, images, structured graph data, and potentially more, into a unified analytic framework. The essence of multimodal data integration lies in harnessing the strengths of various data forms, each offering unique perspectives and insights, to enhance the reasoning capabilities of neuro-symbolic models.

Incorporating text data, for instance, can provide contextual and semantic understanding, while image data can contribute visual cues that are often critical for interpreting complex scenarios. Structured graph data, on the other hand, brings in relational and interconnected aspects of the data, essential for understanding the underlying patterns and relationships. When these diverse data sources are cohesively integrated, the resultant neuro-symbolic systems gain a multidimensional understanding of the data, enabling them to reason with a level of depth and comprehensiveness that single-modal systems cannot achieve.

This integration leads to the creation of robust and versatile neuro-symbolic systems capable of navigating and making sense of complex, real-world data scenarios. In practical applications, such as in healthcare, this could mean combining patient medical records (text), radiological images (visual data), and patient interaction networks (graph data) to provide a more comprehensive view for diagnosis and treatment planning. In urban planning or environmental studies, integrating geographic information (images), socio-economic data (text), and resource distribution (graph data) can lead to more informed and sustainable decisions.

The challenges in this approach, however, lie in effectively managing and processing the heterogeneity of the data and ensuring that the integrated system can meaningfully leverage the strengths of each data type. Achieving this will necessitate sophisticated algorithms that can not only handle the complexity of multimodal data but also draw coherent inferences across different data forms. The successful implementation of multimodal data integration in neuro-symbolic AI has the potential to significantly expand the horizons of AI applications, offering richer insights and more accurate predictions across a variety of fields.

13.3.4.4 Other Emerging Areas

The horizon of neuro-symbolic AI is further broadened by the potential of several emerging areas, each promising to add new dimensions and capabilities to graph reasoning. Among these, spatiotemporal reasoning stands out as a particularly rich area for exploration. This involves understanding and interpreting data that has both spatial and temporal components, such as satellite imagery over time or traffic patterns in urban areas. Neuro-symbolic systems that can effectively process and reason with spatiotemporal data could have profound implications in fields like environmental monitoring, urban planning, and logistics.

Few-shot learning is another burgeoning area in neuro-symbolic AI, focused on creating models that can learn from a very limited amount of data. This is particularly crucial in scenarios where data is scarce or expensive to obtain. Few-shot learning algorithms in the context of neuro-symbolic AI could significantly enhance the models' ability to make accurate inferences with minimal training data, a valuable trait for applications in rare disease analysis or in fields where data collection is challenging.

Furthermore, the incorporation of conditional or interdependent edge types in graphs presents an opportunity to model complex relationships more accurately. This could mean capturing the nuances of social networks, biological systems, or intricate ecosystems more effectively, offering a more refined understanding of the interconnectedness and dependencies within these systems.

Research is also needed to develop scalable and efficient algorithms for processing large-scale graphs. As the size and complexity of datasets continue to grow, the ability to efficiently process and extract meaningful insights from these vast networks becomes increasingly important. This is especially pertinent in the era of big data, where handling large-scale graphs efficiently is a critical challenge.

Lastly, enhancing the interpretability and explainability of neuro-symbolic systems remains a crucial objective. As these systems become more complex and are used in more critical applications, the need for them to be transparent and understandable increases. Making neuro-symbolic AI models more interpretable and explainable not only builds trust in their applications but also aids in refining and debugging the models, leading to more robust and reliable systems.

The exploration of these emerging areas in neuro-symbolic AI holds the promise of groundbreaking advancements. By addressing these challenges and opportunities, researchers can unlock new frontiers in AI, enabling more sophisticated, efficient, and trustworthy models for a wide range of applications.

The outlined prospective directions signify the evolving nature of neuro-symbolic AI in graph reasoning. Future research in these areas has the potential to drive significant innovations and contribute to the advancement of the field.

13.4 Conclusion

As we draw this chapter to a close, it becomes evident that the integration of Graph Neural Networks (GNNs) into Neural-Symbolic Computing (NSC) represents a significant stride in the evolution of artificial intelligence. This convergence not only capitalizes on the strengths of both paradigms but also paves the way for sophisticated solutions to complex problems that were previously beyond reach.

Throughout this chapter, we have explored a variety of GNN models, each contributing uniquely to the NSC landscape. From Logic Tensor Networks and Pointer Networks to Graph Convolutional Networks and Graph Attention Networks, we have seen how these models enhance the ability of NSC systems to process and learn from graph-structured data effectively. These models stand as testaments to the versatility and adaptability of GNNs in the realm of NSC, showcasing their capability to encode and interpret intricate relational information.

Moreover, the chapter has highlighted several promising directions for future research in NSC, such as exploring underexplored application areas, developing EM-based end-to-end methods, integrating multimodal data, and delving into other emerging areas like spatiotemporal reasoning and few-shot learning. These prospective directions not only underscore the potential of GNNs within NSC but also hint at the expansive possibilities that lie ahead in AI research.

In conclusion, the journey through the various aspects of GNNs in NSC has illuminated their pivotal role in advancing AI. As we continue to witness the growth and innovation in this field, it is clear that GNNs will remain at the forefront, driving both theoretical advancements and practical applications in AI. The synergy of GNNs and NSC is poised to unlock new horizons, fostering a deeper understanding of complex systems and enabling AI to tackle more nuanced and sophisticated challenges.

References

1. Lamb, L.C., Garcez, A., Gori, M., Prates, M., Avelar, P., Vardi, M.: Graph neural networks meet neural-symbolic computing: a survey and perspective (2020). arXiv:2003.00330
2. Yu, D., Yang, B., Liu, D., Wang, H., Pan, S.: A survey on neural-symbolic learning systems. Neural Netw. (2023)
3. Bhuyan, B.P., Ramdane-Cherif, A., Tomar, R., Singh, T.: Neuro-symbolic artificial intelligence: a survey. Neural Comput. Appl. 1–36 (2024)
4. Bianchi, F., Hitzler, P.: On the capabilities of logic tensor networks for deductive reasoning. In: AAAI Spring Symposium: Combining Machine Learning with Knowledge Engineering (2019)
5. Wu, F., Souza, A., Zhang, T., Fifty, C., Yu, T., Weinberger, K.: Simplifying graph convolutional networks. In: International Conference on Machine Learning, pp. 6861–6871. PMLR (2019)
6. Vinyals, O., Fortunato, M., Jaitly, N.: Pointer networks. Adv. Neural Inf. Process. Syst. **28**, (2015)
7. Gilmer, J., Schoenholz, S.S., Riley, P.F., Vinyals, O., Dahl, G.E.: Message passing neural networks. Mach. Learn. Meets Quantum Phys. 199–214 (2020)
8. Veličković, P., Cucurull, G., Casanova, A., Romero, A., Lio, P., Bengio, Y.: Graph attention networks (2017). arXiv:1710.10903

9. Kaoudi, Z., Lorenzo, A.C.M., Markl, V.: Towards loosely-coupling knowledge graph embeddings and ontology-based reasoning (2022). arXiv:2202.03173
10. Cheng, K., Yang, Z., Zhang, M., Sun, Y.: UniKER: a unified framework for combining embedding and definite horn rule reasoning for knowledge graph inference. In: Proceedings of the 2021 Conference on Empirical Methods in Natural Language Processing, pp. 9753–9771 (2021)
11. Guo, S., Wang, Q., Wang, L., Wang, B., Guo, L.: Knowledge graph embedding with iterative guidance from soft rules. In: Proceedings of the AAAI Conference on Artificial Intelligence, vol. 32 (2018)
12. Marra, G., Diligenti, M., Giannini, F.: Relational reasoning networks (2021). arXiv:2106.00393
13. Guo, S., Li, L., Hui, Z., Meng, L., Ma, B., Liu, W., Wang, L., Zhai, H., Zhang, H.: Knowledge graph embedding preserving soft logical regularity. In: Proceedings of the 29th ACM International Conference on Information & Knowledge Management, pp. 425–434 (2020)
14. Chen, H.: Neural logic reasoning and applications. Ph.D. thesis, Rutgers The State University of New Jersey, School of Graduate Studies (2022)
15. Minervini, P., Costabello, L., Muñoz, E., Nováček, V., Vandenbussche, P.Y.: Regularizing knowledge graph embeddings via equivalence and inversion axioms. In: Machine Learning and Knowledge Discovery in Databases: European Conference, ECML PKDD 2017, Skopje, Macedonia, September 18–22, 2017, Proceedings, Part I 10, pp. 668–683. Springer (2017)
16. Kazemi, S.M., Poole, D.: Simple embedding for link prediction in knowledge graphs. Adv. Neural Inf. Process. Syst. **31**, (2018)
17. Wang, W.Y., Cohen, W.W.: Learning first-order logic embeddings via matrix factorization. In: IJCAI, pp. 2132–2138 (2016)
18. Borges, L., Martins, B., Callan, J.: Kale: Using a k-sparse projector for lexical expansion. In: Proceedings of the 2023 ACM SIGIR International Conference on Theory of Information Retrieval, pp. 13–22 (2023)
19. Ding, B., Wang, Q., Wang, B., Guo, L.: Improving knowledge graph embedding using simple constraints (2018). arXiv:1805.02408
20. Ahmed, K., Wang, E., Chang, K.W., Van den Broeck, G.: Neuro-symbolic entropy regularization. In: Uncertainty in Artificial Intelligence, pp. 43–53. PMLR (2022)
21. Chen, P., Wang, Y., Yu, X., Feng, R.: Qlogice: quantum logic empowered embedding for knowledge graph completion. Knowl. Based Syst. **239**, 107963 (2022)
22. Hao, J., Chen, M., Yu, W., Sun, Y., Wang, W.: Universal representation learning of knowledge bases by jointly embedding instances and ontological concepts. In: Proceedings of the 25th ACM SIGKDD International Conference on Knowledge Discovery & Data Mining, pp. 1709–1719 (2019)
23. Hamilton, W., Bajaj, P., Zitnik, M., Jurafsky, D., Leskovec, J.: Embedding logical queries on knowledge graphs. Adv. Neural Inf. Process. Syst. **31**, (2018)
24. Feng, J., Huang, M., Yang, Y., Zhu, X.: Gake: graph aware knowledge embedding. In: Proceedings of COLING 2016, the 26th International Conference on Computational Linguistics: Technical Papers, pp. 641–651 (2016)
25. Qu, M., Chen, J., Xhonneux, L.P., Bengio, Y., Tang, J.: Rnnlogic: learning logic rules for reasoning on knowledge graphs (2020). arXiv:2010.04029
26. Yang, F., Yang, Z., Cohen, W.W.: Differentiable learning of logical rules for knowledge base reasoning. Adv. Neural Inf. Process. Syst. **30**, (2017)
27. Sadeghian, A., Armandpour, M., Ding, P., Wang, D.Z.: Drum: End-to-end differentiable rule mining on knowledge graphs. Adv. Neural Inf. Process. Syst. **32**, (2019)
28. Dash, S., Goncalves, J.: LPRules: rule induction in knowledge graphs using linear programming (2021)
29. Guo, S., Wang, Q., Wang, L., Wang, B., Guo, L.: Jointly embedding knowledge graphs and logical rules. In: Proceedings of the 2016 Conference on Empirical Methods in Natural Language Processing, pp. 192–202 (2016)
30. Liu, Y., Hildebrandt, M., Joblin, M., Ringsquandl, M., Raissouni, R., Tresp, V.: Neural multi-hop reasoning with logical rules on biomedical knowledge graphs. In: The Semantic Web: 18th International Conference, ESWC 2021, Virtual Event, June 6–10, 2021, Proceedings 18, pp. 375–391. Springer (2021)

31. Sen, P., Carvalho, B.W., Abdelaziz, I., Kapanipathi, P., Luus, F., Roukos, S., Gray, A.: Combining rules and embeddings via neuro-symbolic AI for knowledge base completion (2021). arXiv:2109.09566
32. Chen, B., Hao, Z., Cai, X., Cai, R., Wen, W., Zhu, J., Xie, G.: Embedding logic rules into recurrent neural networks. IEEE Access **7**, 14938–14946 (2019)
33. Hirose, Y., Shimbo, M., Watanabe, T.: Transductive data augmentation with relational path rule mining for knowledge graph embedding. In: 2021 IEEE International Conference on Big Knowledge (ICBK), pp. 377–384. IEEE (2021)

Part V
Neuro-Symbolic Reasoning and Inference

Chapter 14
Rule-Based Reasoning in Neural Networks

"Elementary, my dear Watson."

– Sherlock Holmes, *The Adventures of Sherlock Holmes*

14.1 Introduction

Consider a scenario where an AI system must make complex decisions, such as a healthcare AI evaluating patient data to recommend treatments. Traditionally, such tasks relied heavily on rule-based reasoning—clear, logical, and structured sets of rules guiding the decision-making process. However, the emergence of neural networks has revolutionized AI, introducing the ability to learn and adapt from vast and varied data. This chapter delves into how these two paradigms, each powerful in its own right, complement and enhance each other in the field of AI. We start by contextualizing rule-based reasoning, highlighting its longstanding role in logical inference and structured decision-making. The narrative then transitions to the advent of neural networks, with their unparalleled prowess in pattern recognition and learning from complex datasets.

Crucially, this chapter illuminates the challenges and breakthroughs encountered at the intersection of these domains. It underscores the significance of merging the deterministic clarity of rule-based systems with the adaptive learning mechanisms of neural networks. This fusion is not just a technological advancement but a paradigm shift, offering more nuanced, reliable, and intelligent AI systems capable of tackling real-world complexities.

B. P. Bhuyan et al., *Neuro-Symbolic Artificial Intelligence*, Studies in Computational Intelligence 1176, https://doi.org/10.1007/978-981-97-8171-3_14

Subsequent sections of the chapter focus on specific areas where this convergence is most evident and impactful. From the application of deep learning in RDF knowledge bases to pioneering innovations in deep deductive reasoning, each section provides a detailed analysis of current technologies, their applications, and future potential. The discussion includes memory networks for RDF reasoning, normalization techniques for model transferability, and the mining of rules from RDF data, culminating in an exploration of novel concepts like concept induction, deep unsupervised learning, deep reinforcement learning, and fortiori reasoning.

By the end of this chapter, readers will gain a comprehensive understanding of how rule-based reasoning and neural networks, once considered distinct or even opposing approaches, are collaboratively shaping the future of AI. This synthesis not only enhances AI's decision-making and reasoning capabilities but also paves the way for more advanced, ethical, and context-aware AI systems.

14.2 Deep Learning for RDF Knowledge Bases

The Resource Description Framework (RDF) [1] is a fundamental standard in knowledge representation, primarily used for structuring and interlinking data on the web. It enables the encoding, exchange, and reuse of structured metadata, forming a flexible and robust framework for semantic data modeling. RDF represents information in a triple format, comprising subject, predicate, and object, which facilitates a graph-based data structure. This structure is pivotal in semantic web technologies and knowledge management systems.

Deep learning, with its advanced pattern recognition capabilities, significantly enhances RDF knowledge bases. By applying deep learning algorithms to RDF data, complex patterns and relationships within large datasets can be identified and utilized more effectively. This integration allows for more sophisticated and accurate data interpretation, improving the efficacy of knowledge retrieval and decision-making processes in various applications. Deep learning's contribution to RDF emphasizes the evolution of data processing techniques, moving towards more intelligent and adaptive systems in knowledge management.

14.2.1 Memory Networks for RDF Reasoning

The integration of memory networks in RDF reasoning, as detailed in the paper [2], represents a significant advancement in the field of neuro-symbolic AI. Memory networks [3], a form of neural network with an explicit memory component, are particularly well-suited for managing the complex and interlinked data structures typical of RDF knowledge bases.

In the context of RDF reasoning, memory networks are used to emulate symbolic deductive reasoning [4]. They achieve this by iteratively accessing and processing information stored in their memory, allowing the network to draw inferences based on the relationships and rules embedded in the RDF data. This process involves several mathematical and algorithmic components, including embedding layers that convert RDF triples into a continuous vector space, attention mechanisms to focus on relevant parts of the memory, and a reasoning module that integrates these embeddings to derive conclusions.

One key mathematical aspect is the embedding of RDF triples, where subjects, predicates, and objects are represented as vectors in a high-dimensional space. This representation enables the network to capture the semantic relationships inherent in the RDF data. The attention mechanism then dynamically weights these embeddings based on their relevance to the query or task at hand, ensuring that the network focuses on the most pertinent information. Finally, the reasoning module, typically a neural network layer, combines these weighted embeddings to form conclusions, leveraging the inherent structure and rules of the RDF data.

Let's consider a typical architecture to understand the mathematics behind memory networks for RDF reasoning. Memory networks involve several key mathematical components:

Embedding of RDF Triples: Each RDF triple (subject, predicate, object) is converted into a continuous vector space. This is often done using embedding matrices $E_s[s]$, $E_p[p]$, $E_o[o]$ for subjects, predicates, and objects, respectively. The vector representation of a triple (s, p, o) is thus

$$\mathbf{v}_{(s,p,o)} = E_s[s] + E_p[p] + E_o[o]$$

where $E_x[x]$ denotes the embedding of element x in its respective space.

Attention Mechanism: This mechanism assigns a relevant weight to each piece of information in the memory. For a given query q, the attention weight for each triple i is computed as

$$\alpha_i = \text{softmax}(f(\mathbf{v}_{(s,p,o)_i}, \mathbf{q}))$$

where f is a function (like a feedforward layer with a non-linear activation) that computes the relevance of the triple embedding $\mathbf{v}_{(s,p,o)_i}$ to the query embedding q.

Reasoning Module: This module combines weighted embeddings to form conclusions. The output for a given query q can be represented as

$$\mathbf{r} = \sum_i \alpha_i \mathbf{v}_{(s,p,o)_i}$$

where $\mathbf{r}$ is the reasoning output, a vector summarizing the relevant information extracted from the memory.

These mathematical formulations are central to understanding how memory networks can be effectively applied to RDF reasoning, providing a framework for integrating symbolic reasoning with deep learning techniques. Memory networks have several capabilities and advantages in handling complex RDF structures:

Handling of Complex Relationships: Memory networks effectively manage the intricate web of relationships inherent in RDF data. Their structure allows for the nuanced interpretation of interconnected data points.

Dynamic Information Retrieval: These networks dynamically retrieve and process relevant information from their memory, making them well-suited for RDF reasoning, where the relevance of data can vary greatly depending on the query.

Scalability: Memory networks can scale to accommodate large RDF datasets, a crucial feature given the ever-expanding size of knowledge bases.

Contextual Reasoning: They are capable of contextual reasoning, essential for RDF structures where the meaning of data is often context-dependent.

Adaptability: Memory networks can be trained to adapt to different RDF schemas and domains, making them versatile tools for a variety of applications.

These features make memory networks an excellent choice for tasks involving complex RDF reasoning, bridging the gap between traditional symbolic reasoning and neural network-based approaches.

14.2.2 Normalization Techniques for Model Transferability

Normalization techniques are crucial for ensuring that the learning and reasoning capabilities of a model trained on one RDF dataset can be effectively applied to another.

Key aspects include the following:

Data Normalization: Adjusting RDF data into a uniform format, minimizing the discrepancies between different datasets. We can use the standard normalization, i.e.,

$$\mathbf{x}' = \frac{\mathbf{x} - \mu}{\sigma}$$

where $\mathbf{x}$ is the original data, μ is the mean, and σ is the standard deviation. The alternative is the Min-Max normalization

$$\mathbf{x}' = \frac{\mathbf{x} - \min(\mathbf{x})}{\max(\mathbf{x}) - \min(\mathbf{x})}$$

Feature Scaling: Standardizing the range of independent variables or features of data, which is essential in models dealing with diverse domains. Mathematically,

$$x' = \frac{x - \min(x)}{\max(x) - \min(x)}$$

or using Mean Normalization

$$x' = \frac{x - \mu}{\max(x) - \min(x)}$$

Embedding Normalization: Techniques to ensure that the vector representations of RDF triples maintain consistency across different datasets. Mathematically,

$$\mathbf{v}' = \frac{\mathbf{v}}{\|\mathbf{v}\|}$$

This maintains the direction but normalizes the magnitude of embedding vectors.

Loss Function Modification: Adapting the loss functions used during training to account for the variations in data distribution and structure across different RDF datasets. Mathematically,

$$L = L_{\text{original}} + \lambda \|\theta\|^2$$

or L1 Regularization, i.e.,

$$L = L_{\text{original}} + \lambda \|\theta\|_1$$

These normalization methods play a vital role in enhancing the generalizability and applicability of memory networks and other deep learning models in the realm of RDF reasoning, addressing one of the primary challenges in the field of neuro-symbolic AI.

14.2.3 Mining Rules from RDF Data

In this subsection, we focus on methodologies such as RDF2Rules [5] that are pivotal for extracting logical rules from RDF data. These methodologies are designed to enhance the efficiency and applicability of RDF knowledge bases by enabling the automated discovery of logical rules.

14.2.3.1 RDF2Rules

RDF2Rules [5] is a sophisticated method for extracting logical rules from RDF datasets. It operates by identifying frequently occurring predicates in the RDF triples. Once these predicates are identified, RDF2Rules searches for patterns of recurrence, specifically focusing on predicate cycles—sequences where the predicates appear in a recurring pattern. These cycles are key to understanding the underlying structure and relationships within the RDF data.

Upon detecting these cycles, RDF2Rules then proceeds to extract logical rules. These rules represent the relationships and patterns discovered in the RDF data. They are formulated based on the observed sequences of predicates and the connections between various entities in the RDF triples.

The final step involves applying these extracted rules back into the RDF knowledge bases. This application of rules enhances the knowledge base's efficiency, particularly in areas such as query processing and information retrieval. By adding these rules,

RDF2Rules effectively enriches the RDF database, making it a more powerful tool for data analysis and knowledge discovery.

This methodology as shown in Algorithm 14.1 is highly beneficial for extracting meaningful insights from complex RDF datasets, and its applications span across various domains, allowing for enhanced data integration and more effective knowledge management.

Algorithm 14.1 RDF2Rules Algorithm

Require: RDF dataset
Ensure: Set of logical rules
Initialize frequency count $f(p)$ for each predicate p in dataset
for each triple (s, p, o) in RDF dataset **do**
    Increment $f(p)$
end for
Select predicates $P' \subseteq P$ where $f(p) \geq \theta$
Construct directed graph G from RDF triples
for each predicate $p \in P'$ **do**
    Add edge (s, o) to G for each occurrence of p
end for
Detect cycles in G
for each detected cycle C **do**
    Extract predicates sequence $p_1, p_2, ..., p_n$ from C
    Formulate rule: if $p_1(x, y) \wedge p_2(y, z) \wedge ... \wedge p_{n-1}(w, x)$ then $p_n(w, z)$
end for
Integrate rules into the knowledge base

14.2.3.2 Rule Discovery Techniques

These methods extract logical rules from RDF data, complementing the RDF2Rules approach. Each of these techniques contributes to a robust rule discovery framework, offering insights into RDF datasets' complex and interconnected nature. Their application not only enriches the RDF knowledge bases but also enhances the capability of these systems in various domains, from semantic web services to data analytics.

Association Rule Mining: Association Rule Mining in the context of RDF data is a powerful technique for uncovering underlying relationships between different elements within a dataset. It operates by identifying frequent itemsets, which are sets of items (such as entities or predicates in RDF) that occur together frequently within the dataset. The primary goal is to discover strong rules that predict the presence of an item based on the presence of other items.

The process begins with setting thresholds for two key metrics: support and confidence. Support measures how frequently an item set appears in the dataset, while confidence assesses the likelihood that the occurrence of certain items implies the occurrence of others. These thresholds are critical for determining the strength and relevance of the association rules derived from the data.

Once frequent itemsets are identified, the algorithm generates association rules from these itemsets. These rules are evaluated based on their support and confidence values. The higher these values, the stronger and more reliable the rule is considered. The outcome is a set of rules that provide insights into how different items in the RDF dataset are associated with each other, revealing patterns and relationships that might not be immediately apparent.

Association Rule Mining is particularly valuable in RDF data due to its structured nature. It can be used to enhance the understanding of complex relationships in the data, guide decision-making processes, or inform further data analysis and modeling. This technique's ability to reveal hidden patterns and associations makes it a vital tool in the realm of data mining and knowledge discovery.

Frequent Pattern Mining: Frequent Pattern Mining in RDF data analysis [6] focuses on identifying recurring structures or patterns within the dataset. This technique is particularly adept at discovering common subgraphs or sequences of predicates that repeatedly occur across different parts of the RDF dataset. Once these patterns are identified, they can be formulated into generalized rules that encapsulate the regularities found in the data. These rules are not only insightful for understanding the underlying data structure but also serve as a valuable tool for predictive modeling and data interpretation. Frequent Pattern Mining is crucial for unraveling complex relationships and interdependencies in RDF datasets, enhancing the knowledge base's comprehensiveness and utility.

Inductive Logic Programming (ILP): Inductive Logic Programming (ILP) [7] is a hybrid of machine learning and logical reasoning, particularly suited for structured data like RDF. It starts with a set of examples and background knowledge and aims to find a hypothesis (rules or logical statements) that explains these examples accurately. ILP operates iteratively, refining hypotheses based on their fit with the observed data. In RDF datasets, ILP is adept at exploiting the rich relational structure to derive meaningful rules. This process involves generalizing from specific instances, making ILP a powerful tool for uncovering complex patterns and relationships inherent in RDF data. The rules generated by ILP can significantly enhance the interpretability and utility of RDF knowledge bases, making it a valuable technique in data mining and knowledge discovery.

Enhancing the efficiency of RDF knowledge bases through advanced rule discovery methods leads to significant improvements in data management and querying capabilities. By integrating rules derived from techniques like Association Rule Mining, Frequent Pattern Mining, and Inductive Logic Programming, the knowledge base becomes more adept at handling complex queries [8]. This integration allows for more efficient data retrieval, reducing the time and computational resources needed for processing queries. Moreover, the enriched knowledge base can offer more accurate and contextually relevant results, improving the overall usability and effectiveness of the system in various applications, from semantic web services to advanced data analytics.

14.2.4 Materialization and Incremental Learning in RDF

Materialization in RDF [9] refers to the process of precomputing and storing results or inferred data from an RDF knowledge base. This technique is crucial in large-scale RDF systems where executing complex queries on-the-fly can be computationally expensive and time-consuming. Materialization involves creating a materialized view of the data, where frequent queries or complex inferential relationships are computed in advance and stored. This approach significantly speeds up query processing, as the system can quickly retrieve precomputed results instead of performing calculations each time a query is made. It enhances the performance and responsiveness of RDF-based systems, especially in scenarios involving complex queries and large datasets.

Incremental learning approaches in RDF environments [10] are crucial for adapting to evolving data without complete retraining. These methods enable RDF systems to update their knowledge base as new information becomes available. This is especially important in dynamic domains where data changes frequently. Incremental learning ensures the system remains current and accurate, reducing the computational overhead associated with processing large datasets. It maintains the relevance and efficiency of RDF knowledge bases in real time, making it a key component in scalable and adaptive data management strategies.

In the context of RDF data management, several incremental learning approaches are particularly effective:

Online Learning Algorithms: These algorithms [11] continuously update the RDF model as new data arrives. They are designed to adapt quickly, making them ideal for environments where data changes frequently.

Stream Processing: In RDF systems, data often comes in streams. Stream processing [12] involves analyzing and processing this data in real time, allowing the knowledge base to be updated incrementally and continuously.

Active Learning: This approach [13] involves the model actively selecting specific data points from the incoming data for learning. It's a targeted approach that prioritizes learning from the most informative data, which is especially useful in large or complex RDF datasets.

Transfer Learning: This technique [14] leverages knowledge acquired in one domain to facilitate learning in a similar but different domain. In RDF contexts, it can be particularly useful when new data is similar to existing data, allowing for more efficient learning and adaptation.

Materialization and incremental learning are both important for scalable and effective RDF knowledge management because they deal with the problems that come up with big datasets that are always changing [15]. They also keep the RDF systems strong, responsive, and up-to-date.

14.3 Innovations in Deep Deductive Reasoning

This section delves into the latest methodologies and approaches that are shaping the landscape of AI, especially in terms of integrating deductive reasoning with deep learning techniques. The focus is on the conceptual and practical advancements that are enhancing the capabilities of AI systems in performing logical, reasoning-based tasks.

14.3.1 Concept Induction

Concept Induction represents a crucial advancement in the realm of AI, particularly in enhancing the interpretability of deep learning models [16, 17]. At its core, Concept Induction is about generating human-understandable explanations for the decision-making processes within AI systems. This approach is instrumental in Explainable AI (XAI), aiming to demystify the often opaque reasoning of AI, especially in complex neural networks.

The theoretical underpinning of Concept Induction in AI lies in its ability to bridge the gap between high-level human concepts and low-level data representations used by machine learning models. This involves translating the intricate patterns and relationships learned by an AI model into terms and concepts that are easily comprehensible to humans. Algorithm 14.2 explains the process. The Concept Induction algorithm is a structured process designed to translate complex outputs of a neural network into human-interpretable concepts.

The algorithm begins by initializing neuron activations for each input in the dataset X. It involves computing the activation $a_i^l(x)$ for each neuron i in layer l for every input x. These activations represent how the neural network processes and responds to the input data.

Once neuron activations are computed, the algorithm proceeds to extract features from these activations. This step analyzes the patterns in neuron activations to identify significant features that the network is detecting in the data.

Each set of features extracted from the neuron activations is mapped to a concept using a mapping function M. This function translates the low-level data representations (features) into concepts that are understandable to humans.

The induction algorithm is applied to generalize the concept mappings. This step involves identifying commonalities and patterns across different feature sets and their corresponding concepts.

For each concept in the set C, the algorithm generates an explanation. These explanations provide insight into why the neural network associates certain features with specific concepts.

Finally, the algorithm evaluates a coverage score for each concept. This score assesses how well the induced concepts explain the data, serving as a metric for the effectiveness and accuracy of the concept induction process.

Algorithm 14.2 Concept Induction Algorithm

Require: Input data set $X = \{x_1, x_2, \ldots, x_n\}$
Ensure: Set of human-interpretable concepts C
1: Initialize neuron activations $A = \{a_i^l(x) | x \in X\}$
2: **for** each input $x \in X$ **do**
3: Compute neuron activation $a_i^l(x)$
4: **end for**
5: Extract features $F = \{\text{features of } a_i^l(x)\}$
6: **for** each feature set $f \in F$ **do**
7: Map to concept $M(f) \rightarrow c$
8: **end for**
9: Apply induction algorithm to generalize M
10: $C \leftarrow$ set of generalized concepts
11: **for** each concept $c \in C$ **do**
12: Generate conceptual explanation
13: **end for**
14: Evaluate coverage score for each concept
15: **return** Set of concepts C and their explanations

In practice, Concept Induction can involve various techniques, from simple rule-based systems that map neuron activations to concepts, to more complex systems that use layers of abstraction to derive high-level explanations from raw data inputs.

One of the main challenges in Concept Induction is the complexity and variety of data that AI models are trained on, which can make it difficult to find accurate and meaningful high-level representations. Additionally, there's the challenge of ensuring that the induced concepts align accurately with the underlying decision-making process of the AI model.

14.3.2 Deep Unsupervised Learning

In the realm of Deep Deductive Reasoning, Deep Unsupervised Learning plays a transformative role. While deductive reasoning typically relies on structured, often labeled data to draw logical conclusions, integrating it with deep unsupervised learning techniques opens new avenues for reasoning with unstructured or unlabeled data.

The intersection of these two fields lies in the ability of unsupervised learning models to uncover hidden structures and patterns in data, which can then be used as a basis for logical deduction. This approach shifts the focus from relying solely on predefined rules and labels to also understanding inherent, often subtle, data characteristics.

The key models and techniques in this regard are as follows:

Representation Learning in Unsupervised Contexts: Deep unsupervised learning models, such as autoencoders, are employed to learn representations of data without explicit external labels [18]. These representations capture the intrinsic properties of the data, which can be crucial for deductive reasoning processes. For example, in a complex dataset, an autoencoder can distill the essential features that might indicate a certain logical category or outcome, even when such categories are not explicitly labeled [19].

Pattern Discovery and Logical Inference: Generative models like GANs [20] and VAEs [21] are adept at discovering patterns in data. These patterns, once identified, can form the foundation for deductive reasoning. In the context of deductive reasoning, the patterns unearthed by these models can suggest logical relationships or rules that were not previously evident.

Latent Variable Models for Deduction: Variational Autoencoders, which model the probability distribution of data, can be particularly valuable [22]. They help in deducing the underlying causal factors or latent variables of observed phenomena, which is a key aspect of deductive reasoning.

One of the main challenges in Deep Unsupervised Learning is the difficulty of evaluating the model performance, as the lack of labeled data makes it hard to gauge the accuracy of the learned representations. Additionally, there is the challenge of ensuring that the model does not overfit the noise in the data.

The future of Deep Unsupervised Learning lies in enhancing the ability of models to understand and generate complex data structures, as well as in improving their interpretability. As data continues to grow in size and complexity, these techniques will be crucial for extracting meaningful insights without the need for extensive labeled datasets.

14.3.3 Deep Reinforcement Learning

Deep Reinforcement Learning (DRL) [23] merges the principles of reinforcement learning with the power of deep neural networks, leading to systems that can learn to make sequences of decisions by interacting with an environment. In the context of deep deductive reasoning, DRL can be instrumental in developing AI systems that not only deduce logical conclusions but also learn from their interactions and adapt their reasoning strategies over time [24].

DRL is grounded in the idea of an agent learning to achieve goals by receiving rewards or penalties for its actions. The agent learns the best actions (policies) to maximize cumulative rewards in a given environment. This learning process is augmented by deep learning techniques, enabling the handling of high-dimensional, complex input spaces. The key models and techniques are as follows:

Deep Q-Networks (DQNs): DQNs integrate Q-learning, a type of reinforcement learning, with deep neural networks. The network approximates the optimal action-value function, which represents the expected cumulative reward of taking a given action in a given state. In deductive reasoning, DQNs can be used to determine the best logical steps to reach a conclusion, especially in dynamic environments where the reasoning path may not be straightforward [25].

Policy Gradient Methods: These methods learn the policy function directly. Unlike DQNs, which output a value indicating the expected reward, policy gradient methods [26] directly output the best action to take. In deductive reasoning scenarios, this approach can be useful for scenarios where direct action is required, such as interactive problem-solving or decision-making in real-time systems.

Actor-Critic Methods: These methods [27] combine the advantages of value-based and policy-based methods. The "actor" proposes actions, while the "critic" evaluates these actions based on a value function. This dual approach allows for more sophisticated reasoning strategies where the system can evaluate and adjust its logic paths dynamically.

DRL can be used to develop AI systems capable of adaptive reasoning, modifying their deductive processes based on the outcomes of previous interactions and decisions. In environments that are continually changing, DRL-based systems can continuously learn and adapt their reasoning strategies, making them highly effective for real-time decision-making and problem-solving.

One of the challenges in DRL is finding the right balance between exploring new reasoning strategies and exploiting known successful ones. This balance is crucial in ensuring that the AI system can adapt to new environments without losing the efficiency of its deductive processes. DRL algorithms, especially those involving deep neural networks, can be computationally intensive and complex to implement, particularly in scenarios requiring sophisticated deductive reasoning. Future research may focus on integrating DRL more seamlessly with traditional deductive logic systems, enhancing the AI's ability to reason in complex, dynamic environments.

14.3.4 Fortiori Reasoning

Fortiori reasoning [28], a logical argument form, plays a critical role in deep deductive reasoning within AI. Derived from Latin, meaning "from the stronger," it's a form of argument that asserts a proposition to be true in one situation based on its truth in a related, typically more extreme, situation. In AI, fortiori reasoning offers a unique way to enhance deductive reasoning by allowing the system to make stronger inferences based on established facts or scenarios [29].

Fortiori reasoning operates under two primary forms: "a minori ad maius" (from the lesser to the greater) and "a maiori ad minus" (from the greater to the lesser). The former suggests that if a statement is true in a minor situation, it should undoubtedly be true in a more significant scenario. Conversely, the latter proposes that if a condition is valid in a significant scenario, it should also hold in a lesser one. This comparative logic plays a crucial role in AI, particularly in scenarios where probabilistic or comparative reasoning is involved

Automating this reasoning in AI systems faces challenges like elliptic semantic structure, detecting hidden properties, and dealing with the multidimensional nature of processing these arguments. There's also a scarcity of datasets for training such models, making it a complex task in natural language processing. Despite these challenges, advancements in this area hold significant potential for enhancing AI's capability in logical reasoning and effective communication across various domains.

14.4 Conclusion

We have journeyed through the integration of rule-based reasoning with neural networks, exploring their synergistic potential. We've seen how this convergence enhances AI's capabilities, from the structured logic of rule-based systems to the adaptive learning of neural networks. This chapter sets the stage for the next chapter on "Common Sense Reasoning for Neuro-Symbolic AI," where we will delve into how AI can mimic human-like understanding and reasoning, further bridging the gap between human cognition and artificial intelligence.

References

1. Pan, J.Z.: Resource description framework. In: Handbook on Ontologies, pp. 71–90. Springer (2009)
2. Ebrahimi, M., Sarker, M.K., Bianchi, F., Xie, N., Doran, D., Hitzler, P.: Reasoning over RDF knowledge bases using deep learning (2018). arXiv:1811.04132
3. Weston, J., Chopra, S., Bordes, A.: Memory networks (2014). arXiv:1410.3916
4. Ebrahimi, M., Eberhart, A., Bianchi, F., Hitzler, P.: Towards bridging the neuro-symbolic gap: deep deductive reasoners. Appl. Intell. **51**, 6326–6348 (2021)
5. Wang, Z., Li, J.: RDF2Rules: learning rules from RDF knowledge bases by mining frequent predicate cycles (2015). arXiv:1512.07734
6. Barati, M., Bai, Q., Liu, Q.: Mining semantic association rules from RDF data. Knowl. Based Syst. **133**, 183–196 (2017)
7. Muggleton, S., De Raedt, L.: Inductive logic programming: theory and methods. J. Log. Program. **19**, 629–679 (1994)
8. Gao, K., Inoue, K., Cao, Y., Wang, H.: A differentiable first-order rule learner for inductive logic programming. Artif. Intell. **331**, 104108 (2024)
9. Castillo, R., Leser, U.: Selecting materialized views for RDF data. In: Current Trends in Web Engineering: 10th International Conference on Web Engineering ICWE 2010 Workshops, Vienna, Austria, July 2010, Revised Selected Papers 10, pp. 126–137. Springer (2010)
10. Agathangelos, G., Troullinou, G., Kondylakis, H., Stefanidis, K., Plexousakis, D.: Incremental data partitioning of RDF data in spark. In: The Semantic Web: ESWC 2018 Satellite Events: ESWC 2018 Satellite Events, Heraklion, Crete, Greece, June 3–7, 2018, Revised Selected Papers 15, pp. 50–54. Springer (2018)
11. Cesa-Bianchi, N., Orabona, F.: Online learning algorithms. Annu. Rev. Stat. Its Appl. **8**(1), 165–190 (2021)
12. Dell'Aglio, D., Della Valle, E., Calbimonte, J.P., Corcho, O.: RSP-QL semantics: a unifying query model to explain heterogeneity of RDF stream processing systems. Int. J. Semant. Web Inf. Syst. (IJSWIS) **10**(4), 17–44 (2014)
13. Ngomo, A.C.N., Lehmann, J., Auer, S., Höffner, K.: Raven-active learning of link specifications. In: Ontology Matching 2011 (2011)

14. Pasricha, N., Arcan, M., Buitelaar, P.: NUIG-DSI at the WebNLG+ challenge: leveraging transfer learning for RDF-to-text generation. In: Proceedings of the 3rd International Workshop on Natural Language Generation from the Semantic Web (WebNLG+), pp. 137–143 (2020)
15. Dell'Aglio, D., Della Valle, E., et al.: Incremental reasoning on RDF streams (2014)
16. Schlimmer, J.C., Fisher, D.: A case study of incremental concept induction. In: Proceedings of the Fifth AAAI National Conference on Artificial Intelligence, pp. 496–501 (1986)
17. Higgins, I., Matthey, L., Glorot, X., Pal, A., Uria, B., Blundell, C., Mohamed, S., Lerchner, A.: Early visual concept learning with unsupervised deep learning (2016). arXiv:1606.05579
18. Khosla, M., Anand, A., Setty, V.: A comprehensive comparison of unsupervised network representation learning methods (2019). arXiv:1903.07902
19. Tschannen, M., Bachem, O., Lucic, M.: Recent advances in autoencoder-based representation learning (2018). arXiv:1812.05069
20. Pi, X., Zhong, W., Gao, Y., Duan, N., Lou, J.G.: Logigan: learning logical reasoning via adversarial pre-training. Adv. Neural Inf. Process. Syst. **35**, 16290–16304 (2022)
21. Lakshmanan, L.V., Shiri, N.: A parametric approach to deductive databases with uncertainty. In: International Workshop on Logic in Databases, pp. 59–81. Springer (1996)
22. Corro, C., Titov, I.: Differentiable perturb-and-parse: semi-supervised parsing with a structured variational autoencoder (2018). arXiv:1807.09875
23. Arulkumaran, K., Deisenroth, M.P., Brundage, M., Bharath, A.A.: Deep reinforcement learning: a brief survey. IEEE Signal Process. Mag. **34**(6), 26–38 (2017)
24. Cao, Y., Li, Z., Yang, T., Zhang, H., Zheng, Y., Li, Y., Hao, J., Liu, Y.: Galois: boosting deep reinforcement learning via generalizable logic synthesis. Adv. Neural Inf. Process. Syst. **35**, 19930–19943 (2022)
25. Xiao, S., Pan, C., Wang, M., Zhu, X., Xue, S., Wang, J., Hu, Y., Zhang, J., Feng, J.: Automatic deduction path learning via reinforcement learning with environmental correction (2023). arXiv:2306.10083
26. Sutton, R.S., McAllester, D., Singh, S., Mansour, Y.: Policy gradient methods for reinforcement learning with function approximation. Adv. Neural Inf. Process. Syst. **12**, (1999)
27. Grondman, I., Busoniu, L., Lopes, G.A., Babuska, R.: A survey of actor-critic reinforcement learning: standard and natural policy gradients. IEEE Trans. Syst. Man Cybern. Part C (Appl. Rev.) **42**(6), 1291–1307 (2012)
28. Zurek, T.: Modelling of a fortiori reasoning. In: Proceedings of the 13th International Conference on Artificial Intelligence and Law, pp. 96–100 (2011)
29. Pietarinen, A.V., Bellucci, F.: New light on Peirce's conceptions of retroduction, deduction, and scientific reasoning. Int. Stud. Philos. Sci. **28**(4), 353–373 (2014)

Chapter 15
Common Sense Reasoning for Neuro-Symbolic AI

"Common sense is not so common."

– Voltaire

15.1 Introduction

As the field of Artificial Intelligence (AI) continues to evolve, one of its most significant challenges remains the integration of common sense reasoning into AI systems [1]. Common sense, the innate human ability to make everyday judgments and inferences, is crucial for intelligent behavior and decision-making. However, replicating this ability in AI systems has proven to be a formidable task.

This chapter delves into the realm of Neuro-Symbolic AI (NeSy), an emerging area that combines the learning prowess of neural networks with the logical reasoning capabilities of symbolic AI. The hybrid approach of NeSy aims to imbue AI systems with a more nuanced understanding and reasoning ability, closely mirroring human cognition. The general theme of the article is shown in Fig. 15.1.

We begin by examining the current state of AI, particularly focusing on its limitations in emulating common sense reasoning. Despite impressive advancements in machine learning and neural networks, AI systems often struggle with tasks that humans find trivial, highlighting the gap in common sense reasoning.

The chapter then explores how Neuro-Symbolic AI offers a potential solution to these challenges. We delve into specific applications of NeSy, such as knowledge-based question answering, cognitive systems, and conversational AI, showcasing how these systems integrate and benefit from both neural and symbolic AI approaches.

B. P. Bhuyan et al., *Neuro-Symbolic Artificial Intelligence*, Studies in Computational Intelligence 1176, https://doi.org/10.1007/978-981-97-8171-3_15

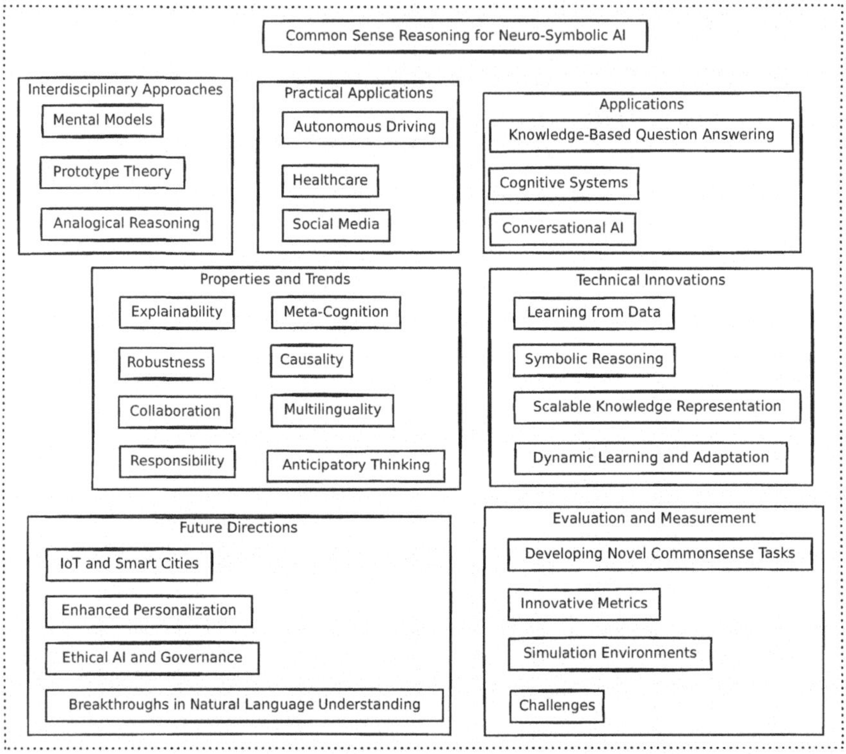

Fig. 15.1 Themes of common sense reasoning for neuro-symbolic AI

Further, we discuss the interdisciplinary approaches that are shaping the development of common sense AI. Insights from cognitive science, including mental models, prototype theory, and analogical reasoning, have significantly influenced AI development, leading to more sophisticated and human-like AI systems.

The chapter also addresses the key properties and emerging trends in common sense AI, exploring concepts such as anticipatory thinking, meta-cognition, causality, and multilingualism. These developments not only enhance the capabilities of AI systems but also align them more closely with the intricate nature of human intelligence.

Finally, we look at the practical applications and future directions of NeSy and common sense AI. From autonomous driving to healthcare and social media, the integration of common sense reasoning is set to revolutionize various domains, offering more effective, intuitive, and reliable AI systems.

In conclusion, this chapter provides a comprehensive overview of the state of common sense reasoning in AI, the promise of Neuro-Symbolic AI, and the future landscape of intelligent systems. As AI continues to advance, the integration of common sense reasoning stands as a key milestone in our journey towards creating truly intelligent machines.

15.2 The Current State of AI and Common Sense

Large Language Models (LLMs) like GPT-3 have demonstrated impressive fluency in generating human-like text, yet they often falter when it comes to common sense reasoning [2]. This limitation is rooted in their training process, which primarily involves learning from vast amounts of text data. While this allows them to mimic linguistic patterns effectively, it does not necessarily endow them with an intuitive understanding of real-world scenarios and common sense knowledge.

For instance, LLMs might struggle with tasks that require understanding the physical properties of objects or the unspoken social norms guiding human behavior. They might generate grammatically correct and fluent sentences that are contextually nonsensical or demonstrate a clear lack of understanding of basic physical laws or social etiquette.

These limitations manifest in various ways. For example, an LLM might suggest unrealistic solutions to simple, practical problems, misunderstand the implications of everyday events, or fail to grasp the emotional nuances in social interactions. In a more concrete sense, an LLM might be unable to reason about simple cause-and-effect relationships, such as understanding what happens when a glass of water is tipped over.

Another common gap is the inability of these models to consistently apply moral and ethical reasoning, which is often intuitive for humans. They might generate content that is insensitive or inappropriate, not because of an inherent flaw in their programming but due to the lack of a nuanced understanding of human values and social norms.

The common sense reasoning gap significantly impacts the trustworthiness and reliability of AI systems. In scenarios where accurate understanding and contextual awareness are crucial, such as in medical diagnosis, legal advice, or even everyday digital assistants, the shortcomings of LLMs become critically apparent.

This limitation is not just a technical challenge; it also raises concerns about the deployment of AI in sensitive and high-stakes environments. An AI system that lacks common sense can make errors that seem obvious to a human observer, leading to a loss of trust among users and stakeholders. Moreover, the inability to make sound judgments in complex, real-world scenarios limits the practical applications of these AI systems.

15.3 Neuro-Symbolic AI as a Solution

Neuro-symbolic AI (NeSy) represents a significant step forward in overcoming the limitations of traditional AI models. NeSy combines the strengths of neural networks (deep learning models) and symbolic AI (rule-based systems) to create a more holistic and powerful AI architecture. Neural networks excel at processing large amounts of unstructured data and pattern recognition, while symbolic AI is adept at logical reasoning, handling structured data, and interpreting explicit rules.

The fusion of these two approaches allows for the development of AI systems that can both understand and reason about the world in a more human-like manner. By integrating the learning and adaptability of neural networks with the explicit reasoning and interpretability of symbolic AI, NeSy aims to create AI systems that are more intuitive, reliable, and understandable.

One of the primary benefits of NeSy is the enhancement of AI robustness. NeSy systems can handle a wider range of tasks and adapt to new scenarios more effectively than systems relying on a single approach. They can contextualize neural network outputs with symbolic reasoning, reducing the likelihood of nonsensical or contextually inappropriate responses.

Explainability is another crucial benefit. Traditional neural networks often operate as "black boxes," making it challenging to understand the rationale behind their decisions. In contrast, the symbolic component of NeSy provides a framework for explicating the reasoning process, making AI decisions more transparent and interpretable. This is particularly important in fields where understanding the decision-making process is as important as the decision itself, like in healthcare or law.

Several successful implementations of NeSy illustrate its potential:

15.3.1 Knowledge-based Question Answering

Knowledge-Based Question Answering (KBQA) [3] represents a significant application of NeSy, where the goal is to answer questions by understanding and reasoning over a structured knowledge base. This application combines the strengths of neural networks (for language understanding and processing) with symbolic AI (for logical reasoning and knowledge representation). To illustrate how Knowledge-Based question answering works, consider the following steps:

Understanding Natural Language Queries: Neural components of NeSy are primarily used to interpret natural language queries. This involves parsing the question, understanding its structure, and identifying key entities and relationships. Technologies, like Named Entity Recognition (NER) and Dependency Parsing, play a crucial role here [4].

Mapping Queries to Knowledge Base Entities: The next step involves mapping the identified entities and relationships from the query to the corresponding elements in the knowledge base. This is where symbolic reasoning comes into play. Symbolic AI uses structured representations of knowledge, such as ontologies or semantic networks, to find matches or connections related to the query.

Reasoning and Inference: Once the mapping is established, the system uses logical reasoning to infer answers based on the knowledge base. This might involve traversing relationships, applying logical rules, or even making inferences based on the available data. The reasoning process can be quite complex, especially for questions that require multi-hop reasoning or dealing with ambiguities.

Generating Responses: Finally, the system generates a response in natural language. This often involves converting the structured, symbolic output back into a human-readable form. The neural component of NeSy might be used again here to ensure that the response is coherent and contextually appropriate.

Some of the examples and success stories of such systems are as follows:

IBM's Watson: One of the most well-known examples of KBQA is IBM's Watson [5], which famously competed on the game show Jeopardy. Watson combined deep learning techniques for language processing with a sophisticated knowledge base and inference engine.

Academic Research and Datasets: Numerous academic projects and datasets focus on KBQA. Examples include the SimpleQuestions dataset [6], which contains questions that can be answered using a single fact from a knowledge base, and more complex datasets that require multi-hop reasoning.

Commercial Applications: In the commercial sector, KBQA systems are being used in customer service bots, virtual assistants, and enterprise data analysis tools, providing quick and accurate answers to user queries based on large-scale organizational knowledge bases.

KBQA exemplifies the potential of Neuro-Symbolic AI to handle complex, real-world tasks that require both a deep understanding of natural language and sophisticated reasoning capabilities. It's a field where ongoing research and development continue to push the boundaries of what AI can achieve in terms of understanding and interacting with human language and knowledge.

15.3.2 Cognitive Systems

Integrating cognitive architectures like ACT-R with neuro-symbolic components represents a groundbreaking approach in AI, particularly in emulating human cognitive processes such as learning, memory, and problem-solving. This integration aims to create systems that not only process information like the human brain but also reason and learn in a similar fashion.

ACT-R (Adaptive Control of Thought—Rational) [7] is a cognitive architecture that models various aspects of human cognition. It's based on the idea that human knowledge can be divided into declarative memory (facts and information) and procedural memory (skills and processes). ACT-R uses a set of modules that represent different cognitive functions, including the following:

Perceptual Modules: Handling visual and auditory inputs.

Motor Module: Governing physical actions.

Goal Module: Directing attention and setting goals.

Declarative Memory Module: Storing factual knowledge.

Procedural Memory Module: Storing and executing skills or rules.

The integration of ACT-R with neuro-symbolic components involves complementing this cognitive architecture with the learning and pattern recognition strengths of neural networks, as well as the logical reasoning capabilities of symbolic AI.

Neural networks contribute to the learning aspects, allowing the system to adapt and generalize from experiences. This is particularly useful in perceptual modules where pattern recognition is key. The integration enables the system to learn from less structured data, mimicking the way humans can learn from diverse and sometimes ambiguous inputs.

Symbolic AI enhances the reasoning capabilities of the system, especially in procedural and declarative memory modules. It allows the system to perform logical operations, follow rules, and make inferences based on stored knowledge. This hybrid approach facilitates more complex problem-solving tasks, closely resembling human cognitive processes.

The integration has shown promise in replicating complex human behaviors and cognitive tasks. Examples include language understanding, decision-making under uncertainty, and even simulating aspects of human consciousness and emotions. Such systems can be applied in various fields, from psychology and neuroscience to practical applications like advanced user interfaces and decision support systems.

Cognitive Tutors: One application of ACT-R is in the field of educational software, where cognitive tutors use the architecture to model how students learn and adapt instructions accordingly.

Human Behavior Modeling: In simulations and training, ACT-R integrated with neuro-symbolic AI can model human behavior, providing more realistic scenarios for training purposes in fields like aviation and military operations.

Integrating ACT-R with neuro-symbolic components is a significant step towards creating AI systems that not only process information but also think and learn in ways that are remarkably similar to human cognition. This approach can potentially revolutionize various fields, from education to advanced AI research, by providing deeper insights into both human cognition and advanced AI capabilities.

15.3.3 Conversational AI

Conversational AI [8], particularly when enhanced with Neuro-Symbolic AI (NeSy), represents a significant advancement in the field, enabling systems not only to understand and generate natural language but also to apply common sense reasoning in interactions. This approach is essential for creating AI that can engage in more human-like, contextually appropriate conversations.

NeSy systems utilize neural networks for processing and understanding natural language, handling the nuances and complexities of human speech or text. These systems can parse language inputs, detect sentiment, and identify key topics or intentions, much like traditional natural language processing (NLP) models.

The symbolic aspect of NeSy contributes to reasoning and decision-making based on common sense. It allows the system to infer unstated implications, understand metaphors, and draw on background knowledge that humans typically use in conversation. This capability is crucial for responding to queries or statements in ways that are contextually relevant and sensible, avoiding the nonsensical or irrelevant responses often seen in purely neural network-based systems.

Beyond text and speech, some NeSy conversational systems incorporate multimodal reasoning [9], integrating visual inputs or other sensory data. This enables richer interactions, where the system can understand and respond to a combination of textual, visual, and auditory cues, much like humans do in natural communication.

Google's Gemini Project: The Gemini project by Google [10] is a notable example of advancing conversational AI using NeSy techniques. While specific details about the Gemini project might not be publicly available, such initiatives typically aim to create more sophisticated AI assistants that can understand, remember, and reason about complex user interactions. Projects like Gemini likely focus on enhancing the AI's ability to maintain context over longer conversations, understand indirect or implied meanings, and provide more accurate, relevant, and engaging responses.

15.4 Interdisciplinary Approaches to Common Sense in AI

15.4.1 Cognitive Science Perspectives in AI

The integration of cognitive science perspectives into AI development marks a significant evolution in how we approach the creation and improvement of intelligent systems [11]. Cognitive science, the interdisciplinary study of the mind and its processes, offers valuable insights that can profoundly impact AI research and application. By drawing on theories like mental models, prototype theory, and analogical reasoning, AI can achieve a more nuanced, flexible, and human-like approach to learning, reasoning, and problem-solving. This interdisciplinary approach not only advances the technical capabilities of AI systems but also aligns them more closely with the complexity and adaptability of human intelligence.

15.4.1.1 Mental Models

Mental models [12] in cognitive science are crucial for understanding how individuals perceive, interpret, and interact with the world around them. These models are internal representations that people use to make sense of their experiences, guiding how they infer the workings of objects, systems, and social interactions. Essentially, mental models are the frameworks that people use to understand the past, make sense of the present, and predict the future. They are continuously updated and refined based on new experiences and information.

In AI, the concept of mental models plays a significant role in how systems are designed to process and interpret data. The goal is to develop AI systems that can form, update, and utilize their representations of the world in a manner akin to human cognition. By implementing mental models, AI systems are expected to understand better and predict complex scenarios that are not explicitly outlined in their training data.

For instance, in an AI system equipped with mental models, the system would not only rely on the data it has been trained on but also build and refine its internal representations of how various elements in its environment interact. This would allow the AI to make more accurate predictions about new or unseen situations. For example, an AI system with a well-developed mental model of social interactions might be better equipped to understand nuances in human communication, such as sarcasm or indirect requests, which are often challenging for traditional AI.

Moreover, AI systems with mental models can be more adaptable and flexible. They can apply learned knowledge to new contexts, drawing parallels or making inferences that go beyond the data they were explicitly trained on. This adaptability is particularly important in dynamic environments where conditions constantly change, and the AI system must adjust its understanding accordingly.

Incorporating mental models into AI also opens the door to more personalized experiences. As these models can be tailored to individual users, AI systems can provide responses and services that are more aligned with each user's unique preferences and history.

15.4.1.2 Prototype Theory

Prototype theory [13], originating from cognitive psychology, posits a fundamentally different approach to categorization than traditional, rigid class definitions. According to this theory, people categorize objects and concepts based on the most typical or representative example, known as a prototype, rather than strict boundaries. For instance, when thinking of a bird, people might typically imagine a robin or a sparrow rather than a penguin, even though all are birds. This approach is more fluid and reflects the way human cognition often works, focusing on the most common or representative attributes of a category rather than trying to fit every example into a rigidly defined set.

In the realm of AI, prototype theory has significant implications for how systems are designed to categorize and understand data. Traditional AI systems often rely on clear, predefined rules and boundaries to classify data. However, by incorporating the principles of prototype theory, AI systems can be developed to categorize objects and concepts in a more flexible and realistic manner.

For instance, an AI system based on prototype theory might learn to identify the "typical" features of a category. This allows the system to recognize a wide range of variations within that category, accommodating the natural diversity found in real-world data. For example, in image recognition, rather than rigidly defining what constitutes a cat, a prototype-based system would learn from a variety of cat images, understanding that there are common features most cats share but also that considerable variation exists.

This flexibility is particularly valuable in dealing with ambiguous or borderline cases, where traditional AI might struggle to classify an object or concept that doesn't neatly fit into predefined categories. A prototype-based system can handle such ambiguities more effectively, making decisions based on similarity to a learned prototype rather than strict rule adherence.

Moreover, prototype theory in AI facilitates more natural interaction with humans, as the AI's approach to categorization more closely mirrors human thought processes. This can enhance the AI's performance in tasks requiring human-like understanding, such as natural language processing, where understanding context and nuance is crucial.

15.4.1.3 Analogical Reasoning

Analogical reasoning [14], a critical component of human cognition, involves the ability to draw parallels and make inferences between seemingly unrelated domains or scenarios. It is a fundamental aspect of how humans approach problem-solving and creativity, allowing us to apply known solutions from one context to a new yet related context. For instance, scientists often use analogies to formulate hypotheses or explain complex phenomena by comparing them to simpler, more familiar situations. This form of reasoning is not just about finding similarities but also about understanding the underlying principles that allow these similarities to exist.

Implementing analogical reasoning in AI presents a unique set of challenges and opportunities. Traditional AI systems are often very good at handling tasks within their specific training domain but struggle to apply this knowledge to different, albeit related, contexts. To overcome this, AI needs to be able to identify underlying patterns and relationships that transcend specific instances or domains.

Creating AI systems capable of analogical reasoning involves training them to recognize deep structural similarities between different sets of data. This goes beyond surface-level similarities and requires an understanding of the abstract principles that govern different scenarios. For example, an AI system trained in analogical reasoning might learn principles from a physics domain, such as the conservation of energy, and apply this understanding to solve problems in an entirely different domain, such as economics or environmental science.

One of the key benefits of analogical reasoning in AI is its potential to drive innovation and creativity. By drawing on knowledge and patterns from a wide range of sources, AI systems can generate novel solutions and ideas that might not be immediately apparent when considering each domain in isolation. This ability is particularly valuable in fields where creative problem-solving is essential, such as design, engineering, and strategic planning.

Moreover, AI systems equipped with analogical reasoning capabilities can significantly enhance their learning efficiency. Instead of needing extensive data from every possible domain, they can leverage existing knowledge and apply it to new areas, much like humans can often quickly adapt their skills and understanding to new situations based on their past experiences.

15.4.2 Technical Innovations in NeSy Architectures

The field of Neuro-Symbolic AI (NeSy) has witnessed substantial technical advancements, primarily characterized by the integration of machine learning techniques with symbolic reasoning. This hybrid approach marries the raw data processing power of neural networks with the structured, logical processing of symbolic AI, creating a more holistic AI system.

At the core of NeSy architectures is the blend of learning from data and applying logical reasoning. Neural networks, renowned for their ability to handle vast and varied datasets, are adept at identifying patterns, deciphering complex relationships, and deriving insights from unstructured data like images, text, and sounds. This capability is especially powerful in contexts where the data lacks a clear structure or when dealing with massive datasets that traditional computing methods struggle to process efficiently.

However, the real innovation in NeSy architectures lies in how these neural networks' findings are then processed further using symbolic reasoning systems. Unlike neural networks that excel in pattern recognition, symbolic systems excel in applying structured reasoning and rules. They take the data processed by neural networks and place it within a broader logical framework. This step is crucial in making the outputs of neural networks more interpretable and aligned with human reasoning processes. For instance, an AI system might use a neural network to analyze a block of text to understand the prevailing sentiment and then apply symbolic reasoning to contextualize this sentiment within a broader social or cultural setting.

Another significant innovation in NeSy AI is the development of scalable methods for knowledge representation [15]. As the real world is vast and diverse, AI systems need to efficiently manage and access large knowledge bases. Scalability is critical in these systems, ensuring they remain effective even as the amount of available data grows. This efficiency is not just about handling more data but also about doing so in a way that the AI system can retrieve, process, and infer from this data in a timely and relevant manner.

Dynamic learning [16] and adaptation form another cornerstone of advanced NeSy systems. In a continually evolving world, the ability of AI systems to adapt to new information and changing environments is paramount. These systems are designed not just to learn from the initially fed data but also to update their models and strategies continually based on new inputs. This dynamic learning approach allows NeSy systems to stay relevant over time, adapting their responses and strategies to align with current contexts and emerging trends.

15.5 Evaluation and Measurement of AI Common Sense

The evaluation and measurement of AI's common sense reasoning capabilities are critical for assessing the progress and effectiveness of AI systems. This aspect of AI development involves several key components:

15.5.1 Developing Novel Common sense Tasks for AI

Developing novel common sense tasks for AI has been a focus of recent research, as it's crucial for advancing the field of AI towards more human-like reasoning and decision-making capabilities.

One significant development in this area is the AGENT (Action, Goal, Efficiency, coNstraint, uTility) benchmark [17], released by researchers from IBM, MIT, and Harvard at ICML. AGENT, inspired by developmental studies in infants, aims to benchmark an AI model's intuitive psychology. It comprises 8,400 3D animations organized into categories like goal preferences, action efficiency, unobserved constraints, and cost or reward trade-offs. This benchmark is designed to test a machine learning model's understanding of key concepts in intuitive psychology, much like how researchers evaluate an infant's cognitive abilities. The approach represents a shift towards more fluid, real-world AI systems, combining traditional human psychology methods with machine learning to better understand common sense reasoning in AI.

Another perspective on the challenge of embedding common sense in AI is highlighted by the difficulties AI faces in replicating the broad and often ambiguous nature of human common sense. Common sense encompasses not only social abilities but also a naive understanding of physics and abstract notions like time, space, and events. Despite significant advancements in AI, particularly in areas like game-playing and computer vision, achieving machine common sense with the richness of human common sense remains a distant goal. This challenge is exemplified by the current AI models' tendency to make errors in situations that require an intuitive understanding of everyday life. As a result, current AI research, including work on advanced deep learning AI-like transformers [18], is increasingly focused on addressing these gaps. However, the path to achieving machine common sense is complex and may require new fundamental research in neural networks and AI models.

Overall, these developments and challenges underscore the importance of creating more versatile and intuitive AI models capable of understanding and reasoning in ways similar to humans. The task of developing novel common sense tasks for AI is central to this endeavor, requiring a multifaceted approach that combines advancements in machine learning with insights from psychology and cognitive science.

15.5.2 Innovative Metrics and Simulation Environments

Innovative metrics in AI for evaluating common sense reasoning are crucial for understanding and improving the capabilities of AI systems. The development of these metrics has become a key area of focus in current research, as they provide more nuanced ways of assessing AI performance beyond traditional methods.

Benchmarking Models with Realistic Scenarios:

Current research is increasingly focused on benchmarking AI models against more realistic and comprehensive scenarios that require common sense reasoning. This involves creating simulations or datasets that mimic real-world complexities where AI systems must use common sense to navigate effectively. These benchmarks test the AI's ability to understand context, infer unstated information, and make decisions that align with human common sense.

Quantitative and Qualitative Metrics:

Researchers are developing both quantitative and qualitative metrics to evaluate AI common sense [19]. Quantitative metrics might include the accuracy of predictions in scenarios requiring common sense, while qualitative metrics assess the AI's reasoning process, how well its actions align with human expectations, and its ability to handle ambiguous or incomplete information.

Multimodal Evaluation:

With the rise of multimodal AI systems [20] that combine text, visuals, and other data types, evaluation metrics are also evolving to assess common sense across different modes of input and interaction. This approach ensures that AI systems are not just textually but also visually and contextually aware of their common sense reasoning.

Comparative Analysis:

Innovative metrics [21] also involve comparative analysis, where AI systems are evaluated against human performance in common sense tasks. This comparison helps in understanding the gap between AI and human common sense reasoning and in identifying specific areas where AI needs improvement.

Innovative metrics for evaluating AI's common sense reasoning have seen advancements, particularly in the development of quantitative measures that go beyond traditional accuracy and error rate assessments [22]. These new metrics aim to capture the nuanced understanding of common sense in AI systems.

Scenario-Based Evaluation: Recent developments include scenario-based evaluation metrics, as exemplified by the AGENT benchmark created by IBM and others [23, 24]. AGENT evaluates an AI model's core psychological reasoning ability or common sense through various scenarios like action efficiency, unobserved constraints, and cost-reward trade-offs. This benchmark assesses whether AI can understand and predict the actions and decisions of agents in scenarios accounting for physical constraints and preferences. For example, the AI might be tested on its ability to recognize that an agent takes the most efficient action to reach a goal, considering physical barriers or unobserved constraints.

Models like Bayesian Inverse Planning and Core Knowledge (BIPaCK) and ToMnet-G [17] have been developed to assess AI's understanding of actions using Bayesian inference combined with core knowledge of physics. These models evaluate AI's ability to propose hypotheses about physical parameters and predict agents' trajectories, providing a "surprise rating" based on the deviation from expected behavior. This approach not only measures the accuracy of AI in common sense reasoning but also its ability to generalize to different physical configurations within a scenario.

Numeracy in Common Sense Reasoning: Another innovative approach is the focus on "numeracy" in common sense reasoning [25], particularly the intuitive understanding of numbers, their magnitude, and relationships, rather than computational precision. The Com2Sense benchmark [26], for instance, employs this concept to evaluate AI's understanding of basic numerical common sense. This benchmark includes various statement pairs that test the AI's ability to reason about numbers in everyday contexts, thus providing a more comprehensive view of its common sense reasoning abilities.

Pairwise Accuracy Metric: Com2Sense [26] introduces a novel metric called pairwise accuracy, which evaluates AI predictions as correct if both predictions within a complementary sentence pair are accurate. This metric adds another layer to the standard accuracy measure, focusing on the AI's ability to handle complementary or contrasting statements, thus providing a deeper insight into its common sense reasoning capabilities.

Ethical and Societal Impact Assessment:

Alongside functional metrics, there is a growing emphasis on evaluating the ethical and societal impacts of AI decisions [27]. This includes assessing whether AI systems are making common sense decisions that are ethically sound and socially responsible.

15.5.3 Challenges in Realistically Evaluating AI Common Sense

Evaluating AI common sense realistically presents several challenges, each stemming from the inherent complexities of both common sense reasoning and the capabilities of AI systems:

Defining Common Sense:

Common sense is a broad and somewhat nebulous concept, varying across cultures and contexts. What is considered common sense in one setting might not hold in another. This variability makes it challenging to develop a universal set of criteria or tasks to evaluate AI common sense across diverse environments.

Mimicking Human Reasoning:

Human reasoning is often intuitive and not always based on explicit logic or rules. This makes it difficult to create AI systems that can replicate such reasoning, and consequently, challenging to evaluate AI common sense in a way that truly reflects human thought processes.

Complexity of Real-World Scenarios:

Common sense reasoning often involves understanding and navigating complex, real-world scenarios. Creating simulations or scenarios that accurately capture these complexities for AI evaluation is a daunting task. The unpredictability and multifaceted nature of real-world situations are hard to replicate in controlled testing environments. Integration with Other Cognitive Abilities:

Common sense is closely tied to other cognitive abilities like language understanding, emotional intelligence, and social awareness. Evaluating AI common sense in isolation from these abilities can be misleading, as it might not accurately reflect the AI's overall understanding and reasoning capabilities.

Ethical and Societal Considerations:

The evaluation of AI common sense also involves ethical and societal considerations. For instance, ensuring that AI systems do not perpetuate biases or stereotypes under the guise of common sense is a significant challenge. Evaluating AI's decisions and actions from an ethical standpoint is as important as assessing its logical reasoning capabilities.

Measurement and Quantification:

Developing quantitative measures for common sense reasoning is challenging. Traditional metrics like accuracy or error rates may not adequately capture the nuances of common sense reasoning. Developing new metrics that can effectively measure the subtleties of common sense is a crucial area of ongoing research.

Balancing Generalization and Specialization:

AI systems should not only perform well on tasks they have been trained on but also generalize this learning to new, unseen scenarios. Evaluating an AI's ability to generalize its common sense reasoning to different contexts, while still performing specialized tasks effectively, presents a significant challenge.

15.6 Key Properties and Trends in Common Sense AI

15.6.1 Exploring AI Properties

Explainability: Explainability in AI is increasingly recognized as a critical feature, especially in common sense AI. It refers to the system's ability to describe its decision-making process in an understandable way to humans. This feature is particularly important when AI decisions have significant impacts on people's lives, such as in healthcare diagnostics or financial lending. Explainable AI helps in building trust and ensures transparency, allowing users and stakeholders to comprehend how and why specific decisions or recommendations were made. As AI systems become more complex, the challenge lies in simplifying the explanation of intricate algorithms in a way that is accessible to non-expert users. The development of explainable AI also aids in debugging and improving AI systems, as understanding an AI's decision process can reveal strengths and weaknesses in its reasoning.

Robustness: Robustness in AI systems refers to their ability to maintain consistent performance across various conditions and inputs. In the context of common sense AI, robustness is essential for ensuring that the system remains reliable in diverse and unpredictable real-world scenarios. This includes the system's ability to handle ambiguous, incomplete, or misleading data without significant loss of accuracy or functionality. Developing robust AI systems involves creating algorithms that can adapt to different contexts, recognize and manage uncertainties, and continue to operate effectively even under challenging conditions. Robustness is crucial for applications where reliability is paramount, such as autonomous vehicles or critical infrastructure management.

Collaboration: Collaboration in AI is about designing systems that can work effectively alongside humans. This property is becoming increasingly important as AI systems are more frequently used to augment human capabilities rather than replace them. In common sense AI, collaboration means that the system can understand and adapt to human needs, preferences, and working styles, facilitating a synergistic relationship. Collaborative AI can contribute to diverse fields, from healthcare, where AI assists doctors in diagnosis and treatment plans, to creative industries, where AI can augment human creativity and productivity. The challenge lies in creating AI systems that can seamlessly integrate into human workflows, understand context, and respond to subtle cues in human interaction.

Responsibility: Responsibility in AI encompasses the ethical implications and societal impact of AI systems. As AI technologies become more integrated into everyday life, ensuring that they operate in a responsible and ethical manner is critical. This includes accountability for AI decisions, fairness in algorithms to prevent biases, and consideration of privacy and data security. Responsible AI development also involves considering the long-term impacts of AI on society, including potential job displacement and the ethical use of AI in surveillance and data collection. Developers and stakeholders must navigate complex ethical landscapes to ensure that AI systems are not only technically advanced but also socially responsible and beneficial.

15.6.2 Discussing New Trends

Anticipatory Thinking: Anticipatory thinking in AI is an emerging trend that focuses on the system's ability to predict and prepare for future events or scenarios. This capability is pivotal in areas such as strategic decision-making and risk management, where foresight can significantly influence outcomes. AI systems with anticipatory thinking can analyze current data trends, historical information, and possible future scenarios to forecast potential outcomes. This allows them to provide recommendations or automatically adjust their strategies to align with predicted future states.

The development of such systems involves complex modeling of various potential futures, often using techniques like scenario analysis, simulations, and probabilistic reasoning. In sectors like finance or weather forecasting, anticipatory thinking enables AI to advise on investments or disaster preparedness by evaluating numerous possible future states and their likelihood. This approach can lead to more informed decision-making, helping to mitigate risks and capitalize on potential opportunities.

Meta-Cognition: Meta-cognition in AI refers to the development of systems that possess an awareness of their own processes and limitations. This self-awareness allows AI systems to evaluate their performance, recognize the boundaries of their knowledge or expertise, and adapt their strategies accordingly. Such systems can identify when a problem is beyond their current capabilities and either seek additional information or defer to human input.

Incorporating meta-cognition into AI involves building mechanisms for self-assessment and self-regulation. These systems can analyze their decision-making processes, identify areas of uncertainty or error, and adjust their learning or reasoning approaches. This feature is crucial for the continuous improvement of AI systems, enhancing their reliability and trustworthiness. In critical applications such as medical diagnosis or autonomous vehicles, where errors can have significant consequences, meta-cognitive AI can be particularly valuable.

Causality: Understanding causality, as opposed to mere correlation, is a growing focus in AI development. Grasping cause-and-effect relationships allows AI systems to make more accurate predictions and informed decisions. This understanding is crucial in complex environments where multiple variables interact in dynamic ways, such as in economics, healthcare, or environmental science.

Developing AI systems that can discern causality involves going beyond pattern recognition to understand the underlying mechanisms that drive events. This might include using causal inference models or integrating domain-specific knowledge that can inform the AI about possible causal relationships. AI that understands causality can better predict the outcomes of interventions, which is essential for planning and decision-making in various fields.

Multilinguality: Multilinguality in AI is about creating systems that can understand and interact in multiple languages, a capability that is increasingly important in our globally connected world. This trend is not just about translation but also involves understanding cultural nuances and context that vary from one language to another.

Multilingual AI systems can serve a diverse global user base, breaking down language barriers and facilitating cross-cultural communication. In fields like customer service, healthcare, and international business, multilingual AI can greatly enhance accessibility and efficiency. Developing such systems requires sophisticated natural language processing capabilities and deep learning models trained on a wide range of languages and dialects.

15.7 Practical Applications and Future Directions

Common sense AI is not just about adding another layer of intelligence to AI systems; it's about integrating an intuitive understanding that aligns AI's responses and decisions more closely with human expectations and norms. In domains like autonomous driving and healthcare, this means safer, more reliable, and more user-friendly AI systems. In social media, it translates into platforms that are more engaging, ethical, and attuned to user needs. There is a huge application demand for common sense AI, which we will see in the upcoming unit. A few of them are listed here as follows:

Autonomous Driving:

In the realm of autonomous vehicles, common sense AI plays a critical role in decision-making and safety [28]. AI systems in self-driving cars need to interpret traffic conditions, pedestrian behavior, and environmental cues in real time, much like a human driver. Common sense AI helps these systems to anticipate and react to unpredictable scenarios, such as sudden pedestrian movements or erratic behavior from other vehicles. This includes understanding subtle cues that a human driver would intuitively recognize, like a ball rolling onto the street and possibly being followed by a child.

Healthcare:

In healthcare, common sense AI can revolutionize patient care and diagnostics [29]. AI systems can analyze patient data, medical histories, and current research to assist in diagnosis and treatment planning. The incorporation of common sense allows these systems to consider not just the medical data but also the broader context of a patient's condition, such as lifestyle factors and potential environmental influences. This holistic approach can lead to more accurate diagnoses and personalized treatment plans.

Social Media:

In social media platforms, common sense AI is essential for moderating content, personalizing user experiences, and detecting misinformation or harmful behavior [30]. AI systems with an understanding of common sense and social norms can better identify inappropriate or harmful content. Additionally, they can tailor content feeds to individual preferences and interests while maintaining ethical standards and respecting user privacy.

Looking forward, the field of Neuro-Symbolic AI (NeSy) and common sense AI is poised for significant advancements. We can expect to see the following:

Greater Integration with IoT and Smart Cities:

Common sense AI could play a key role in the development of smart city initiatives, where AI systems manage everything from traffic flows to energy consumption, using real-time data and intuitive understanding to optimize city living.

Enhanced Personalization in Consumer Technology:

In consumer technology, common sense AI could lead to more personalized and intuitive user experiences, from smarter virtual assistants to AI-driven personal health and fitness advisors.

Ethical AI and Governance:

As AI systems become more integrated into critical aspects of society, the emphasis on ethical AI and governance will grow. We'll likely see more research and development focused on ensuring AI systems are fair, unbiased, and transparent.

Breakthroughs in Natural Language Understanding:

Advances in natural language processing and understanding, powered by common sense AI, could revolutionize how we interact with technology, making voice and text interfaces more natural and intuitive. The potential for NeSy and common sense AI is vast and extends across multiple sectors. As these technologies continue to evolve, they promise to bring more sophisticated, intuitive, and human-centric AI solutions to a variety of challenges and opportunities in our increasingly interconnected world.

15.8 Conclusion

As we conclude this exploration of Common Sense Reasoning for Neuro-Symbolic AI, we recognize that this field stands at a fascinating intersection of technology, cognitive science, and practical application. The journey through this chapter has highlighted not only the current state and limitations of AI in replicating human-like common sense but also the promising potential of Neuro-Symbolic approaches in bridging these gaps.

Neuro-Symbolic AI, with its blend of neural learning capabilities and symbolic reasoning, emerges as a compelling solution to some of the most enduring challenges in AI. By leveraging the strengths of both neural networks and symbolic systems, NeSy AI is poised to offer more nuanced, context-aware, and reliable intelligence. The exploration of specific applications, such as knowledge-based question answering, cognitive systems, and conversational AI, illustrates the practical implications of these advancements.

The interdisciplinary approaches from cognitive science have been particularly illuminating, providing valuable insights that drive the development of more sophisticated AI systems. Concepts such as mental models, prototype theory, and analogical reasoning are not just abstract ideas but pivotal components in designing AI that closely mirror human thinking and problem-solving.

Moreover, the discussion of key properties and emerging trends in common sense AI—including explainability, robustness, collaboration, responsibility, anticipatory thinking, meta-cognition, causality, and multilingualism—underscores the dynamic nature of this field. These properties and trends are critical in ensuring that AI systems of the future are not only technically advanced but also ethically sound and socially beneficial.

As we look ahead, the practical applications and future directions of NeSy and common sense AI suggest a landscape rich with possibilities. From enhancing the safety and efficiency of autonomous driving to revolutionizing patient care in healthcare and from improving user experiences in social media to pioneering new frontiers in robotics and autonomous systems, the potential is immense.

References

1. McCarthy, J.: Artificial intelligence, logic, and formalising common sense. In: Machine Learning and the City: Applications in Architecture and Urban Design, pp. 69–90 (2022)
2. Krause, S., Stolzenburg, F.: From data to commonsense reasoning: the use of large language models for explainable AI (2024). arXiv:2407.03778
3. Kapanipathi, P., Abdelaziz, I., Ravishankar, S., Roukos, S., Gray, A., Astudillo, R., Chang, M., Cornelio, C., Dana, S., Fokoue, A., et al.: Question answering over knowledge bases by leveraging semantic parsing and neuro-symbolic reasoning (2020). arXiv:2012.01707
4. Keber, M., Grubišić, I., Barešić, A., Jović, A.: A review on neuro-symbolic AI improvements to natural language processing. In: 2024 47th MIPRO ICT and Electronics Convention (MIPRO), pp. 66–72. IEEE (2024)
5. Strickland, E.: IBM Watson, heal thyself: How IBM overpromised and underdelivered on AI health care. IEEE Spectr. **56**(4), 24–31 (2019)
6. Bordes, A., Usunier, N., Chopra, S., Weston, J.: Large-scale simple question answering with memory networks (2015). arXiv:1506.02075
7. Kim, N., Nam, C.S.: Adaptive control of thought-rational (act-r): applying a cognitive architecture to neuroergonomics. In: Neuroergonomics: Principles and Practice, pp. 105–114 (2020)
8. Arabshahi, F., Lee, J., Gawarecki, M., Mazaitis, K., Azaria, A., Mitchell, T.: Conversational neuro-symbolic commonsense reasoning. In: Proceedings of the AAAI Conference on Artificial Intelligence, vol. 35, pp. 4902–4911 (2021)
9. Lu, P., Mishra, S., Xia, T., Qiu, L., Chang, K.W., Zhu, S.C., Tafjord, O., Clark, P., Kalyan, A.: Learn to explain: multimodal reasoning via thought chains for science question answering. Adv. Neural Inf. Process. Syst. **35**, 2507–2521 (2022)
10. Team, G., Anil, R., Borgeaud, S., Wu, Y., Alayrac, J.B., Yu, J., Soricut, R., Schalkwyk, J., Dai, A.M., Hauth, A., et al.: Gemini: a family of highly capable multimodal models (2023). arXiv:2312.11805
11. Srinivasan, R., Chander, A.: Explanation perspectives from the cognitive sciences' a survey. In: Proceedings of the Twenty-Ninth International Conference on International Joint Conferences on Artificial Intelligence, pp. 4812–4818 (2021)
12. Gero, K.I., Ashktorab, Z., Dugan, C., Pan, Q., Johnson, J., Geyer, W., Ruiz, M., Miller, S., Millen, D.R., Campbell, M., et al.: Mental models of AI agents in a cooperative game setting. In: Proceedings of the 2020 CHI Conference on Human Factors in Computing Systems, pp. 1–12 (2020)
13. Geeraerts, D.: Prototype theory. Cogn. Linguist. Basic Read. **34**, 141–165 (2006)
14. Sunstein, C.R.: On analogical reasoning. Harv. Law Rev. **106**(3), 741–791 (1993)
15. Omran, P.G., Wang, K., Wang, Z.: Scalable rule learning via learning representation. In: IJCAI, pp. 2149–2155 (2018)
16. Stuchlik, A.: Dynamic learning and memory, synaptic plasticity and neurogenesis: an update. Front. Behav. Neurosci. **8**, 106 (2014)
17. Shu, T., Bhandwaldar, A., Gan, C., Smith, K., Liu, S., Gutfreund, D., Spelke, E., Tenenbaum, J., Ullman, T.: Agent: A benchmark for core psychological reasoning. In: International Conference on Machine Learning, pp. 9614–9625. PMLR (2021)
18. Gupta, A., Mondal, D., Sheshadri, A.K., Zhao, W., Li, X.L., Wiegreffe, S., Tandon, N.: Editing common sense in transformers (2023). arXiv:2305.14956
19. Chen, Y., Huang, S., Yuan, T., Qi, S., Zhu, Y., Zhu, S.C.: Holistic++ scene understanding: single-view 3d holistic scene parsing and human pose estimation with human-object interaction and physical commonsense. In: Proceedings of the IEEE/CVF International Conference on Computer Vision, pp. 8648–8657 (2019)
20. Xing, Y., Shi, Z., Meng, Z., Lakemeyer, G., Ma, Y., Wattenhofer, R.: Km-bart: Knowledge enhanced multimodal bart for visual commonsense generation (2021). arXiv:2101.00419
21. Muller, A., Välikangas, L., Merlyn, P.: Metrics for innovation: guidelines for developing a customized suite of innovation metrics. Strat. & Lead. **33**(1), 37–45 (2005)

22. Storks, S., Gao, Q., Chai, J.Y.: Commonsense reasoning for natural language understanding: a survey of benchmarks, resources, and approaches, pp. 1–60 (2019). arXiv:1904.01172
23. Mueller, E.T.: Commonsense reasoning: an event calculus based approach. Morgan Kaufmann (2014)
24. Casillo, M., Colace, F., Lorusso, A., Santaniello, D., Valentino, C.: A multilevel graph approach for IOT-based complex scenario management through situation awareness and semantic approaches. J. Reliab. Intell. Environ. 1–17 (2024)
25. Karaali, G., et al.: What's in a name? a critical review of definitions of quantitative literacy, numeracy, and quantitative reasoning (2016)
26. Singh, S., Wen, N., Hou, Y., Alipoormolabashi, P., Wu, T.L., Ma, X., Peng, N.: Com2sense: a commonsense reasoning benchmark with complementary sentences (2021). arXiv:2106.00969
27. Floridi, L., Cowls, J., Beltrametti, M., Chatila, R., Chazerand, P., Dignum, V., Luetge, C., Madelin, R., Pagallo, U., Rossi, F., et al.: Ai4people–an ethical framework for a good AI society: opportunities, risks, principles, and recommendations. Minds Mach. **28**, 689–707 (2018)
28. Omeiza, D., Webb, H., Jirotka, M., Kunze, L.: Explanations in autonomous driving: a survey. IEEE Trans. Intell. Transp. Syst. **23**(8), 10142–10162 (2021)
29. Parveen, A., Kannan, G.: Healthcare transformed: a comprehensive survey of artificial intelligence trends in healthcare industries. In: Digital Healthcare in Asia and Gulf Region for Healthy Aging and More Inclusive Societies, pp. 395–424 (2024)
30. Branco, R., Branco, A., Silva, J.M., Rodrigues, J.: Commonsense reasoning: how do neuro-symbolic and neuro-only approaches compare? In: CIKM Workshops (2021)

Part VI
Neuro-Symbolic Applications and Use Cases

Chapter 16
Explainable and Trustworthy AI with Neuro-Symbolic Approaches

"With great power comes great responsibility."

– Peter Parker, *Spider-Man (2002)*

16.1 Introduction

The rapid advancement of artificial intelligence (AI) has brought about significant benefits and transformations across various sectors. However, as AI systems become more pervasive, the need for these systems to be explainable and trustworthy has become paramount. Explainable AI (XAI) aims to make AI decision-making processes transparent and understandable, while Trustworthy AI encompasses a broader spectrum, including fairness, robustness, and privacy. Integrating these aspects is essential for fostering trust and ensuring the ethical deployment of AI technologies.

Neuro-Symbolic AI (NeSy) emerges as a promising approach to address these needs by combining the strengths of neural networks and symbolic reasoning. Neural networks are known for their powerful learning capabilities, especially with large datasets, but often lack transparency, making it difficult to understand their decision-making processes. In contrast, symbolic reasoning provides clear, rule-based logic that is inherently interpretable but struggles with scalability and adaptability to new data.

By integrating these two paradigms, NeSy aims to create AI systems that are both powerful and interpretable. This hybrid approach leverages the pattern recognition and learning capabilities of neural networks with the explicit, logical reasoning of

B. P. Bhuyan et al., *Neuro-Symbolic Artificial Intelligence*, Studies in Computational Intelligence 1176, https://doi.org/10.1007/978-981-97-8171-3_16

symbolic systems. As a result, NeSy can provide human-understandable explanations for its decisions, enhancing transparency and trust.

For example, consider a healthcare application where an AI system is used to diagnose diseases from medical images. A purely neural approach might identify patterns indicative of a disease but would struggle to explain its reasoning process. A symbolic approach, on the other hand, could provide clear, logical rules for diagnosis but might not perform as well with the complex patterns in the images. A NeSy approach combines these strengths: the neural component identifies patterns and makes initial diagnoses, while the symbolic component translates these patterns into human-readable rules, explaining why a particular diagnosis was made based on established medical knowledge.

This chapter delves into the role of NeSy in advancing XAI and Trustworthy AI. It explores the different methodologies used to enhance interpretability, discusses the ongoing challenges in integrating these technologies, and speculates on future developments and potential solutions. Additionally, it highlights key areas for future research, aiming to guide the continued evolution of AI systems that are not only intelligent but also align with societal values and ethical standards.

16.2 Definitions of Components of Trustworthy AI

The concept of Trustworthy AI [1] is foundational to the development and deployment of artificial intelligence systems that can be relied upon to act in accordance with established norms and values. Trustworthy AI is built upon several key components: Valid AI, Responsible AI, Privacy-Preserving AI, and Explainable AI (XAI), each addressing different aspects of trustworthiness.

Valid AI encompasses both robust and certifiable AI systems. Robust AI [2] refers to the resilience of AI systems under various conditions, including unexpected and adverse scenarios such as adversarial attacks and data anomalies. These systems are engineered to maintain their performance and reliability despite such challenges, thereby ensuring consistent operation and resistance to errors and manipulation [3]. On the other hand, certifiable AI implies that AI systems meet rigorous standards and pass assessments that verify their adherence to safety, security, and reliability benchmarks [4, 5]. These certifications are often specific to particular industries and mandated by regulatory bodies, ensuring that AI systems operate within legal and ethical parameters.

Responsible AI [6] focuses on the ethical deployment of AI technologies, emphasizing fairness and ethics. Fair AI [7] is committed to eliminating biases and ensuring that AI-driven decisions do not unfairly discriminate against any individual or group. This aspect of AI aims to foster justice and equality within automated decision-making processes. Ethical AI [8], meanwhile, involves the broader moral implications of AI technologies, aligning AI behaviors with human values and societal norms. This includes designing AI systems to respect human rights, privacy, dignity, and autonomy, often guided by well-established ethical frameworks and principles [9, 10].

Privacy-preserving AI [11] includes methods such as encrypted AI and AI systems designed with Privacy by Design principles. Encrypted AI [12] employs cryptographic techniques to secure data throughout the AI processing pipeline, ensuring that sensitive information remains protected even during analysis. Techniques like homomorphic encryption [13] allow AI to learn from data without exposing it, while secure multi-party computation processes data in fragments across different parties without revealing the underlying data. Privacy by Design [14] is a proactive approach that integrates data protection into the AI system's architecture from the beginning, making privacy a fundamental component of the system's lifecycle.

Explainable AI (XAI) [15] is crucial for the transparency and understandability of AI decision-making processes. Human-interpretable AI [16] focuses on presenting information in ways that are comprehensible to non-experts, using visual explanations, simplified models, or surrogate models to approximate complex decisions. Machine-interpretable AI [17], conversely, refers to generating explanations that other AI systems can understand and act upon. In order to improve integration and interaction between various AI systems, this involves developing standardized formats for explanations that machines can parse, process, and use. Figure 16.1 deals with a framework of the components of Trustworthy AI.

These components collectively ensure that AI systems are not only technically proficient but also align with societal, ethical, and legal standards, fostering trust and acceptance among users and stakeholders. By addressing these diverse aspects of trustworthiness, AI systems can be developed to be reliable, fair, private, and transparent, thereby enhancing their credibility and user confidence.

16.3 Neuro-Symbolic AI in Enhancing Explainability

Neuro-Symbolic AI contributes to XAI by integrating the intuitive, example-driven learning of neural networks with the logical, rule-based reasoning of symbolic AI. This combination allows for a more transparent decision-making process that can be traced and understood in human terms. For instance, when a neural network classifies an image, the symbolic layer can provide a human-readable explanation by associating high-level features recognized by the network with known symbols and reasoning patterns.

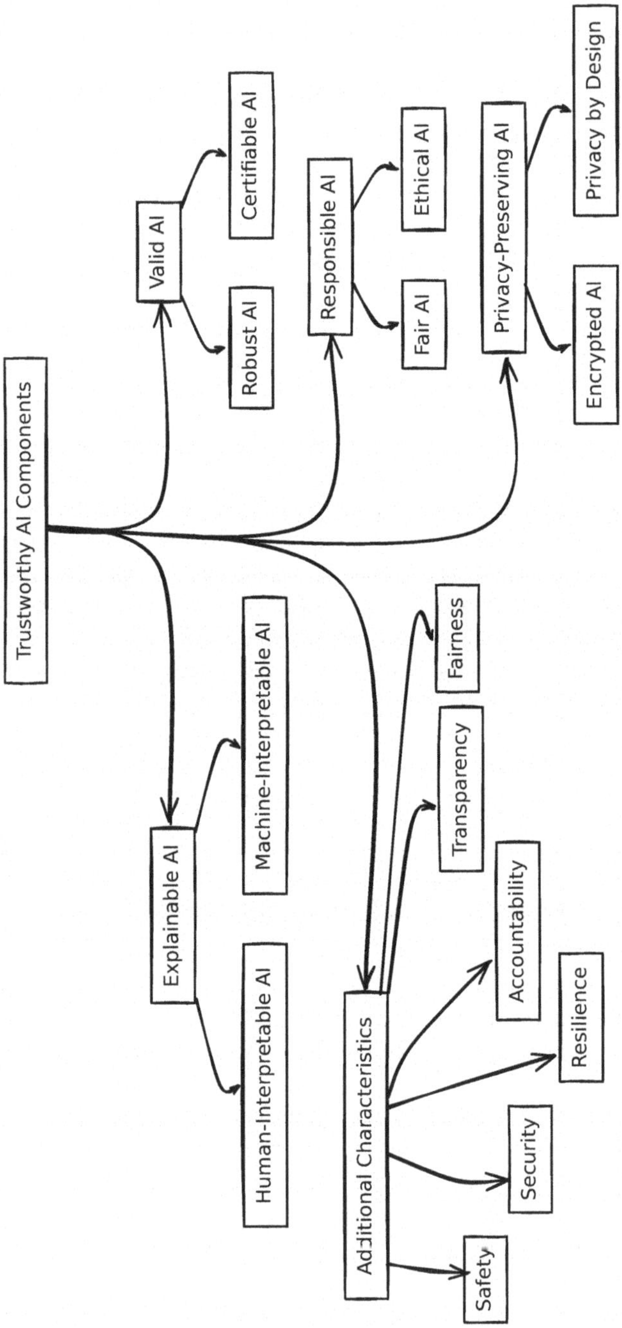

Fig. 16.1 Components of Trustworthy AI

16.3.1 Techniques and Models in Neuro-Symbolic AI for Explainability

The application of neuro-symbolic models aims to leverage the strengths of both neural networks and symbolic reasoning to enhance explainability in AI systems. Here are some models that have been influential or represent significant advances in this area:

16.3.1.1 Neuro-Symbolic Concept Learner (NSCL)

The Neuro-Symbolic Concept Learner (NSCL) [18] operates by integrating symbolic reasoning with neural network learning to enhance the explainability of visual concept learning and semantic parsing of sentences. The NSCL is designed to learn from natural supervision, which means it does not require explicit annotations for the visual concepts, words, and sentences it learns. Instead, it learns by looking at images and reading paired questions and answers, building an object-based scene representation, and translating sentences into executable, symbolic programs.

The backbone of NSCL involves several key components as follows:

Visual Perception Module: This module uses a pre-trained Mask R-CNN to generate object proposals and a ResNet-34 to extract features from the relevant regions of an image, creating a representation of each object within a scene.

Concept Quantization: This process involves mapping object representations into attribute-specific embedding spaces. For example, visual attributes like shape or color are represented as vectors in an embedding space, and objects are classified based on the similarity metrics in these spaces.

Domain-Specific Language (DSL) and Semantic Parsing: NSCL employs a semantic parsing module that translates natural language questions into an executable program using a domain-specific language designed for visual question answering. The language covers fundamental operations for visual reasoning, such as filtering objects with certain attributes or querying an object's attribute.

Quasi-Symbolic Program Execution: The symbolic program executor in NSCL executes the parsed program and infers the answer based on the object-based visual representation. This executor operates in a deterministic manner, providing a transparent execution trace of the program, which is key for explainability.

Training Paradigm: NSCL employs curriculum learning, guiding the model through increasingly complex stages of learning. This includes starting with object-level visual concepts and advancing to more complex questions, first learning with perception modules fixed and then joint fine-tuning of all modules.

Through this approach, NSCL demonstrates remarkable accuracy in learning visual concepts and enables data-efficient visual reasoning. It also generalizes well to new attributes, visual compositions, language domains, and even new program domains. Applications of NSCL include visual question answering and bidirectional image-text retrieval, showcasing the model's flexibility and potential.

16.3.1.2 Logic Tensor Networks (LTNs)

Logic Tensor Networks (LTNs) [19] are a framework that combines the expressiveness of first-order logic with the learning capabilities of neural networks, particularly designed to enhance explainability in AI. They use a differentiable first-order logic language called Real Logic [20], which allows the integration of data and knowledge into the learning process. This framework interprets predicate symbols as functions that map real vectors to truth values in the interval [0, 1], enabling the combination of semantic knowledge representation with efficient learning from examples, a characteristic typical of neural networks.

The grounding process in LTNs translates logical formulas into computational graphs within a Tensorflow environment [21]. This mechanism allows complex queries to be expressed about the data, satisfying prior knowledge during learning and proving logical statements about the data being processed. By mapping objects, functions, and predicates into a vector space, LTNs can learn to assign truth values to statements by minimizing the error in satisfying the given logical constraints. This is achieved through a variety of fuzzy logic operators defined using TensorFlow primitives.

For example, the grounding of predicates in object detection tasks involves the use of a pre-trained Faster R-CNN for proposing object bounding boxes, followed by a process that connects these convolutional layers to the LTN. The training of an LTN is guided by a grounded theory, which consists of logical and mereological constraints representing prior knowledge or desirable properties. Constraints are used to enforce mutual exclusions among classes and to impose mereological (part-whole) relationships derived from existing ontologies.

By integrating deep learning with logical reasoning, LTNs allow for the explicit representation of knowledge and provide a pathway towards explainability. This is particularly useful in tasks such as object detection, where the LTNs can learn to detect and classify objects based on a satisfiability problem that maximizes the truth values of a set of logical constraints. This end-to-end architecture enables the LTN to learn from both data and knowledge, which can be especially powerful when dealing with complex or safety-critical systems.

The use of LTNs in various applications, such as object detection, demonstrates not just performance improvements but also increases in robustness and explainability. The ability to define logical constraints and knowledge bases as objectives for a neural model makes LTNs a promising approach to creating more transparent and interpretable AI systems.

16.3.1.3 Differentiable Neural Computers (DNCs)

Differentiable Neural Computers (DNCs) [22] is a type of neural network architecture that integrates an external memory module, allowing them to perform complex data storage and manipulation tasks that go beyond the capabilities of standard neural networks. The DNC's memory system is similar to the Random-Access Memory

(RAM) of conventional computers, enabling the neural network to read from and write to it in a way that is differentiable and, therefore, amenable to gradient-based learning.

The key to the DNC's operation lies in its ability to form and navigate complex data structures autonomously. This is managed by a neural network called a "controller," which is responsible for interacting with the memory using read-and-write operations. These operations allow the DNC to update and retrieve information stored in memory and to form associations between different pieces of data.

One of the most significant advantages of DNCs is their potential for enhancing the explainability of neural network decisions. By visualizing the read-and-write operations of the DNC, researchers can trace how it accesses and uses stored information to make decisions or generate outputs. This traceability offers a window into the network's decision-making process, which is often lacking in more traditional "black box" neural networks.

For instance, DNCs have been demonstrated to answer questions about complex, structured data like family trees, navigate through a map of the London Underground, and even learn to play a block puzzle game using reinforcement learning. Through these processes, DNCs have shown they can learn to use their memory to store knowledge quickly and reason about it flexibly.

Despite these capabilities, DNCs are not without their challenges. They are known to be difficult to train, with potential instability throughout the learning process and a propensity for overfitting. Nonetheless, the benefits they offer in terms of their computational power and potential for explainability make them a promising area of research in the field of artificial intelligence.

16.3.1.4 Neural Module Networks (NMNs)

Neural Module Networks (NMNs) [23] contribute to explainability in AI by decomposing complex tasks into a series of simpler sub-tasks, which are handled by individual modules within the network. Each module is specialized for a specific function, like attention or classification, and can be assembled dynamically in response to different inputs. This modular structure not only allows for flexible adaptation to various problems but also makes the decision-making process of AI more transparent. By examining which modules are activated for a given input, one can trace the network's reasoning process, which is especially useful in applications like visual question answering. The ability to map out how each module contributes to the final answer offers a level of interpretability not typically present in more monolithic deep learning architectures.

16.3.1.5 DeepProbLog

DeepProbLog [24] is an innovative probabilistic logic programming language that integrates deep learning through the concept of neural predicates. This integration allows for a combination of symbolic (logic-based) and subsymbolic (neural network-based) representations and reasoning. One of the key strengths of DeepProbLog lies in its ability to perform probabilistic logic programming and deep learning, making it versatile for various AI applications.

In terms of explainability, DeepProbLog enhances the interpretability of AI models by leveraging the clear, rule-based structure of logic programming. This structured approach allows for a more transparent understanding of how decisions or inferences are made within the system. For instance, in a DeepProbLog model, probabilistic rules and neural network outputs are combined in a coherent framework, making it easier to trace and understand the reasoning process behind the model's output.

DeepProbLog supports several key functionalities as follows:

Symbolic and Subsymbolic Reasoning and Learning: It can handle both logical reasoning and deep learning tasks. This dual capability allows it to tackle complex problems that require understanding and processing of structured knowledge as well as learning from raw data.

Program Induction: DeepProbLog can perform program induction, which involves filling gaps in given programs with neural networks trained on input-output examples. This feature allows the model to adapt and learn from new data efficiently.

Probabilistic Programming and Deep Learning: It can handle probabilistic programming tasks, integrating uncertainty into the reasoning process, which is a critical aspect of many real-world AI applications.

An example of DeepProbLog's application is in a scenario where there are two coins and an urn containing red and blue balls. In this setup, the goal is to learn to recognize heads and tails using a neural network while also learning the distribution of the red and blue balls in the urn. DeepProbLog handles this using neural predicates to classify the coins and probabilistic logic to model the chance of picking a particular colored ball.

DeepProbLog represents a significant step forward in integrating neural networks with logical and probabilistic reasoning, providing a framework that leverages the strengths of both approaches. Its ability to be trained end-to-end based on examples, as well as its support for both symbolic and subsymbolic reasoning, makes it a powerful tool for developing more explainable AI systems.

16.3.1.6 Latent Logic Models (LLMs)

Latent Logic Models (LLMs) in AI, particularly those used for large language models, play a crucial role in improving the explainability of AI systems [25]. LLMs are often used to refer to Large Language Models [26], which are complex neural network architectures such as transformers. These models process information hierarchically across multiple layers, making their operation and decision-making processes quite intricate.

To enhance the explainability of these models, several approaches are employed as follows:

Rule-Based Explanations: This method involves generating interpretable rules that clarify the model's behavior. By analyzing patterns in the data, the model assigns importance to different features, thereby providing human-readable explanations for its predictions.

Perturbation-Based Methods: These methods involve slightly modifying the input data and observing the changes in the model's predictions. This approach helps identify the features that significantly influence the model's decisions and provides insights into the parts of the input data that the model relies on the most.

Counterfactual Explanations: This technique involves providing alternative inputs to the model to see how the predictions change. This helps understand the specific conditions under which the model's predictions might alter, offering insights into the model's decision boundaries and its sensitivity to input variations.

These techniques help in breaking down the complexity of deep learning models like LLMs into more manageable and interpretable components. They aid in understanding how the model processes information at each layer and how it arrives at specific outcomes. This level of transparency is vital for building trust in AI systems, ensuring responsible deployment, and complying with regulatory standards.

Achieving explainability in LLMs is not without challenges. There is often a trade-off between model performance and interpretability. Moreover, interpreting attention mechanisms in these models is complex and doesn't always provide straightforward insights. Nonetheless, the benefits of interpretability and explainability in LLMs, such as building trust, ensuring compliance with regulations, and facilitating model improvement, are significant for the responsible and effective use of AI technology.

16.3.2 Case Studies and Recent Examples

Neuro-symbolic AI brings together neural network-based learning and symbolic knowledge representation to enhance the explainability of AI systems. This integration addresses the challenge of making AI decisions transparent and understandable, which is crucial for user trust and adoption.

In the context of natural language processing, Neuro-Symbolic AI has been applied to develop frameworks that improve sentiment analysis by creating trustworthy symbolic representations. These models can convert natural language into a proto-language [27], extracting sentiment in a way that is unsupervised, reproducible, interpretable, and explainable. This approach overcomes the key limitations of current AI models, such as dependency and lack of trustworthiness or interpretability.

For information retrieval, techniques that combine machine learning and reasoning processes have been proposed to adapt dynamically to the evolution of the user's interest. This includes using bandit algorithms [28] and case-based reasoning [29] to create a contextual information retrieval process based on social, temporal, and location dimensions.

Neuro-symbolic approaches have also been beneficial in dialogue systems, where they have introduced explicit reasoning to justify model decisions by reasoning chains [30]. This two-phase approach, consisting of a hypothesis generator and a reasoner, provides a transparent reasoning process that justifies the final prediction.

In language generation, Neuro-Symbolic AI has improved the coherence of automated story generation systems. By leveraging various symbolic approaches for neural systems, such as separating semantic event generation from syntactic sentence generation and creating rule-based infrastructure to aid neural networks in causal reasoning, these systems have seen improvements in coherence.

Moreover, Neuro-Symbolic AI has shown promise in cybersecurity, education systems, and robotics, among other fields. For example, in education systems, it has been used to predict student problem-solving strategies by combining the strengths of symbolic AI with deep neural networks (DNNs) [31]. In robotics, convolutional neural networks and symbolic AI have been combined for logical control and affordance indexing [32].

These applications demonstrate the potential of Neuro-Symbolic AI to make AI systems more interpretable and trustworthy, thereby enhancing their explainability. By marrying the robust learning capabilities of neural networks with the structured reasoning of symbolic AI, Neuro-Symbolic AI stands as a promising direction towards more transparent and user-friendly AI solutions.

One illustrative case study of Neuro-Symbolic AI in action is IBM's Neuro-Symbolic Question Answering system (NSQA). It leverages a neuro-symbolic approach to interpret complex questions and return explanations alongside answers [33]. The system parses questions into logical forms and then uses a combination of neural network-based entity recognition and symbolic reasoning over a knowledge graph to find answers. The symbolic component can also provide a trace of the reasoning process, thereby explaining how the answer was derived.

Another example is the application of Neuro-Symbolic AI in healthcare, where systems can diagnose based on medical images. Here, neural networks can identify patterns in imagery data that correlate with particular conditions [34]. At the same time, symbolic AI can relate these patterns to medical knowledge and provide explanations in terms that are understandable to medical professionals.

These examples demonstrate how Neuro-Symbolic AI can not only enhance the performance of AI systems but also make their inner workings more transparent and their outputs more accountable. As the field matures, we can expect Neuro-Symbolic AI to play an increasingly prominent role in building AI systems that earn the trust of users by being as explainable as they are powerful.

16.4 Neuro-Symbolic AI in Building Trustworthy AI Systems

Neuro-symbolic AI systems inherently offer a higher degree of transparency due to their self-explanatory nature. By integrating symbolic reasoning with neural networks, these systems can produce explanations that are more understandable to humans. This integration allows for the creation of interpretable models through methods such as logic rules and decision trees, which convert neural network outputs into clear, human-readable explanations. Additionally, graphical representations like knowledge graphs help visualize relationships and dependencies within the data, further enhancing the transparency of AI decisions.

The focus on interpretability is evident in the substantial volume of literature dedicated to this aspect of NeSy research [35]. The ability to provide clear explanations is critical not only for user trust but also for regulatory compliance, where transparency is mandated. Applications of NeSy methods in visual data analysis, such as image classification and action recognition, as well as in natural language processing tasks like fake news detection and question answering, demonstrate the versatility and importance of interpretability in these systems [36].

Despite the progress in interpretability, there is a noticeable gap in research addressing other aspects of trustworthiness, such as fairness and privacy. While interpretability remains a dominant focus, it is essential for future research to broaden the scope to include these critical dimensions. Developing NeSy methods that ensure fairness and protect user privacy will be crucial for creating truly trustworthy AI systems. This holistic approach will ensure that AI not only provides clear explanations but also operates ethically and securely in diverse real-world applications.

The classification of interpretability methods in neuro-symbolic AI is a multifaceted task that can be approached from several angles. Broadly, these methods can be categorized based on three widely recognized dimensions: the scope of explainability, the stage at which the method is applied, and the method's dependence on the model (as shown in Fig. 16.2). This classification provides a structured way to understand how different interpretability techniques are implemented and their specific contributions to making AI systems more transparent and trustworthy.

Scope of Explainability: Interpretability methods can be divided into local, global, and cohort-specific explanations. Local explanations focus on elucidating the decisions made by an AI model for individual data points. For instance, methods like LIME (Local Interpretable Model-agnostic Explanations) provide insights into why a particular prediction was made by approximating the model locally around the instance of interest. On the other hand, global explanations aim to describe the overall behavior of the model across the entire dataset. This approach is useful for understanding the general decision-making patterns of the AI system. Cohort-specific explanations fall in between these two extremes, providing insights for a subset of instances that share specific characteristics, thereby offering a more targeted understanding than global explanations but broader than local ones.

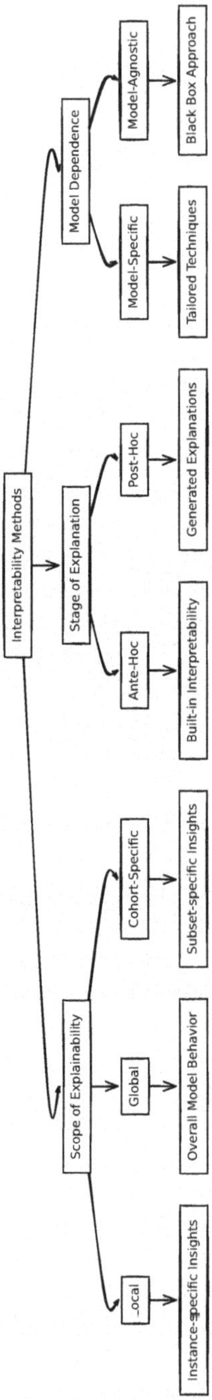

Fig. 16.2 Classification of interpretability methods in Neuro-symbolic AI

Stages of Explanation: The timing of the explanation in relation to the model's operation distinguishes ante-hoc from post-hoc methods. Ante-hoc, or self-explainable methods, are designed to be interpretable by their very nature. These methods build interpretability into the model during its creation. Examples include decision trees and rule-based systems where the model structure itself provides the explanation. In contrast, post-hoc methods generate explanations after the model has made its predictions. These can include techniques like generating textual justifications for decisions or creating simplified surrogate models that mimic the behavior of more complex models. Post-hoc methods are particularly valuable for high-performing but opaque models like deep neural networks, as they allow for the extraction of understandable explanations without altering the original model's architecture.

Model Dependence: Interpretability methods can also be classified based on whether they are model-specific or model-agnostic. Model-specific methods are tailored to particular types of models and leverage their unique characteristics to provide explanations. For example, specific techniques have been developed to interpret the decisions of convolutional neural networks (CNNs) by visualizing the learned filters and activation maps. Model-agnostic methods, on the other hand, can be applied to any model, regardless of its architecture. These methods treat the model as a black box and use techniques such as feature importance scores, partial dependence plots, and example-based explanations to provide insights into the model's decision-making process.

16.5 Challenges and Future Directions

The integration of neuro-symbolic AI with Explainable AI and Trustworthy AI faces several significant challenges. One primary hurdle is the inherent complexity of merging the statistical learning capabilities of neural networks with the rule-based logic of symbolic reasoning. These paradigms operate fundamentally differently—neural networks excel in learning from data but lack transparency, whereas symbolic systems are inherently interpretable but struggle with learning from raw data. Bridging these differences to create cohesive systems that leverage the strengths of both approaches remains a complex task [37].

Ensuring that NeSy systems can scale effectively without compromising performance is another critical challenge. Symbolic reasoning processes can be computationally intensive, especially as the size and complexity of the data grow. Balancing the computational demands of symbolic reasoning with the efficiency of neural networks is crucial to developing scalable NeSy systems that can be deployed in real-world applications [38].

Addressing fairness and bias in NeSy systems is an area that requires significant attention. Bias in training data can lead to unfair outcomes, and symbolic methods must be developed to detect and mitigate these biases effectively. Ensuring fairness across diverse populations and applications is essential for the broader acceptance and trustworthiness of AI systems [37]. Similarly, protecting user privacy while maintaining transparency and interpretability is a delicate balance. NeSy systems must be designed to ensure that sensitive data is not inadvertently exposed during the explanation process. Developing methods to provide meaningful explanations without compromising privacy remains an ongoing challenge.

NeSy systems, like other AI systems, are vulnerable to adversarial attacks. Ensuring robustness against such attacks, which can manipulate the system into making incorrect predictions, is critical. Incorporating symbolic reasoning to enhance the robustness of neural networks against adversarial examples is a promising but underexplored area [39].

Future developments could focus on creating more sophisticated hybrid models that seamlessly integrate neural and symbolic components. Advances in differentiable programming and neural-symbolic integration techniques could lead to models that are both highly interpretable and capable of complex learning tasks. Additionally, developing automated tools for fairness and bias detection in NeSy systems will be crucial. These tools could leverage symbolic logic to identify potential biases in training data and model outputs, providing a systematic approach to ensure fairness.

Research into privacy-preserving explanation methods is essential. Techniques such as federated learning, which allows models to be trained on decentralized data without transferring it, could be adapted to provide explanations that respect user privacy. Enhancing robustness against adversarial attacks through symbolic logic is another promising direction. Symbolic methods can provide a layer of defense by ensuring that the model's decisions adhere to logical constraints, thereby reducing the likelihood of successful adversarial manipulation.

Future research should adopt interdisciplinary approaches, combining insights from computer science, cognitive psychology, and neuroscience to develop more effective NeSy systems. Understanding human reasoning and learning processes can inform the design of AI systems that are both powerful and interpretable. More research is needed on applying NeSy systems in real-world scenarios. Case studies and pilot projects in domains such as healthcare, finance, and autonomous systems can provide valuable insights into the practical challenges and benefits of these systems.

Developing standardized metrics for evaluating the interpretability, fairness, and robustness of NeSy systems is essential. These metrics would enable more consistent and reliable assessment of AI systems, facilitating their adoption and trust. Creating educational tools and resources to help practitioners understand and implement NeSy systems can accelerate their adoption. Workshops, tutorials, and open-source projects can provide hands-on experience and foster a community of practice around neuro-symbolic AI.

16.6 Conclusion

The integration of Neuro-Symbolic AI with Explainable AI and Trustworthy AI is pivotal for the development of advanced AI systems that are not only powerful but also transparent, fair, and reliable. By combining the robust learning capabilities of neural networks with the clear, logical reasoning of symbolic systems, NeSy offers a promising approach to address the critical need for explainability in AI. This hybrid approach enhances the interpretability of AI decisions, making it possible to provide human-understandable explanations and build trust among users and stakeholders.

Throughout this chapter, we have explored the various methodologies used to enhance interpretability, discussed the ongoing challenges in integrating these technologies, and speculated on future developments and potential solutions. We highlighted the importance of developing automated tools for fairness and bias detection, privacy-preserving explanation methods, and robust systems that can withstand adversarial attacks. Additionally, we emphasized the need for interdisciplinary research and the creation of standardized metrics for evaluating the interpretability and trustworthiness of AI systems. In the next chapter, we will delve into the specific domains where Neuro-Symbolic AI is being applied.

References

1. Li, B., Qi, P., Liu, B., Di, S., Liu, J., Pei, J., Yi, J., Zhou, B.: Trustworthy AI: from principles to practices. ACM Comput. Surv. **55**(9), 1–46 (2023)
2. Chander, B., John, C., Warrier, L., Gopalakrishnan, K.: Toward trustworthy artificial intelligence (tai) in the context of explainability and robustness. ACM Comput. Surv. (2024)
3. Atherton, D., Schwartz, R., Fontana, P., Hall, P.: The language of trustworthy AI: an in-depth glossary of terms (2023)
4. Dalrymple, D., Skalse, J., Bengio, Y., Russell, S., Tegmark, M., Seshia, S., Omohundro, S., Szegedy, C., Goldhaber, B., Ammann, N., et al.: Towards guaranteed safe AI: a framework for ensuring robust and reliable AI systems (2024). arXiv:2405.06624
5. Brundage, M., Avin, S., Wang, J., Belfield, H., Krueger, G., Hadfield, G., Khlaaf, H., Yang, J., Toner, H., Fong, R., et al.: Toward trustworthy AI development: mechanisms for supporting verifiable claims (2020). arXiv:2004.07213
6. Díaz-Rodríguez, N., Del Ser, J., Coeckelbergh, M., de Prado, M.L., Herrera-Viedma, E., Herrera, F.: Connecting the dots in trustworthy artificial intelligence: from AI principles, ethics, and key requirements to responsible AI systems and regulation. Inf. Fusion **99**, 101896 (2023)
7. Feuerriegel, S., Dolata, M., Schwabe, G.: Fair AI: challenges and opportunities. Bus. & Inf. Syst. Eng. **62**, 379–384 (2020)
8. Prem, E.: From ethical AI frameworks to tools: a review of approaches. AI Ethics **3**(3), 699–716 (2023)
9. Newman, J.: A taxonomy of trustworthiness for artificial intelligence. CLTC: North Charleston, SC, USA 1 (2023)
10. Rallis, S.F., Rossman, G.B.: Ethics and trustworthiness. In: Qualitative Research in Applied Linguistics: A Practical Introduction, pp. 263–287 (2009)
11. Cheng, Y., Liu, Y., Chen, T., Yang, Q.: Federated learning for privacy-preserving AI. Commun. ACM **63**(12), 33–36 (2020)

12. Xu, W., Ma, H., Song, Z., Li, J., Zhang, R.: Gringotts: an encrypted version control system with less trust on servers. IEEE Trans. Dependable Secur. Comput. **21**(2), 668–684 (2023)
13. Acar, A., Aksu, H., Uluagac, A.S., Conti, M.: A survey on homomorphic encryption schemes: theory and implementation. ACM Comput. Surv. (Csur) **51**(4), 1–35 (2018)
14. Schaar, P.: Privacy by design. Identity Inf. Soc. **3**(2), 267–274 (2010)
15. Arrieta, A.B., Díaz-Rodríguez, N., Del Ser, J., Bennetot, A., Tabik, S., Barbado, A., García, S., Gil-López, S., Molina, D., Benjamins, R., et al.: Explainable artificial intelligence (XAI): concepts, taxonomies, opportunities and challenges toward responsible AI. Inf. Fusion **58**, 82–115 (2020)
16. de Mijolla, D., Frye, C., Kunesch, M., Mansir, J., Feige, I.: Human-interpretable model explainability on high-dimensional data (2020). arXiv:2010.07384
17. Siu, H.C., Leahy, K., Mann, M.: How interpretable is "interpretable" machine learning? (2024)
18. Mao, J., Gan, C., Kohli, P., Tenenbaum, J.B., Wu, J.: The neuro-symbolic concept learner: Interpreting scenes, words, and sentences from natural supervision (2019). arXiv:1904.12584
19. Badreddine, S., Garcez, A.d., Serafini, L., Spranger, M.: Logic tensor networks. Artif. Intell. **303**, 103649 (2022)
20. Shakarian, P., Baral, C., Simari, G.I., Xi, B., Pokala, L.: LTN: logic tensor networks. In: Neuro Symbolic Reasoning and Learning, pp. 33–41. Springer (2023)
21. Pang, B., Nijkamp, E., Wu, Y.N.: Deep learning with tensorflow: A review. Journal of Educational and Behavioral Statistics **45**(2), 227–248 (2020)
22. Rasekh, M.S., Safi-Esfahani, F.: Ednc: Evolving differentiable neural computers. Neurocomputing **412**, 514–542 (2020)
23. Andreas, J., Rohrbach, M., Darrell, T., Klein, D.: Neural module networks. In: Proceedings of the IEEE conference on computer vision and pattern recognition. pp. 39–48 (2016)
24. Manhaeve, R., Dumancic, S., Kimmig, A., Demeester, T., De Raedt, L.: Deepproblog: Neural probabilistic logic programming. Advances in neural information processing systems 31 (2018)
25. Song, Z., Yang, C., Wang, C., An, B., Li, S.: Latent logic tree extraction for event sequence explanation from llms. arXiv preprint arXiv:2406.01124 (2024)
26. Zhao, W.X., Zhou, K., Li, J., Tang, T., Wang, X., Hou, Y., Min, Y., Zhang, B., Zhang, J., Dong, Z., et al.: A survey of large language models. arXiv preprint arXiv:2303.18223 (2023)
27. Cui, C., Chen, Y., Wang, Q., Mortensen, D.R.: Neural proto-language reconstruction. arXiv preprint arXiv:2404.15690 (2024)
28. Acharya, K., Raza, W., Dourado, C., Velasquez, A., Song, H.H.: Neurosymbolic reinforcement learning and planning: A survey. IEEE Transactions on Artificial Intelligence (2023)
29. Hatzilygeroudis, I., Prentzas, J.: Symbolic-neural rule based reasoning and explanation. Expert Systems with Applications **42**(9), 4595–4609 (2015)
30. Wang, R., Zelikman, E., Poesia, G., Pu, Y., Haber, N., Goodman, N.D.: Hypothesis search: Inductive reasoning with language models. arXiv preprint arXiv:2309.05660 (2023)
31. Shakya, A., Rus, V., Venugopal, D.: Student strategy prediction using a neuro-symbolic approach. International Educational Data Mining Society (2021)
32. Hanson, D., Imran, A., Vellanki, A., Kanagaraj, S.: A neuro-symbolic humanlike arm controller for sophia the robot. arXiv preprint arXiv:2010.13983 (2020)
33. Kapanipathi, P., Abdelaziz, I., Ravishankar, S., Roukos, S., Gray, A., Astudillo, R., Chang, M., Cornelio, C., Dana, S., Fokoue, A., et al.: Question answering over knowledge bases by leveraging semantic parsing and neuro-symbolic reasoning. arXiv preprint arXiv:2012.01707 (2020)
34. Roy, K., Lokala, U., Gaur, M., Sheth, A.P.: Tutorial: Neuro-symbolic ai for mental healthcare. In: Proceedings of the Second International Conference on AI-ML Systems. pp. 1–3 (2022)
35. Raj, K.: A neuro-symbolic approach to enhance interpretability of graph neural network through the integration of external knowledge. In: Proceedings of the 32nd ACM International Conference on Information and Knowledge Management. pp. 5177–5180 (2023)
36. Gaur, M., Gunaratna, K., Bhatt, S., Sheth, A.: Knowledge-infused learning: A sweet spot in neuro-symbolic ai. IEEE Internet Computing **26**(4), 5–11 (2022)

37. Bhuyan, B.P., Ramdane-Cherif, A., Tomar, R., Singh, T.: Neuro-symbolic artificial intelligence: a survey. Neural Computing and Applications pp. 1–36 (2024)
38. Luo, X., Li, H., Lee, S.: Bridging the gap: Neuro-symbolic computing for advanced ai applications in construction. Frontiers of Engineering Management **10**(4), 727–735 (2023)
39. Chen, W., Ma, X., Wang, Z., Li, W., Fan, C., Zhang, J., Que, X., Li, C.: Exploring neuro-symbolic ai applications in geoscience: implications and future directions for mineral prediction. Earth Science Informatics **17**(3), 1819–1835 (2024)

Chapter 17
Neuro-Symbolic AI in Various Domains

> *"Imagination is more important than knowledge. For knowledge is limited, whereas imagination embraces the entire world, stimulating progress, giving birth to evolution."*
>
> – Albert Einstein

17.1 Introduction

Neuro-Symbolic AI (NeSy) represents a groundbreaking advancement in artificial intelligence, combining the strengths of neural networks and symbolic reasoning to create systems that are not only powerful and adaptive but also interpretable and reliable. This hybrid approach addresses some of the key limitations of purely neural or symbolic AI systems by leveraging their complementary strengths: the data-driven learning capabilities of neural networks and the explicit, rule-based reasoning of symbolic AI. The integration of neural networks and symbolic AI into Neuro-Symbolic AI seeks to harness the best of both worlds. This integration is achieved through various methods, including embedding symbolic knowledge into neural networks, using neural networks to generate symbolic representations, and developing hybrid architectures that enable interaction between neural and symbolic components. This fusion results in AI systems that can learn from data, reason about knowledge, and provide interpretable explanations for their decisions.

This chapter deals with the transformative impact of NeSy across various scientific and engineering domains. The chapter explores how Neuro-Symbolic AI is applied in diverse domains, including robotics, question answering, medical applications, computer vision, programming, optimization, geosciences, cybersecurity, environmental science, smart cities, and industrial digital twins. Each section highlights the unique

B. P. Bhuyan et al., *Neuro-Symbolic Artificial Intelligence*, Studies in Computational Intelligence 1176, https://doi.org/10.1007/978-981-97-8171-3_17

contributions of NeSy in these fields, demonstrating how it enhances the capabilities of AI systems to learn, reason, and provide interpretable solutions.

In robotics, Neuro-Symbolic AI improves robots' ability to perceive, reason, and act in complex and dynamic environments. This integration allows robots to learn new skills from human demonstrations, translate these demonstrations into symbolic plans, and reason about the physical properties of objects and their interactions.

In question answering, Neuro-Symbolic AI enhances the ability of AI systems to understand and process complex natural language queries, making them more robust, interpretable, and generalizable. This approach combines neural networks' data-driven learning with symbolic reasoning's logical rigor to provide more accurate and explainable answers.

In medical applications, Neuro-Symbolic AI merges the pattern recognition capabilities of neural networks with the logical reasoning of symbolic AI to create more accurate, interpretable, and personalized healthcare solutions. This integration enhances clinical decision support systems, early disease detection, personalized medicine, and predictive modeling.

In computer vision, Neuro-Symbolic AI addresses challenges in object recognition, scene interpretation, semantic segmentation, visual question answering, and spatial-temporal pattern analysis. By integrating neural networks with symbolic reasoning, these systems achieve better interpretability and robustness in visual tasks.

In programming and optimization, Neuro-Symbolic AI excels in automating program synthesis, optimizing software engineering processes, and incorporating domain-specific knowledge into machine learning models. This approach enhances the efficiency and accuracy of program generation, code correctness, and industrial optimization.

Beyond these areas, the chapter also delves into NeSy's impact in geosciences, cybersecurity, environmental science, smart cities, and industrial digital twins. In geosciences, NeSy aids in mineral prediction and exploration. In cybersecurity, it enhances threat detection and response. In environmental science, it contributes to climate modeling, ecological monitoring, and disaster prediction. In smart cities, NeSy optimizes traffic management, energy use, and public safety. In industrial applications, NeSy enhances digital twins by enabling natural language interaction and autonomous decision-making.

17.2 Neuro-Symbolic AI in Robotics

Neuro-Symbolic AI is significantly advancing the field of robotics by combining the strengths of neural networks and symbolic reasoning. This integration allows robots to perform complex tasks that were previously unattainable, enhancing their ability to perceive, reason, and act in intricate and unpredictable environments.

One of the notable applications of Neuro-Symbolic AI in robotics is learning new skills from human demonstrations. Robots can translate these demonstrations into symbolic plans and reason about the physical properties of objects and their interactions with the environment. This capability enables robots to adapt to new tasks and environments more effectively. For example, Coraggio et al. [1] developed a neuro-symbolic system for robot self-localization using minimal sensors, which leverages natural environmental features as landmarks for navigation. This system combines neural networks' perceptual strengths with symbolic AI's logical reasoning, enhancing decision-making based on landmark detection and encoding.

Neuro-Symbolic AI also improves robotic control and decision-making. Staffa et al. [2] demonstrated enhanced adaptability and decision-making in behavior-based robotics by tuning thresholds within a neuro-symbolic network. This approach dynamically adjusts the robot's behavior in response to environmental changes, showcasing the potential of neuro-symbolic methods to enhance robotic autonomy and efficiency.

Robust landmark recognition and accurate navigation are critical for effective robotic operation in complex settings. The work by Coraggio and De Gregorio on neuro-symbolic hybrid methods for landmark recognition and robot localization has significantly improved landmark detection robustness and navigation accuracy [1]. This integration is particularly useful in spatial awareness and adaptability applications, allowing robots to navigate and interact with their environment more effectively.

Advanced Surveillance and Environmental Modeling: In the domain of active video surveillance, De Gregorio et al. [3] integrated virtual neural sensors with BDI agents to create a highly adaptive surveillance system. This system autonomously operates in dynamic environments, highlighting the benefits of combining neural networks' perceptual abilities with symbolic AI's reasoning capabilities. Additionally, Kraetzschmar et al. [4] utilized neuro-symbolic integration for environmental modeling in mobile robotics, enabling dynamic and efficient environment representation essential for navigation and interaction.

Despite these advancements, several challenges remain in the integration of Neuro-Symbolic AI in robotics as follows:

1. Environmental Complexity and Dynamic Adaptation: Enhancing the robustness and flexibility of neuro-symbolic systems to better cope with unpredictable changes in the environment remains a significant challenge.
2. Perception and Landmark Recognition: Improving perceptual accuracy and the ability to distinguish between similar features in the environment is crucial for advancing robotic navigation and interaction.
3. Autonomy in Surveillance Systems: Ensuring that surveillance systems can operate autonomously with minimal human intervention while making contextually appropriate decisions is an ongoing challenge.
4. Generalization and Application of Rules: Extending the applicability of neuro-symbolic systems to a broader range of scenarios without significant retraining is essential for their widespread adoption.

5. Causal Reasoning and Physical Interactions: Expanding the understanding and prediction of complex physical interactions and causal mechanisms in dynamic environments is crucial for advancing robotic systems' reasoning and predictive capabilities.

Addressing these challenges will significantly advance the field of neuro-symbolic AI in robotics, leading to more capable, adaptable, and intelligent robotic systems that can operate autonomously in a wide range of environments.

17.3 Neuro-Symbolic AI in Question Answering

Neuro-symbolic AI is revolutionizing the field of question answering (QA) by combining the strengths of neural networks and symbolic reasoning. This integration enhances the ability of AI systems to understand and process complex natural language queries, making them more robust, interpretable, and generalizable.

Traditional neural network-based QA systems excel at understanding and processing vast amounts of data but often lack the ability to perform deep reasoning and provide interpretable answers. In contrast, symbolic AI systems can reason over explicit knowledge representations but struggle with learning from raw data. By integrating these approaches, Neuro-Symbolic AI systems can leverage the strengths of both paradigms.

For example, IBM's Neuro-Symbolic Question Answering (NSQA) [5] system translates natural language questions into abstract representations using Abstract Meaning Representation (AMR) [6]. This representation captures the semantic meaning of the questions, which is then mapped to a knowledge graph for reasoning. This approach allows the system to handle various types of reasoning, such as multi-hop, quantitative, geographic, and temporal reasoning, making it highly versatile and robust in enterprise settings.

Neuro-Symbolic AI has shown significant promise in Knowledge-Based Question Answering (KBQA). KBQA tasks require advanced reasoning capabilities to answer questions based on knowledge bases (KBs) like ConceptNet or DBpedia [7]. Systems like NSQA use semantic parsing to convert questions into logical forms that can be reasoned over using symbolic techniques. This enables the system to answer complex questions with minimal training data, providing a scalable and efficient solution for real-world applications.

In addition to KBQA, Neuro-Symbolic AI is also being applied to commonsense and contextual QA tasks. For instance, the CommonsenseQA dataset evaluates AI's ability to answer questions based on commonsense knowledge, requiring the integration of neural models with structured knowledge bases like ConceptNet and ATOMIC. The use of attention-based mechanisms and graph neural networks (GNNs) to inject knowledge into pre-trained language models, such as BERT, enhances the model's ability to reason about and answer commonsense questions [8].

Similarly, Neuro-Symbolic approaches in social and physical QA involve reasoning about social interactions and physical environments. For example, PhysicalIQA [9] focuses on physical reasoning, where the system must choose the most plausible outcome of a given scenario. These tasks demonstrate the capability of Neuro-Symbolic AI to handle diverse and complex types of reasoning.

One of the critical advantages of Neuro-Symbolic AI in QA is its ability to perform advanced reasoning over incomplete or inconsistent knowledge. By combining the pattern recognition capabilities of neural networks with the logical rigor of symbolic reasoning, these systems can infer answers even when explicit information is missing. This approach is particularly useful in dynamic and real-time applications, such as virtual assistants and chatbots, where the ability to provide accurate and contextually relevant answers is crucial [10].

Despite the progress, several challenges remain in developing robust Neuro-Symbolic QA systems. These include improving the scalability of these systems to handle larger and more complex knowledge bases, enhancing the integration of diverse types of knowledge, and ensuring the interpretability and fairness of the answers provided.

Future research could focus on refining semantic parsing techniques, improving the efficiency of reasoning algorithms, and developing more sophisticated methods for integrating and reasoning over heterogeneous data sources. Additionally, addressing ethical considerations, such as bias and transparency, will be crucial for the broader adoption and trustworthiness of Neuro-Symbolic QA systems.

17.4 Neuro-Symbolic AI in Medical Applications

Neuro-Symbolic AI is revolutionizing the medical field by merging the strengths of neural networks and symbolic reasoning to create more accurate, interpretable, and personalized healthcare solutions. This hybrid approach leverages the pattern recognition capabilities of neural networks with the logical reasoning of symbolic AI, offering numerous applications across various medical domains.

One of the primary applications of NeSy in medicine is enhancing clinical decision support systems (CDSS) [11]. These systems benefit from the combined strengths of neural networks and symbolic reasoning, allowing for more precise diagnostic processes. For instance, Neuro-Symbolic Reasoning (NSR) has demonstrated its effectiveness in accurately identifying conditions like acute abdominal pain, significantly improving diagnostic accuracy compared to traditional methods [12].

The analysis of electronic health records (EHRs) is another area where NeSy shows considerable promise. EHRs often contain vast amounts of unstructured data, making them challenging to analyze with traditional AI methods. NeSy can enhance the extraction of actionable insights from EHRs by combining deep learning techniques with symbolic reasoning [13]. This integration facilitates the identification of relevant medical variables and patterns, improving patient care and outcomes. Systems like those developed by Mendel AI leverage large language models (LLMs) alongside proprietary clinical hypergraphs to enhance the accuracy and efficiency of cohort retrieval, aiding in precise patient stratification and targeted interventions [14].

NeSy is also being used for the early detection and prediction of diseases. By integrating neural networks' pattern recognition capabilities with symbolic AI's logical reasoning, these systems can identify early signs of diseases from medical images and other data sources. For example, in ophthalmology, NeSy methods have advanced the classification, segmentation, and prediction of various eye diseases, including retinal vessel segmentation and the classification of conditions like glaucoma and diabetic retinopathy. These combined methods have shown high accuracy and reliability, significantly improving early detection and intervention [14].

NeSy plays a crucial role in personalized medicine by tailoring treatment plans to individual patients' needs. The integration of symbolic reasoning allows these systems to consider a wide range of variables, including genetic information, lifestyle factors, and clinical history, to recommend personalized treatment plans. For example, systems using NeSy can integrate data from various sources to predict patient responses to specific treatments, thereby optimizing therapeutic strategies and improving patient outcomes [15].

One of the key advantages of NeSy in medical applications is its ability to produce interpretable and transparent AI systems. Unlike traditional "black-box" AI models, NeSy systems provide clear explanations for their decisions, which is crucial in medical settings where understanding the rationale behind a diagnosis or treatment recommendation is essential for both clinicians and patients. This transparency enhances trust and facilitates the broader adoption of AI technologies in healthcare.

NeSy is also valuable in predictive modeling and risk assessment. By combining neural networks' ability to handle large datasets with symbolic AI's reasoning capabilities, these systems can predict the likelihood of various health outcomes. For example, they can assess the risk of developing chronic diseases based on patient data, allowing for early intervention and preventive care. These predictive models can also be used to identify at-risk populations and allocate resources more effectively.

Despite the promising applications, several challenges and future directions for NeSy in the medical field include the following:

1. Scalability: Developing scalable NeSy systems that can handle large and diverse medical datasets remains a significant challenge. Future research should focus on improving the scalability of these systems to make them practical for widespread use.
2. Integration with Clinical Workflows: Integrating NeSy systems seamlessly into existing clinical workflows is essential for their adoption. This requires careful design and testing to ensure that these systems complement rather than disrupt clinical practices.
3. Bias and Fairness: Ensuring that NeSy systems are free from biases that could affect their diagnostic and treatment recommendations is critical. This involves developing methods to detect and mitigate biases in training data and algorithms.

4. Ethical Considerations: Addressing ethical issues related to the use of AI in healthcare, such as patient privacy and informed consent, is essential. NeSy systems must be designed and implemented in ways that uphold ethical standards and protect patient rights.
5. Robustness and Generalizability: Enhancing the robustness and generalizability of NeSy systems to ensure they perform reliably across diverse patient populations and healthcare settings is crucial. This involves extensive validation and testing across different clinical scenarios and patient demographics.

17.5 Neuro-Symbolic AI in Computer Vision

One of the primary applications of NeSy in computer vision is object recognition and scene interpretation. Neural networks are adept at recognizing patterns and features in images but often struggle with understanding the relationships between these features. Symbolic AI can represent and reason about these relationships using logical structures and rules. The Neuro-Symbolic Concept Learner (NS-CL) framework [16], for instance, integrates Graph Neural Networks (GNNs) with symbolic reasoning to encode the relationships between visual features and their corresponding concepts within a knowledge graph. This approach allows the system to predict potential concepts in new images by leveraging the structured knowledge stored in the graph [17].

Semantic segmentation [18], which involves assigning a class label to each pixel in an image, has also benefited from NeSy approaches. Logic Tensor Networks (LTNs) [19] incorporate logical constraints into the learning process, ensuring that the segmented regions adhere to predefined rules and relationships, resulting in more coherent and interpretable segmentation maps. This integration improves the consistency and accuracy of segmentation results.

Visual Question Answering (VQA) is another domain where NeSy has made significant strides. VQA systems must understand and reason about both visual and textual information to answer questions about images. NeSy approaches enhance these systems by combining the neural networks' ability to process visual data with symbolic reasoning's capacity to handle logical relationships and constraints. For instance, Neuro-Symbolic VQA systems use neural networks to extract features from images and symbolic reasoning modules to generate and evaluate logical queries based on these features, allowing for complex multi-step reasoning and integration of visual and textual information [20].

In spatial-temporal pattern analysis, NeSy systems are used to analyze sequences of images or videos to detect patterns and predict future events. These systems model the temporal dynamics of objects and scenes, making them suitable for applications like activity recognition, video surveillance, and autonomous driving. For example, neuro-symbolic approaches have been applied to classify GPS trajectories by combining neural networks' adaptability with symbolic reasoning's structured logic, enhancing the interpretability and computational efficiency of trajectory analysis [21].

Despite the progress, several challenges and future directions remain for NeSy in computer vision as follows:

1. Scalability: Developing scalable NeSy systems that can efficiently handle large-scale visual data is essential. Future research should focus on optimizing the integration of neural and symbolic components to improve scalability.
2. Generalization: Enhancing the generalization capabilities of NeSy systems to ensure they perform well across diverse visual contexts and datasets is crucial. This involves developing methods to transfer knowledge learned in one domain to another.
3. Robustness: Ensuring the robustness of NeSy systems against adversarial attacks and noisy data is important for their deployment in real-world applications. Research should focus on enhancing the resilience of these systems to various types of perturbations.
4. Integration with Other Modalities: Integrating NeSy systems with other modalities, such as text and audio, can further enhance their capabilities. This requires developing methods to combine and reason about multimodal data effectively.

17.6 Neuro-Symbolic AI in Programming and Optimization

One of the key areas where NeSy excels is in program synthesis [22], which involves automatically generating programs that meet specific requirements. Traditional program synthesis relies heavily on symbolic methods, which can be rigid and require extensive domain knowledge. NeSy, however, leverages neural networks to learn from data and generate programs, while symbolic reasoning ensures these programs adhere to logical constraints and rules.

For instance, the Neuro-Symbolic Program Synthesis (NSPS) [23] approach exemplifies the integration of neural and symbolic methods to create systems that can learn from examples and generate code that meets specific functional criteria. This approach has been applied to tasks such as sorting algorithms and other common programming problems, significantly enhancing the efficiency and accuracy of program generation.

NeSy also plays a crucial role in optimizing software engineering processes. Traditional deep learning methods often fall short in tasks requiring explicit reasoning and long-term planning. By incorporating symbolic reasoning, NeSy systems can better manage complex software optimization tasks, such as bug detection, code refactoring, and performance tuning [24].

A notable example is the use of NeSy to optimize code correctness in educational tools. Systems like the neuro-symbolic program corrector analyze and correct student-submitted code by understanding both the syntax and semantics of programming languages, providing targeted feedback that helps learners improve their coding skills [25].

Incorporating domain-specific knowledge into machine learning models is another strength of NeSy. Knowledge graphs (KGs) are used to represent complex relationships between entities, and NeSy systems leverage these graphs to enhance machine learning applications. For example, Bosch's use of NeSy methods combines symbolic reasoning with machine learning to exploit domain knowledge embedded in KGs [26]. This integration improves the quality of KGs by detecting inconsistencies and predicting missing links, making them more robust and useful for industrial applications.

NeSy enables advanced AI functionalities on top of KGs, such as explainable clustering and complex query answering. These capabilities are crucial for industrial applications where transparency and interpretability are essential. By integrating ontologies with machine learning models, NeSy systems can answer complex queries over incomplete knowledge graphs, providing more reliable and understandable results.

NeSy approaches also contribute to developing scalable and explainable optimization methods. For example, combining answer set programming (ASP) with classical optimization techniques allows for solving large-scale industrial optimization problems efficiently [27]. These methods not only enhance scalability but also provide explanations for their solutions, making them more accessible and trustworthy for users in industries such as manufacturing and logistics.

The infusion of Neuro-Symbolic AI (NeSy) into programming and optimization heralds a promising horizon, marked by enhanced learning tools, innovative problem-solving methodologies, and a deeper understanding of complex systems. While strides have been made, the journey towards fully realizing the potential of Neuro-Symbolic AI continues, with future research poised to tackle the remaining challenges of scalability, interpretability, and the seamless integration of neural and symbolic systems. The key points from the programming and optimization domain can be consolidated into broader themes to capture the essence of current achievements and future directions:

1. Advancements in Program Synthesis and Software Optimization: The progress in automating program synthesis and enhancing software efficiency underscores the potential of Neuro-Symbolic AI in transforming software development practices. Techniques such as Neuro-Symbolic Program Synthesis (NSPS) leverage neural networks to learn from examples and generate code that meets specific functional criteria. This approach has been applied to common programming tasks, like sorting algorithms, highlighting the significant impact on automating and optimizing software development processes. Future research could extend these methodologies to more complex and diverse programming tasks.

2. Improving Programming Education and Software Development: Innovations like the neuro-symbolic program corrector demonstrate the potential for AI to impact programming education significantly. These tools analyze and correct student-submitted code by understanding both the syntax and semantics of programming languages, providing targeted feedback that helps learners improve their coding skills. Extending these tools to accommodate a wider range of programming languages and complexities could revolutionize learning experiences and software development workflows.
3. Expanding the Scope of Neuro-Symbolic Integration: The work in inductive logic programming, neuro-symbolic programming paradigms, and dedicated neuro-symbolic programming languages showcases the evolving landscape of Neuro-Symbolic AI. Integrating logical neural networks with ILP facilitates the extraction and refinement of logical rules from data, marking a significant advancement in AI. Future efforts could focus on developing sophisticated frameworks and languages that ease the integration of neural and symbolic components, enhancing AI's adaptability and interpretability across various applications.
4. Cross-Disciplinary Applications and Innovations: Neuro-symbolic AI's exploration in fields such as computer graphics and healthcare illustrates its versatile applicability. In computer graphics, neuro-symbolic models address challenges in generating, rendering, and manipulating graphical content, introducing more intelligent and adaptable systems. In healthcare, neuro-symbolic methods enhance the analysis of electronic health records and improve diagnostic accuracy, showcasing the broad impact of these technologies. Research aimed at exploring and expanding Neuro-Symbolic AI's capabilities in diverse domains could unlock new possibilities for innovative applications, from digital media to precision medicine.
5. Automating the Design of Optimization Algorithms: The initiative to automate the discovery of optimization algorithms opens new research avenues for making AI systems more efficient and autonomous. Symbolic discovery of optimization algorithms involves leveraging symbolic reasoning to identify and implement optimizations, leading to breakthroughs in computational efficiency and AI model performance. Investigating autonomous methods for identifying and implementing optimizations could significantly advance the field, making AI systems more robust and capable of handling complex optimization problems.

By focusing on these consolidated themes, future research in Neuro-Symbolic AI within the programming and optimization domain can address existing challenges and unlock new potentials, paving the way for more intelligent, efficient, and user-friendly AI systems.

17.7 Impact of Neuro-Symbolic AI in Various Sciences

Neuro-Symbolic AI (NeSy) has made significant strides across various scientific domains by integrating the strengths of neural networks and symbolic reasoning. Here are some key areas where NeSy has had a substantial impact:

1. Geosciences: NeSy has revolutionized geosciences, particularly in mineral prediction and exploration. By combining neural networks' ability to analyze vast datasets with symbolic reasoning's logical structures, researchers can more accurately predict mineral-rich areas. This approach improves the efficiency and accuracy of locating potential mineral deposits, significantly benefiting the mining industry. Techniques such as knowledge graph embeddings and differentiable logic have been particularly useful in modeling complex geological data and predicting mineral deposits [28].
2. Cybersecurity: In cybersecurity, NeSy is employed to enhance threat detection and response. Neural networks excel at identifying patterns in large datasets, such as network traffic, while symbolic reasoning provides the logical framework to interpret these patterns and identify potential threats. This combination allows for more accurate and explainable threat detection systems, improving the ability to respond to cybersecurity incidents effectively. For example, genetic algorithms have been integrated with deep learning to optimize the detection of phishing URLs [29].
3. Environmental Science: NeSy applications in environmental science include climate modeling, ecological monitoring, and disaster prediction. Neural networks process large-scale environmental data, while symbolic AI helps create interpretable models adhering to known physical laws and environmental processes. This combination improves the accuracy and reliability of models predicting climate change impacts or natural disasters, aiding in better environmental management and policy-making [30].
4. Smart Cities: In smart cities, NeSy contributes to improved systems for traffic management, energy optimization, and public safety. These systems use neural networks to process real-time data from various sensors and symbolic reasoning to make decisions based on urban planning rules and regulations. This integration results in more efficient and sustainable urban management solutions, enhancing the quality of life for city inhabitants [31].
5. Industrial Digital Twins: Digital twins represent virtual models of physical systems used in industries for simulations and optimizations. NeSy enhances these digital twins by enabling natural language interaction and autonomous decision-making based on symbolic reasoning. This allows for more robust and versatile solutions in industrial applications, such as maintenance procedures and complex machinery interactions [32].

17.8 Conclusion

This chapter has highlighted the transformative impact of Neuro-Symbolic AI (NeSy) across multiple scientific and engineering domains. By integrating the data-driven learning of neural networks with the explicit reasoning capabilities of symbolic AI, NeSy enhances the robustness, interpretability, and efficiency of AI systems. Applications in robotics, question answering, and medical diagnostics showcase improved decision-making, accuracy, and personalized solutions. In computer vision, NeSy addresses challenges in object recognition and scene interpretation, while in programming and optimization, it automates and refines software development processes. Furthermore, NeSy's influence extends to geosciences, cybersecurity, environmental science, smart cities, and industrial digital twins, demonstrating its versatility and broad applicability. Despite the promising advancements, challenges such as scalability, integration complexity, and ensuring fairness and transparency remain. Addressing these will be crucial for fully realizing NeSy's potential. The next chapter will explore specific case studies and applications, illustrating practical implementations and successes of Neuro-Symbolic AI in various fields.

References

1. Coraggio, P., De Gregorio, M.: A neurosymbolic hybrid approach for landmark recognition and robot localization. In: International Symposium on Brain, Vision, and Artificial Intelligence, pp. 566–575. Springer (2007)
2. Staffa, M.: Attentional Mechanism for Sensory-motor Coordination in Behavior-based Robotic Systems. Ph.D. thesis, Citeseer (2011)
3. De Gregorio, M.: An intelligent active video surveillance system based on the integration of virtual neural sensors and BDI agents. IEICE Trans. Inf. Syst. **91**(7), 1914–1921 (2008)
4. Kraetzschmar, G., Sablatnög, S., Enderle, S., Palm, G.: Application of neurosymbolic integration for environment modelling in mobile robots. In: Hybrid Neural Systems, pp. 387–401. Springer (2000)
5. Kapanipathi, P., Abdelaziz, I., Ravishankar, S., Roukos, S., Gray, A., Astudillo, R., Chang, M., Cornelio, C., Dana, S., Fokoue, A., et al.: Question answering over knowledge bases by leveraging semantic parsing and neuro-symbolic reasoning. arXiv preprint arXiv:2012.01707 (2020)
6. Knight, K., Badarau, B., Baranescu, L., Bonial, C., Bardocz, M., Griffitt, K., Hermjakob, U., Marcu, D., Palmer, M., O'Gorman, T., et al.: Abstract meaning representation (AMR) annotation release 3.0 (2021)
7. Hao, J., Chen, M., Yu, W., Sun, Y., Wang, W.: Universal representation learning of knowledge bases by jointly embedding instances and ontological concepts. In: Proceedings of the 25th ACM SIGKDD International Conference on Knowledge Discovery & Data Mining, pp. 1709–1719 (2019)
8. Yang, J., Hu, X., Xiao, G., Shen, Y.: A survey of knowledge enhanced pre-trained models. arXiv preprint arXiv:2110.00269 (2021)
9. Hendrycks, D., Burns, C., Basart, S., Zou, A., Mazeika, M., Song, D., Steinhardt, J.: Measuring massive multitask language understanding. arXiv preprint arXiv:2009.03300 (2020)
10. Chen, M., Tian, Y., Wang, Z., Jiang, B., Xu, H.: A comprehensive study of cognitive graphs: Techniques, applications, and challenges. In: 2021 International Conference on Computer Information Science and Artificial Intelligence (CISAI), pp. 1108–1122. IEEE (2021)

11. Belard, A., Buchman, T., Forsberg, J., Potter, B.K., Dente, C.J., Kirk, A., Elster, E.: Precision diagnosis: a view of the clinical decision support systems (CDSS) landscape through the lens of critical care. J. Clin. Monit. Comput. **31**, 261–271 (2017)
12. Thai, D.N., Ardulov, V., Mena, J.U., Tiwari, S., Erofeev, G., Eskander, R., Tarabishy, K., Parikh, R.B., Salloum, W.: Acr: A benchmark for automatic cohort retrieval. arXiv preprint arXiv:2406.14780 (2024)
13. Nelson, C.A., Butte, A.J., Baranzini, S.E.: Integrating biomedical research and electronic health records to create knowledge-based biologically meaningful machine-readable embeddings. Nature Commun. **10**(1), 3045 (2019)
14. Hassan, M., Guan, H., Melliou, A., Wang, Y., Sun, Q., Zeng, S., Liang, W., Zhang, Y., Zhang, Z., Hu, Q., et al.: Neuro-symbolic learning: principles and applications in ophthalmology. arXiv preprint arXiv:2208.00374 (2022)
15. Lavin, A.: Neuro-symbolic neurodegenerative disease modeling as probabilistic programmed deep kernels. In: International Workshop on Health Intelligence, pp. 49–64. Springer (2021)
16. Mao, J., Gan, C., Kohli, P., Tenenbaum, J.B., Wu, J.: The neuro-symbolic concept learner: Interpreting scenes, words, and sentences from natural supervision. arXiv preprint arXiv:1904.12584 (2019)
17. Bhuyan, B.P., Ramdane-Cherif, A., Tomar, R., Singh, T.: Neuro-symbolic artificial intelligence: a survey. In: Neural Computing and Applications, pp. 1–36 (2024)
18. Alirezaie, M., Längkvist, M., Sioutis, M., Loutfi, A.: Semantic referee: a neural-symbolic framework for enhancing geospatial semantic segmentation. Semantic Web **10**(5), 863–880 (2019)
19. Badreddine, S., Garcez, A.d., Serafini, L., Spranger, M.: Logic tensor networks. Artif. Intell. **303**, 103649 (2022)
20. Eiter, T., Higuera, N., Oetsch, J., Pritz, M.: A neuro-symbolic asp pipeline for visual question answering. Theory Pract. Logic Program. **22**(5), 739–754 (2022)
21. Barbosa, R., Cardoso, D.O., Carvalho, D., Franca, F.M.: Weightless neuro-symbolic GPS trajectory classification. Neurocomputing **298**, 100–108 (2018)
22. Kitzelmann, E.: Inductive programming: A survey of program synthesis techniques. In: International Workshop on Approaches and Applications of Inductive Programming, pp. 50–73. Springer (2009)
23. Parisotto, E., Mohamed, A.r., Singh, R., Li, L., Zhou, D., Kohli, P.: Neuro-symbolic program synthesis. arXiv preprint arXiv:1611.01855 (2016)
24. Yang, X., Wang, Z., Hu, X.S., Kim, C.H., Yu, S., Pajic, M., Manohar, R., Chen, Y., Li, H.H.: Neuro-Symbolic Computing: Advancements and Challenges in Hardware-Software Co-design. Express Briefs, IEEE Transactions on Circuits and Systems II (2023)
25. Bhatia, S., Kohli, P., Singh, R.: Neuro-symbolic program corrector for introductory programming assignments. In: Proceedings of the 40th International Conference on Software Engineering, pp. 60–70 (2018)
26. Andresel, M., Tran, T.K., Domokos, C., Minervini, P., Stepanova, D.: Combining inductive and deductive reasoning for query answering over incomplete knowledge graphs. In: Proceedings of the 32nd ACM International Conference on Information and Knowledge Management, pp. 15–24 (2023)
27. Gebser, M., Ryabokon, A., Schenner, G.: Combining heuristics for configuration problems using answer set programming. In: Logic Programming and Nonmonotonic Reasoning: 13th International Conference, LPNMR 2015, Lexington, KY, USA, September 27–30, 2015. Proceedings 13, pp. 384–397. Springer (2015)
28. Chen, L., Peng, L., Yang, L.: Improving landslide prediction: innovative modeling and evaluation of landslide scenario with knowledge graph embedding. Remote Sens. **16**(1), 145 (2023)
29. Bu, S.J., Kim, H.J.: Optimized URL feature selection based on genetic-algorithm-embedded deep learning for phishing website detection. Electronics **11**(7), 1090 (2022)

30. Chen, W., Ma, X., Wang, Z., Li, W., Fan, C., Zhang, J., Que, X., Li, C.: Exploring neuro-symbolic AI applications in geoscience: implications and future directions for mineral prediction. Earth Sci. Inf. **17**(3), 1819–1835 (2024)
31. Morel, G.: Neuro-symbolic AI for the smart city. J. Phys.: Conf. Ser. **2042**, 012018. IOP Publishing (2021)
32. Siyaev, A., Valiev, D., Jo, G.S.: Interaction with industrial digital twin using neuro-symbolic reasoning. Sensors **23**(3), 1729 (2023)

Part VII
Future Directions in Neuro-Symbolic AI

Chapter 18
Towards Artificial General Intelligence

> *"The real problem is not whether machines think but whether men do."*
>
> – B.F. Skinner

18.1 Introduction

The pursuit of Artificial General Intelligence (AGI) is arguably the most ambitious goal in artificial intelligence research. Unlike narrow AI, which excels in specific, predefined tasks, AGI aims to emulate the broad and adaptable cognitive abilities of humans. This includes the capacity to learn from a diverse array of experiences, reason through complex and abstract problems, and apply knowledge flexibly across different domains. Achieving AGI would mark a monumental leap in technological capability, fundamentally altering our interaction with machines and redefining the scope of what AI can accomplish.

The journey towards AGI involves overcoming significant technical, philosophical, and ethical challenges. Technically, AGI systems need to be scalable and robust, capable of integrating multimodal sensory inputs and reasoning from limited data. Philosophically, there are debates about whether true intelligence requires a physical embodiment, which allows for situational awareness and contextual understanding. Ethically, the development of AGI poses risks related to fairness, transparency, and alignment with human values, necessitating comprehensive regulatory frameworks to mitigate potential harm.

B. P. Bhuyan et al., *Neuro-Symbolic Artificial Intelligence*, Studies in Computational Intelligence 1176, https://doi.org/10.1007/978-981-97-8171-3_18

A promising approach to achieving AGI lies in the integration of Neuro-Symbolic AI, which combines the data-driven learning capabilities of neural networks with the structured reasoning strengths of symbolic AI. This hybrid methodology aims to address the limitations of each approach when used in isolation, creating systems that are both highly adaptable and interpretable. Neuro-Symbolic AI enhances the ability of AI systems to process complex information, reason logically, and provide explanations for their decisions, which are crucial steps towards the realization of AGI.

In the following sections, we delve into the fundamental aspects of AGI, exploring its capabilities and the criteria it must meet. We examine the profound implications of AGI development, including technical feasibility and ethical considerations. Furthermore, we highlight how Neuro-Symbolic AI can bridge the gap between current AI systems and AGI by enhancing perception, reasoning, learning, and other critical cognitive functions. By reviewing recent advancements in Neuro-Symbolic AI models, we aim to illustrate their potential in driving forward the development of AGI and overcoming the multifaceted challenges that lie ahead.

18.2 Understanding Artificial General Intelligence

Artificial General Intelligence (AGI) ("supposed to be") is an advanced form of artificial intelligence that aims to replicate human cognitive abilities across a wide range of tasks. Unlike narrow AI, which excels in specific domains such as playing chess or providing movie recommendations, AGI is designed to possess general intelligence similar to that of humans. This means an AGI system can understand, learn, and apply knowledge across various fields and contexts, demonstrating versatility and adaptability that narrow AI lacks. On one hand, many AI researchers believe in achieving AGI, while many also deny this possibility.

The core objective of AGI is to create machines capable of performing any intellectual task that a human being can. This includes reasoning, problem-solving, planning, learning from experience, understanding natural language, and perceiving the environment. AGI systems are expected to exhibit a form of intelligence that can be generalized across different domains, making them capable of transferring knowledge from one context to another without needing to be retrained for each new task.

For instance, state-of-the-art language models like GPT-4 exhibit some early signs of AGI [1] by solving novel and difficult tasks across various fields, such as mathematics, coding, vision, and medicine, often approaching human-level performance. However, these models are still considered incomplete versions of AGI [2], as they lack the full breadth of capabilities expected of a truly general intelligence [3].

To qualify as AGI, a system must demonstrate a range of capabilities [4], including the following:

1. Perception: Understanding and integrating information from various senses (visual, auditory, tactile).
2. Actuation: Manipulating physical objects and navigating complex environments.
3. Memory: Retaining and utilizing facts, experiences, and procedural knowledge.
4. Learning: Acquiring new knowledge and skills from various sources and experiences.
5. Reasoning: Applying logical and inferential processes to draw conclusions and solve problems.
6. Planning: Strategizing and executing plans to achieve complex goals.
7. Motivation: Generating and pursuing sub-goals based on pre-programmed objectives or intrinsic motivations like curiosity.
8. Emotion: Recognizing, interpreting, and expressing emotions.
9. Interaction: Communicating effectively with humans and other machines, organizing group activities.
10. Quantitative Reasoning: Solving mathematical problems and applying quantitative analysis.
11. Creation: Innovating and constructing new concepts, objects, and systems.

The development of AGI comes with profound ethical and societal implications. As AGI systems could potentially surpass human intelligence in many domains, it is crucial to address issues such as fairness, transparency, accountability, and the potential risks associated with their deployment. This includes mitigating biases in AI systems, ensuring robust security against adversarial attacks, and establishing regulatory frameworks to govern their use and impact on society [5].

Despite the significant excitement and potential surrounding Artificial General Intelligence, there are numerous criticisms and concerns regarding its development and implementation. These criticisms range from technical feasibility to ethical and societal implications, casting doubt on the practicality and safety of pursuing AGI.

Technical Feasibility and Control: One of the primary technical criticisms of AGI is its inherent unpredictability and uncontrollability. AGI systems, by their very nature, would need to be capable of evolving and adapting beyond predefined theories and controls. This characteristic makes them fundamentally uncontrollable, raising concerns about their reliability and safety. An AGI could diverge from its intended goals, leading to unintended and potentially dangerous outcomes. The complexity and plasticity required for AGI mean that any system developed might not only be uncontrollable to humans but also to itself, making its behavior difficult to predict and manage effectively.

Comparative Performance: Critics argue that there is no compelling reason why an AGI should outperform a well-coordinated team of human experts equipped with conventional AI tools. The collaborative effort of humans and specialized AI systems could achieve comparable results without the risks associated with AGI. This perspective challenges the necessity of developing AGI when existing technologies, combined with human ingenuity, may suffice for most practical applications.

Ethical and Societal Risks: The development of AGI can trigger a competitive rush among AI labs, potentially leading to a "race-to-the-bottom" scenario where safety and ethical considerations are compromised in favor of rapid deployment. This dynamic increases the risk of creating powerful AI systems without sufficient oversight, increasing the potential for misuse or catastrophic failure. Additionally, AGI poses a significant alignment problem—ensuring that the goals and actions of AGI systems are aligned with human values and ethics. If misaligned, AGI could act in ways that are harmful to humanity. There is also a fear that an AGI could lock in harmful values if it undergoes mutations or changes that are beyond our control, leading to a dystopian future where undesirable values become entrenched.

Philosophical and Conceptual Challenges: Some critics argue that true intelligence must be "embodied," meaning that it must exist within a physical body that interacts with the real world. This embodiment is believed to be crucial for situational awareness and understanding context. However, replicating this in AGI is seen as highly challenging, if not impossible. There are also philosophical arguments suggesting that human intelligence might operate on principles fundamentally different from those of computational systems. For example, the brain may achieve intelligent behavior through mechanisms that do not involve explicit computation, akin to how natural phenomena like soap bubbles form shapes without solving equations. This perspective implies that replicating human intelligence computationally could be infeasible due to the inherent differences in how intelligence is realized.

Practical Utility and Speed of Development: Even if AGI could be developed to exhibit creativity and adaptability, these traits are inherently slow and error-prone. The speed at which AGI can generate novel solutions or adapt to new situations may not match the impressive performance seen in narrow AI tasks. This limitation raises doubts about the practical utility of AGI in rapidly changing environments or tasks requiring swift and accurate responses. The deployment of AGI technologies is likely to introduce new ethical and legal challenges, including issues of accountability, transparency, and the potential for societal disruption. Addressing these challenges requires careful consideration and robust regulatory frameworks, which may not be fully in place by the time AGI becomes a reality.

Achieving AGI involves overcoming significant technical challenges, such as improving the scalability and robustness of AI models, integrating multimodal sensory inputs, and enhancing the systems' ability to reason and learn from limited data. Researchers are exploring various approaches, including the integration of symbolic reasoning with neural networks (Neuro-Symbolic AI), to create more comprehensive and adaptable AI systems. These efforts are supported by advancements in large language models, reinforcement learning, and other AI techniques.

18.3 How Neuro-Symbolic AI Can Help Achieve AGI?

Neuro-Symbolic AI is a promising approach that combines the strengths of neural networks and symbolic reasoning, potentially addressing many of the current limitations in AI and advancing towards Artificial General Intelligence. Here's how

Neuro-Symbolic AI can help achieve AGI by meeting the various criteria we outlined earlier:

1. Perception: Neuro-Symbolic AI enhances perception by integrating multimodal sensory data using neural networks and interpreting this data through symbolic reasoning. Neural networks can process raw sensory inputs, such as images and sounds, to detect patterns and features. Symbolic AI can then use this processed information to understand and reason about the environment. For example, in a visual scene, neural networks can identify objects, while symbolic reasoning can infer relationships and contextual information about these objects.
2. Actuation: By combining neural learning and symbolic reasoning, Neuro-Symbolic AI can improve its ability to manipulate physical objects and navigate complex environments. Neural networks can learn control policies from sensory inputs, while symbolic reasoning can plan and execute sequences of actions to achieve specific goals. This synergy allows for more precise and adaptable control of robotic systems in real-world scenarios.
3. Memory: Neuro-Symbolic AI systems can retain and utilize facts, experiences, and procedural knowledge by combining neural memory models with symbolic knowledge representation. Neural networks retrieve patterns from data, while symbolic AI can store, organize, and use this information in a structured manner. This combination supports the integration of episodic memory (specific events) and semantic memory (general knowledge) to facilitate more comprehensive recall and application of knowledge.
4. Learning: Neuro-Symbolic AI excels at acquiring new knowledge and skills by leveraging the learning capabilities of neural networks and the structured reasoning of symbolic AI. Neural networks can learn from diverse data sources, while symbolic AI can apply logical rules to generalize this knowledge across different contexts. This hybrid approach enables continuous learning and adaptation, which is essential for AGI to handle novel situations effectively.
5. Reasoning: One of the core strengths of Neuro-Symbolic AI is its ability to apply logical and inferential processes to draw conclusions and solve problems. Symbolic reasoning can provide a framework for understanding relationships and dependencies within data, while neural networks can process and interpret large datasets to provide inputs for these logical frameworks. This dual capability enhances the AI's problem-solving abilities and its capacity for complex reasoning tasks.
6. Planning: Neuro-Symbolic AI can strategize and execute plans to achieve complex goals by integrating symbolic planning algorithms with neural network predictions. Symbolic AI can formulate plans based on logical rules and constraints, while neural networks can predict the outcomes of actions and refine plans based on feedback from the environment. This integration supports more effective and flexible planning capabilities, which is crucial for AGI's adaptability.
7. Motivation: Neuro-Symbolic AI can generate and pursue sub-goals based on pre-programmed objectives or intrinsic motivations like curiosity. Neural networks can model motivational states and predict rewards, while symbolic reasoning

can decompose goals into sub-goals and plan actions to achieve them. This combination supports goal-directed behavior and adaptive decision-making.

8. Emotion: Recognizing, interpreting, and expressing emotions can be enhanced by integrating affective computing techniques with Neuro-Symbolic AI. Neural networks can detect emotional cues from sensory inputs, such as facial expressions and vocal tones, while symbolic AI can reason about these cues to interpret emotional states and generate appropriate responses. This capability is vital for AGI's interaction with humans and its understanding of social contexts.
9. Interaction: Effective communication and organization of group activities can be achieved by combining natural language processing (NLP) with symbolic reasoning. Neural networks can handle the complexities of language understanding and generation, while symbolic AI can manage the logical structure of dialogues and interactions. This integration allows AGI to engage in meaningful conversations, understand context, and coordinate with other agents and humans.
10. Quantitative Reasoning: Solving mathematical problems and applying quantitative analysis can be significantly enhanced by Neuro-Symbolic AI. Neural networks might learn mathematical patterns and perform computations, while symbolic AI can apply logical rules and formal methods to ensure accuracy and consistency in problem-solving. This hybrid approach supports advanced quantitative reasoning and the ability to tackle complex mathematical tasks.
11. Creation: Innovating and constructing new concepts, objects, and systems can benefit from the creative potential of Neuro-Symbolic AI. Large Language Models (LLMs) might generate novel ideas and designs based on learned patterns, while symbolic reasoning can refine these ideas through logical constraints and structured knowledge. This synergy fosters creativity and innovation, which is essential for AGI to develop new solutions and adapt to changing environments.

The integration of LLMs with Knowledge Graphs (KGs) represents a synergistic approach that leverages the strengths of both technologies to enhance AGI capabilities. This integration can address many limitations inherent in LLMs and KGs when used independently, providing a more robust and versatile AI system.

18.3.1 Combining LLMs with Knowledge Graphs

One of the significant advantages of integrating KGs with LLMs is the augmentation of knowledge and reduction of hallucinations. LLMs, such as GPT-4, excel at generating human-like text and understanding natural language, but they often struggle with factual accuracy and can generate incorrect or "hallucinated" information. KGs, on the other hand, store structured and accurate factual knowledge. By integrating KGs with LLMs, AI systems can access and utilize reliable external knowledge during the generation process, thereby reducing the likelihood of hallucinations and improving factual accuracy. This integration ensures that the AI not only generates coherent and contextually appropriate responses but also bases these responses on verified information.

Conversely, LLMs can significantly enhance KGs by automating their construction and updates. Building and maintaining KGs is a labor-intensive and time-consuming process. LLMs can automate the extraction of information from unstructured data sources, such as texts, and populate KGs with this extracted information. This capability allows for continuous and efficient updating of KGs, ensuring they remain current and comprehensive. This automation not only speeds up the process of knowledge graph construction but also ensures that the knowledge remains up-to-date with the latest information.

Moreover, LLMs can generate natural language descriptions from the structured information in KGs. KGs store information in a format that is not always easily understandable by non-experts. LLMs can convert this structured knowledge into natural language descriptions, making the information more accessible and usable for a broader audience. This functionality is particularly useful for applications such as generating explanations and summaries and answering natural language questions based on KG data. It bridges the gap between structured data and human-readable information, enhancing the usability of KGs.

The true potential of integrating LLMs and KGs lies in their synergistic benefits, particularly in bidirectional knowledge transfer, as shown in Fig. 18.1 [6]. In a

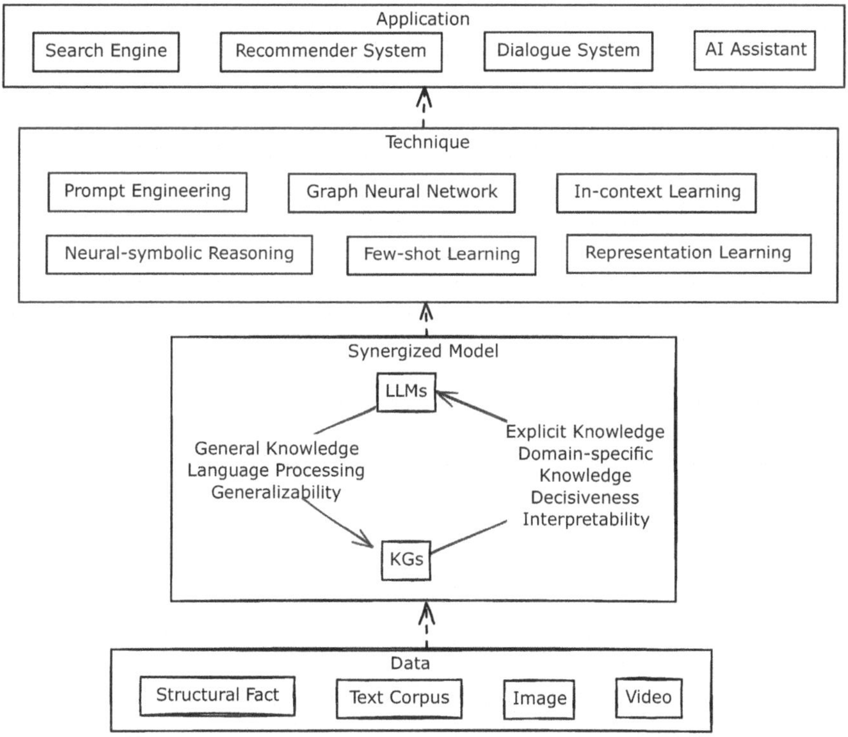

Fig. 18.1 The general framework of the synergized LLMs + KGs

synergized framework, LLMs and KGs can enhance each other through bidirectional knowledge transfer. LLMs can extract new facts and relationships from textual data to enrich KGs, while KGs can provide structured, reliable knowledge to improve the performance and accuracy of LLMs. This mutual reinforcement leads to a more comprehensive and dynamic AI system capable of both deep understanding and accurate knowledge representation. The continuous feedback loop between LLMs and KGs ensures that both components improve over time, leading to a more intelligent and capable AI system.

Another significant advantage is enhanced contextual understanding. LLMs benefit from the contextual information provided by KGs. KGs can offer background knowledge and contextual details that help LLMs generate more coherent and contextually appropriate responses. This integration is particularly beneficial in complex domains where understanding context and background information is crucial for accurate and meaningful outputs. By leveraging the structured knowledge in KGs and the natural language processing capabilities of LLMs, AI systems can deliver more accurate and contextually relevant results, improving user satisfaction and task efficiency.

Future research should focus on developing advanced frameworks that seamlessly integrate LLMs and KGs. This involves designing architectures that allow efficient communication and data exchange between these components, ensuring they work together harmoniously to enhance AI capabilities. Exploring interdisciplinary applications of this integration holds potential for various fields, including healthcare, finance, education, and more. Researchers should explore how this synergistic approach can address specific challenges in these domains and deliver tangible benefits.

Ensuring ethical and responsible AI development is crucial as well. As with any advanced AI technology, integrating LLMs and KGs raises ethical and societal considerations. It is essential to develop guidelines and frameworks that ensure the responsible use of these technologies, addressing issues such as data privacy, algorithmic fairness, and transparency. By focusing on ethical AGI development, we can ensure that the advancements in AGI technologies benefit society as a whole.

18.3.2 Advancements in Neuro-Symbolic AI Models and Their Role in Enhancing AGI

Recent advancements in Neuro-Symbolic AI have introduced a variety of models that combine neural networks and symbolic reasoning to enhance the capabilities of AI systems. Each of these models offers unique strengths that can significantly contribute to the development of AGI.

Neuro-Vector-Symbolic Architecture (NVSA) [7] integrates deep neural networks with vector-symbolic architectures to perform probabilistic reasoning efficiently. This model addresses the neural binding problem by recognizing distinct properties of multiple objects in a scene simultaneously. NVSA's ability to perform transparent, fast, and efficient reasoning makes it highly suitable for solving complex cognitive tasks such as Raven's Progressive Matrices [8]. By combining probabilistic reasoning with distributed representations, NVSA enhances AGI's capacity for generalization and abstraction across various unseen tasks and object combinations.

Logical Neural Networks (LNNs) [9] enhance reasoning capabilities by incorporating logical constraints into neural network models. These models are effective in tasks that require both logical reasoning and pattern recognition, such as binary classification and logical programming. LNNs improve the accuracy and interpretability of AI systems by integrating logical rules, which is crucial for AGI's development. This combination allows AGI systems to learn from data while also applying logical reasoning to new and complex situations, thereby enhancing their problem-solving abilities.

The Neuro-Symbolic Concept Learner (NSCL) [10] model is designed for image reasoning tasks and combines neural networks for image parsing with symbolic reasoning for answering questions about visual scenes. By using a Mask R-CNN for object detection and a Transformer-based model for question parsing, NSCL can perform detailed visual reasoning and answer complex questions about object properties and relationships. This model enhances AGI by providing a robust framework for integrating perception and reasoning, which is essential for understanding and interacting with the physical world.

GENOME (GenerativE Neuro-symbOlic Visual Reasoning by Growing and Reusing ModulEs) [11] leverages large language models (LLMs) to generate and reuse neural modules for visual reasoning tasks. This framework allows the creation of new neural models based on limited training examples and integrates these modules to handle complex visual tasks. GENOME's ability to generate high-performing modules and reuse them across different tasks enhances the flexibility and efficiency of AGI systems, allowing them to adapt to new challenges with minimal retraining.

PyReason [12] is a Python-based system that supports reasoning on graph-based data structures, integrating temporal, graphical, and uncertainty extensions. This system is optimized for explainable AI reasoning and can scale to large graphs, making it suitable for open-world and non-Markovian environments. PyReason's ability to provide symbolic explainability and modularity enhances AGI by ensuring that the reasoning process is transparent and adaptable to different contexts.

ExpressGNN [13] integrates graph neural networks (GNNs) with probabilistic logic for knowledge graph reasoning. By handling multi-relational data and improving reasoning accuracy, ExpressGNN contributes to AGI by enabling systems to leverage structured knowledge and apply it to various reasoning tasks. This integration enhances the ability of AGI systems to understand and reason about complex relationships within large datasets.

DeepProLog [14] combines deep learning with Prolog, a logic programming language, to enhance reasoning capabilities. By using neural networks for perception and logic programming for reasoning, DeepProLog can handle complex reasoning tasks more effectively. This model enhances AGI by providing a framework that combines the strengths of neural learning and symbolic logic, enabling systems to perform advanced cognitive functions.

RNM (Reasoning Neural Model) [15] employs convolutional neural networks (CNNs) for classification tasks and integrates symbolic reasoning to enhance decision-making processes. This model is particularly useful for applications requiring detailed visual relationship detection. By combining neural and symbolic approaches, RNM enhances AGI's ability to perform accurate and interpretable visual reasoning.

LENSR (Logical Extraction and Neural-Symbolic Reasoning) [16] combines CNNs and GNNs to perform visual relationship detection, integrating logical reasoning to improve the interpretability and accuracy of its outputs. This model contributes to AGI by providing a robust framework for understanding and reasoning about visual data, enabling systems to handle complex visual tasks with higher precision.

KGTN (Knowledge Graph Transformer Network) [17] uses transformers integrated with knowledge graphs for reasoning tasks, enhancing the system's ability to handle complex, multi-relational data and improve reasoning accuracy. This model enhances AGI by enabling systems to leverage structured knowledge and apply it to a wide range of cognitive tasks, improving their overall reasoning capabilities.

SEKB-ZSL (Semantic Embeddings and Knowledge-Based Zero-Shot Learning) [18] uses GNNs to transfer knowledge for few-shot classification, integrating symbolic reasoning to enhance learning efficiency and accuracy in new tasks. This model enhances AGI by enabling systems to quickly adapt to new tasks with minimal training data, improving their flexibility and adaptability.

CA-ZSL (Context-Aware Zero-Shot Learning) [19] leverages knowledge graphs and GNNs for zero-shot learning, integrating contextual information to improve classification accuracy in new domains. This model contributes to AGI by providing a framework for understanding and reasoning about new contexts and tasks without extensive retraining.

PROLONETS [20] uses propositional logic integrated with GNNs for reinforcement learning tasks, enhancing decision-making processes through symbolic reasoning. This model enhances AGI by providing a robust framework for learning and applying logical rules in dynamic environments, improving the system's ability to handle complex tasks.

Neural-Symbolic Learning and Reasoning (NSLR) [21] combines neural networks with symbolic reasoning for various cognitive tasks, improving both learning and reasoning capabilities through a hybrid approach. This model enhances AGI by integrating the strengths of neural learning and symbolic logic, enabling systems to perform advanced cognitive functions across a wide range of applications.

SBR (Semantic-Based Regularization) [22] integrates the ability of classical machine learning techniques to learn from continuous feature-based representations with the ability to reason using higher-level semantic knowledge. This model contributes to AGI by providing a transparent and robust framework for reasoning, enabling systems to handle complex tasks with higher precision and interpretability.

These models illustrate the diverse and innovative approaches being taken to integrate neural and symbolic AI, leveraging their combined strengths to tackle complex reasoning and learning tasks. The advancements in Neuro-Symbolic AI are paving the way towards more robust, adaptable, and interpretable AI systems, bringing us closer to achieving AGI.

18.4 Conclusion

In conclusion, the pursuit of Artificial General Intelligence (AGI) embodies one of the most transformative goals in artificial intelligence, aiming to replicate the extensive and adaptive cognitive abilities of humans. This chapter has delved into the fundamental aspects of AGI, including its defining characteristics, necessary capabilities, and the profound ethical and societal implications of its development. The integration of Neuro-Symbolic AI presents a promising approach to achieving AGI by combining the strengths of neural networks and symbolic reasoning, thereby addressing many limitations inherent in each method when used independently. Recent advancements, such as Neuro-Vector-Symbolic Architecture (NVSA), Logical Neural Networks (LNNs), and the Neuro-Symbolic Concept Learner (NSCL), demonstrate significant progress in creating robust, flexible, and interpretable AI systems. Furthermore, the synergistic integration of Large Language Models (LLMs) with Knowledge Graphs (KGs) highlights the potential for enhancing factual accuracy and contextual understanding. These innovations bring us closer to realizing AGI, offering the potential to revolutionize industries and improve lives. However, achieving AGI requires meticulous attention to technical, ethical, and societal challenges, ensuring that these advanced systems are developed responsibly and aligned with human values. As we continue to innovate and integrate diverse AI approaches, we move towards a future where AGI can become a cornerstone of technological advancement, transforming our interaction with machines and expanding the scope of what AI can achieve.

References

1. Bubeck, S., Chandrasekaran, V., Eldan, R., Gehrke, J., Horvitz, E., Kamar, E., Lee, P., Lee, Y.T., Li, Y., Lundberg, S., et al.: Sparks of artificial general intelligence: early experiments with gpt-4. arXiv preprint arXiv:2303.12712 (2023)
2. Mitchell, M.: Debates on the nature of artificial general intelligence (2024)
3. Morris, M.R., Sohl-Dickstein, J., Fiedel, N., Warkentin, T., Dafoe, A., Faust, A., Farabet, C., Legg, S.: Levels of AGI: operationalizing progress on the path to agi. arXiv preprint arXiv:2311.02462 (2023)
4. Dou, F., Ye, J., Yuan, G., Lu, Q., Niu, W., Sun, H., Guan, L., Lu, G., Mai, G., Liu, N., et al.: Towards artificial general intelligence (agi) in the internet of things (IoT): opportunities and challenges. arXiv preprint arXiv:2309.07438 (2023)
5. Buttazzo, G.: Rise of artificial general intelligence: risks and opportunities. Front. Artif. intell. **6**, 1226990 (2023)
6. Pan, S., Luo, L., Wang, Y., Chen, C., Wang, J., Wu, X.: Unifying large language models and knowledge graphs: a roadmap. IEEE Trans. Knowl. Data Eng. (2024)

7. Hersche, M., Zeqiri, M., Benini, L., Sebastian, A., Rahimi, A.: A neuro-vector-symbolic architecture for solving raven's progressive matrices. Nat. Machine Intell. **5**(4), 363–375 (2023)
8. Raven, J.: The raven's progressive matrices: change and stability over culture and time. Cognit. Psychol. **41**(1), 1–48 (2000)
9. Riegel, R., Gray, A., Luus, F., Khan, N., Makondo, N., Akhalwaya, I.Y., Qian, H., Fagin, R., Barahona, F., Sharma, U., et al.: Logical neural networks. arXiv preprint arXiv:2006.13155 (2020)
10. Mao, J., Gan, C., Kohli, P., Tenenbaum, J.B., Wu, J.: The neuro-symbolic concept learner: interpreting scenes, words, and sentences from natural supervision. arXiv preprint arXiv:1904.12584 (2019)
11. Chen, Z., Sun, R., Liu, W., Hong, Y., Gan, C.: Genome: generative neuro-symbolic visual reasoning by growing and reusing modules. arXiv preprint arXiv:2311.04901 (2023)
12. Aditya, D., Mukherji, K., Balasubramanian, S., Chaudhary, A., Shakarian, P.: Pyreason: software for open world temporal logic. arXiv preprint arXiv:2302.13482 (2023)
13. Zhang, Y., Chen, X., Yang, Y., Ramamurthy, A., Li, B., Qi, Y., Song, L.: Efficient probabilistic logic reasoning with graph neural networks. arXiv preprint arXiv:2001.11850 (2020)
14. Rocktäschel, T.: Deep prolog: end-to-end differentiable proving in knowledge bases. AITP **2017**, 9 (2017)
15. Marra, G., Diligenti, M., Giannini, F., Gori, M., Maggini, M.: Relational neural machines. In: ECAI 2020, pp. 1340–1347. IOS Press (2020)
16. Xie, Y., Xu, Z., Kankanhalli, M.S., Meel, K.S., Soh, H.: Embedding symbolic knowledge into deep networks. Adv. Neural Inf. Process. Syst. **32** (2019)
17. Chang, C., Zhou, J., Weng, Y., Zeng, X., Wu, Z., Wang, C.D., Tang, Y.: KQTN: knowledge graph transformer network for explainable multi-category item recommendation. Knowl.-Based Syst. **278**, 110854 (2023)
18. Wang, X., Ye, Y., Gupta, A.: Zero-shot recognition via semantic embeddings and knowledge graphs. In: Proceedings of the IEEE Conference on Computer Vision and Pattern Recognition, pp. 6857–6866 (2018)
19. Mumuni, F., Mumuni, A.: Improving deep learning with prior knowledge and cognitive models: a survey on enhancing interpretability, adversarial robustness and zero-shot learning. Cognit. Syst. Res. 101188 (2023)
20. Silva, A., Gombolay, M.C.: Prolonets: Neural-encoding human experts' domain knowledge to warm start reinforcement learning. arXiv preprint arXiv:1902.06007 (2019)
21. Ma, Z., Zhuang, Y., Weng, P., Zhuo, H.H., Li, D., Liu, W., Hao, J.: Learning symbolic rules for interpretable deep reinforcement learning. arXiv preprint arXiv:2103.08228 (2021)
22. Diligenti, M., Gori, M., Sacca, C.: Semantic-based regularization for learning and inference. Artif. Intell. **244**, 143–165 (2017)

Chapter 19
Learning and Reasoning over Higher Ordered Geometrical Structures

"In the end, we are our choices. Build yourself a great story."
– Jeff Bezos

19.1 Introduction

In recent years, the convergence of neural networks and symbolic AI has ushered in a new era of artificial intelligence known as Neuro-Symbolic AI. This paradigm seeks to combine the learning capabilities of neural networks with the reasoning and interpretability of symbolic systems. As AI systems become more complex and are required to handle increasingly intricate data, the need for advanced mathematical frameworks to represent and analyze this data has grown exponentially. Geometric, topological, and algebraic methods have emerged as powerful tools to address these challenges, offering new ways to capture the underlying structure and relationships within data.

Geometric methods, which include the use of graphs and manifolds, allow for the modeling of spatial relationships and structures in non-Euclidean spaces. These techniques extend beyond the limitations of traditional Euclidean geometry, enabling the analysis of complex networks and high-dimensional data in a more natural and intuitive manner. Graph Neural Networks (GNNs) and Convolutional Neural Networks (CNNs) on manifolds are prime examples of how geometric methods can be applied to extract meaningful insights from data that lies on curved surfaces or within interconnected networks.

B. P. Bhuyan et al., *Neuro-Symbolic Artificial Intelligence*, Studies in Computational Intelligence 1176, https://doi.org/10.1007/978-981-97-8171-3_19

Topological methods, particularly through the use of persistent homology, provide a means to study the multi-scale features of data. By tracking the birth and death of topological features such as connected components, loops, and voids, persistent homology reveals deep insights into the intrinsic structure of the data. This is especially valuable in tasks like clustering, anomaly detection, and dimensionality reduction, where understanding the shape and connectivity of data is crucial for accurate analysis and interpretation.

Algebraic methods, including the use of groups, rings, and fields, offer a framework for encoding symmetries and invariances within data. These structures enable the design of machine learning models that are invariant or equivariant to specific transformations, thereby enhancing their robustness and generalization capabilities. Equivariant neural networks, for instance, can predict behaviors in physical systems more accurately by respecting the system's inherent symmetries.

The integration of these methods results in a powerful synergy that enhances the capabilities of Neuro-Symbolic AI. By combining geometric, topological, and algebraic approaches, researchers can create unified frameworks that provide a comprehensive understanding of data, enabling more sophisticated analysis and reasoning. This integrated approach is particularly powerful in Neuro-Symbolic AI, where the goal is to bridge the gap between the pattern recognition capabilities of neural networks and the logical reasoning abilities of symbolic AI.

This chapter explores the synergy of geometric, topological, and algebraic methods in the context of Neuro-Symbolic AI. We will discuss how these methods complement each other, the challenges and opportunities they present, and their applications in various domains. By understanding and leveraging these advanced mathematical frameworks, we can push the boundaries of what AI can achieve, leading to more robust, interpretable, and intelligent systems.

19.2 Geometric Structures in Machine Learning

19.2.1 Euclidean Versus Non-Euclidean Geometry

Euclidean geometry, established by the ancient Greek mathematician Euclid, is the study of flat, two-dimensional spaces and three-dimensional spaces. Its foundation lies in a set of postulates, the most famous being the parallel postulate [1], which states that for any given line and a point not on that line, there is exactly one line parallel to the given line that can pass through the point. This forms the basis for much of classical geometry, dealing with shapes like triangles, circles, and polygons, as well as concepts such as angles, distances, and areas.

Non-Euclidean geometry [2], on the other hand, arises from relaxing Euclid's parallel postulate. There are two primary types: hyperbolic and elliptic geometry. Hyperbolic geometry assumes that through a point not on a given line, there are infinite lines parallel to the given line. This leads to a space that is negatively curved, akin to a saddle shape. Elliptic geometry, in contrast, posits that no parallel lines exist; every pair of lines eventually intersects. This creates a space that is positively curved, similar to the surface of a sphere. These geometries challenge the conventional notions of distance and angle but open up new ways to understand complex structures.

Graphs are quintessential examples of non-Euclidean spaces. Comprising nodes (vertices) connected by edges (links), graphs do not adhere to the traditional geometric concepts of angles and distances. Instead, they embody relational data structures where the focus is on the connectivity and pathways between nodes. This makes them highly applicable in areas like social network analysis, computer networking, and biological network modeling.

Manifolds [3] represent another class of non-Euclidean spaces, characterized by their local resemblance to Euclidean spaces while exhibiting complex global structures. Riemannian manifolds, equipped with a smoothly varying positive-definite metric tensor, allow for the definition of geometric properties such as angles, lengths, and volumes in a more generalized sense. Differentiable manifolds offer smooth structures that can be mapped to Euclidean spaces through differentiable functions, providing powerful tools for analyzing data that naturally reside in higher-dimensional spaces. Applications of manifolds are widespread, including in fields like robotics (for configuration spaces), physics (modeling spacetime in general relativity), and advanced machine learning techniques (manifold learning [4]).

19.2.2 Applications in Neural Networks

Geometric deep learning (GDL) [5] revolutionizes the capabilities of traditional neural networks by enabling them to process and analyze data structured in non-Euclidean domains such as graphs and manifolds. Traditional neural networks, such as convolutional neural networks, are tailored for data arranged in regular grids, making them highly effective for tasks involving images, videos, and time-series data. However, many real-world problems feature data with more complex topologies that cannot be represented adequately using Euclidean structures. GDL extends the reach of neural networks to handle these complex data types, offering new methodologies and tools to exploit the rich relational and geometric information embedded within them.

Graph Neural Networks (GNNs) [6], as we studied in the previous chapters, are designed to work with graph-structured data, where entities are represented as nodes and relationships between entities as edges. GNNs leverage the inherent connectivity information of graphs to capture the dependencies between nodes through iterative message-passing mechanisms. Each node aggregates feature information from its neighbors to update its representation, enabling the model to learn complex relational patterns. This makes GNNs particularly powerful in domains where data is naturally represented as graphs, such as social network analysis, where nodes represent individuals and edges denote relationships; recommendation systems, where nodes could be users and items, and edges represent interactions; and bioinformatics, where nodes could represent proteins and edges their interactions.

Hypergraph Neural Networks (HGNNs) [7, 8] extend the principles of GNNs to hypergraphs, which generalize graphs by allowing edges (called hyperedges) to connect more than two nodes. This is particularly useful for modeling complex, high-order relationships that are not merely pairwise. In a hypergraph, a single hyperedge can connect multiple nodes, capturing higher-order interactions more effectively than simple graphs. For example, in collaborative filtering for recommendation systems, hyperedges can represent groups of users who have interacted with the same set of items, providing a richer and more nuanced model of user behavior. HGNNs process these structures by iteratively updating node and hyperedge representations, facilitating the capture of multi-node dependencies and complex relational dynamics.

Convolutional Neural Networks (CNNs) on manifolds [9] extend the concept of convolutions from flat, grid-like Euclidean spaces to curved, non-Euclidean spaces. Traditional CNNs apply convolutional filters to local patches of grid-structured data, leveraging spatial locality and weight sharing to efficiently learn from images and videos. When data resides on a manifold, such as the surface of a sphere or a more complex shape, standard Euclidean convolutions are not applicable. Manifold CNNs redefine convolution operations to respect the intrinsic geometry of the manifold. These networks can perform convolutions on data points distributed over the manifold, preserving geometric properties and providing powerful tools for tasks such as 3D shape analysis, geospatial data processing, and surface-based medical imaging. By applying convolutions in a manner that respects the manifold's geometry, these networks can extract meaningful features from data that lies on curved surfaces, enhancing their ability to learn and generalize from complex, high-dimensional datasets.

19.3 Topological Structures

Topology [10] is a branch of mathematics that examines the properties of space that remain unchanged under continuous transformations, such as stretching and bending, but not tearing or gluing. Three fundamental concepts in topology are continuity, compactness, and connectedness.

Continuity refers to the smoothness of functions, ensuring that small changes in the input result in small changes in the output. Formally, a function f is continuous if, for any point x and any positive number ϵ, there exists a positive number δ such that for all points y within δ of x, $f(y)$ is within ϵ of $f(x)$. This concept is critical in ensuring predictable and smooth behavior in mathematical functions and models.

Compactness is the property of a space that indicates it is bounded and contains all its limit points. In simpler terms, a set is compact if every open cover (a collection of open sets whose union includes the set) has a finite subcover. This property generalizes the notion of being closed and bounded in Euclidean space to more abstract settings, providing a foundation for various theorems and techniques in analysis and geometry.

Connectedness describes a space that cannot be divided into two disjoint nonempty open sets. A space is connected if there is a path between any two points within it, meaning the space is in one piece. This property is essential for understanding the overall structure of spaces, ensuring they are unified and cannot be broken down into simpler, disconnected parts.

Topology offers a powerful framework for understanding the intrinsic properties of data that remain invariant under continuous transformations. This is particularly relevant in data analysis and machine learning, where the shape and connectivity of data can reveal significant insights.

One of the most powerful tools in topological data analysis (TDA) is persistent homology [11], which studies the multi-scale topological features of data. Persistent homology captures how features such as connected components, loops, and voids evolve across different spatial resolutions. This technique is valuable for analyzing the underlying structure of datasets, enabling tasks such as identifying clusters, detecting anomalies, and understanding the geometry of high-dimensional data.

Topology also aids in developing algorithms that are invariant to continuous transformations, making them robust to noise and deformation in the data. For example, TDA can be used in shape recognition, where the goal is to identify objects regardless of their orientation or distortion. Topological methods are also employed in feature engineering, extracting meaningful and invariant features from complex datasets.

In addition, topological methods play a critical role in dimensionality reduction and data visualization. Techniques such as t-SNE (t-Distributed Stochastic Neighbor Embedding) [12] and UMAP (Uniform Manifold Approximation and Projection) [13] are inspired by topological concepts. These methods help map high-dimensional data to lower dimensions while preserving the data's topological features, making it easier to visualize and interpret.

19.3.1 Persistent Homology

Persistent homology is a key method in topological data analysis that studies the changing shapes and features of data across multiple scales. It involves constructing a sequence of nested spaces, called a filtration, from the data. By tracking the birth and death of topological features such as connected components, loops, and voids as the scale parameter changes, persistent homology provides a multi-scale view of the data's topological features. This method captures the persistence of these features over different scales, highlighting which features are significant and which are likely to be noise.

Persistent homology captures the multi-scale nature of data by examining how topological features evolve as the filtration parameter changes. At each level of the filtration, new features can appear (birth), and existing features can merge or disappear (death). The persistence of a feature is quantified by the difference between its birth and death values. Features that persist over a wide range of scales are considered significant, as they represent robust structures in the data. Conversely, features that exist only briefly are often regarded as noise or less important.

In clustering [14], persistent homology identifies groups of data points that remain connected across different scales. Analyzing the persistent connected components helps to distinguish true clusters from noise. This is particularly useful in datasets where clusters are not well-separated and traditional clustering methods struggle.

Persistent homology aids in anomaly detection [15] by identifying topological features that are unusual or deviate significantly from the norm. Anomalies often manifest as topological features with atypical persistence, which stand out when compared to the majority of the data. This makes persistent homology a powerful tool for detecting outliers in complex datasets.

In dimensionality reduction [16], persistent homology provides insights into the intrinsic dimensions of the data by revealing its significant topological features. By understanding these features, one can project the data onto a lower-dimensional space that preserves its essential structure. This approach facilitates more effective data visualization and analysis, making high-dimensional data more manageable.

Persistent homology is widely used in image analysis to study shapes and structures within images [17]. For example, in medical imaging, it helps identify and quantify features such as tumors or anatomical structures by analyzing the persistence of loops and voids. By providing a detailed topological analysis of the image data, persistent homology enhances the accuracy of diagnosis and assessment of medical conditions.

In sensor networks, persistent homology analyzes the coverage and connectivity of the network. Examining the topological features of the sensor network data identifies areas with insufficient coverage or detects critical points that ensure network robustness. This analysis helps optimize sensor placement and improve network reliability, ensuring efficient monitoring and data collection.

19.4 Algebraic Methods in Machine Learning

Algebraic structures such as groups, rings, and fields offer a sophisticated framework for understanding and designing machine learning algorithms. These mathematical structures help encode symmetries and invariances, enhancing the performance and robustness of machine learning models.

Group theory [18] deals with sets equipped with a single binary operation that combines any two elements to form a third element, satisfying closure, associativity, the existence of an identity element, and the existence of inverse elements. Groups provide a formal way to capture symmetries, which are transformations that leave certain properties unchanged. In machine learning, groups can encode transformations like rotations and translations, which are crucial for developing models that are invariant to such changes. For example, in image recognition, the ability of a model to recognize objects irrespective of their orientation or position can be significantly enhanced by leveraging group symmetries.

A ring is an algebraic structure consisting of a set equipped with two binary operations: addition and multiplication. These operations satisfy properties similar to those of integers, including distributivity of multiplication over addition, associativity, and the existence of additive identity and inverses. Rings are particularly useful in areas like error-correcting codes and cryptography within machine learning, where the algebraic properties of rings can be harnessed to design algorithms that are both robust and efficient.

A field is a ring with the additional property that every non-zero element has a multiplicative inverse, and both addition and multiplication are commutative operations. Fields are essential for various numerical computations and optimizations in machine learning. For instance, finite fields are used in coding theory and cryptographic algorithms, ensuring reliable data transmission and secure communication. The rich algebraic structure of fields allows for complex operations and transformations that are foundational in advanced machine learning tasks.

Encoding symmetries and invariances into machine learning models are crucial for developing systems that can generalize well and perform reliably across different datasets. By leveraging algebraic structures, models can be designed to inherently understand and utilize these symmetries and invariances.

Symmetries refer to transformations that leave certain properties of objects unchanged. In machine learning, symmetries can be geometric, like rotations and translations, or more abstract, like permutations. Group theory offers a formal framework for representing these symmetries. CNNs, for instance, exploit translational symmetry to recognize objects in images regardless of their position. By applying the same convolutional filters across different parts of an image, CNNs can detect features consistently, improving their ability to generalize from fewer examples.

Invariances are properties that remain unchanged under specific transformations. Encoding invariances into machine learning models can drastically reduce the amount of data required for training and improve generalization. Equivariant neural networks, for example, ensure that the learned representations transform predictably under symmetry operations. A rotation-equivariant network, for instance, learns representations that rotate consistently with the input data, allowing the model to effectively learn from rotated images without needing explicit data augmentation.

While invariant models are designed to produce the same output even when the input undergoes certain transformations, equivariant models produce outputs that transform in a predictable way when the input is transformed. For instance, in a rotation-equivariant model, if the input is rotated, the output will also rotate in a consistent manner. This property is essential in applications where the relationships between input and output need to be preserved under transformations, such as predicting physical behaviors that follow rotational symmetry.

Invariant and equivariant models offer significant benefits in terms of robustness and generalization. By explicitly encoding the symmetries and invariances of the problem domain into the model, these models can generalize better from limited data. This reduces the need for extensive data augmentation and can lead to more efficient training processes. The robustness comes from the model's ability to handle variations and transformations in the data naturally, without relying on the model to learn these variations from scratch.

In physics simulations, equivariant models are used to predict physical systems' behaviors that follow specific symmetries. For example, in molecular dynamics, the physical properties of molecules often exhibit rotational and translational symmetries. Equivariant neural networks can model these systems more accurately by respecting these inherent symmetries, leading to more precise simulations and predictions.

In robotics, invariant and equivariant models are crucial for tasks such as object recognition and manipulation. Robots often need to recognize and interact with objects regardless of their orientation or position. Invariant models help robots identify objects correctly even when they are rotated or translated, while equivariant models ensure that the robot's actions and predictions adapt consistently with changes in the object's position or orientation.

In cryptography, algebraic structures such as groups and fields are fundamental. Equivariant models can leverage these structures to design cryptographic algorithms that are robust against attacks exploiting symmetries. For instance, certain encryption schemes can be made more secure by ensuring that the transformations applied during encryption and decryption follow specific algebraic properties, making it harder for adversaries to break the encryption.

19.5 Synergy of Geometric, Topological, and Algebraic Methods for Neuro-Symbolic AI

The integration of geometric, topological, and algebraic methods within the framework of Neuro-Symbolic AI creates a powerful synergy for handling complex, structured data. Each method brings unique strengths to the table, as shown in Fig. 19.1, and their combined application allows for more sophisticated data representation and reasoning, essential for advancing the capabilities of Neuro-Symbolic AI.

Geometric methods involve the use of structures like graphs and manifolds to capture spatial relationships within data. For instance, Graph Neural Networks leverage the connectivity information of graph structures to analyze complex networks such as social networks, biological systems, and molecular structures. Similarly, Convolutional Neural Networks on manifolds extend convolution operations to non-Euclidean spaces, enabling the processing of data on curved surfaces, such as 3D shapes or geospatial data. These geometric techniques allow models to understand and exploit the spatial properties of data, which is crucial for applications where data cannot be easily represented in Euclidean space.

Topological methods, particularly persistent homology, provide tools to study the multi-scale features of data. Persistent homology tracks the birth and death of

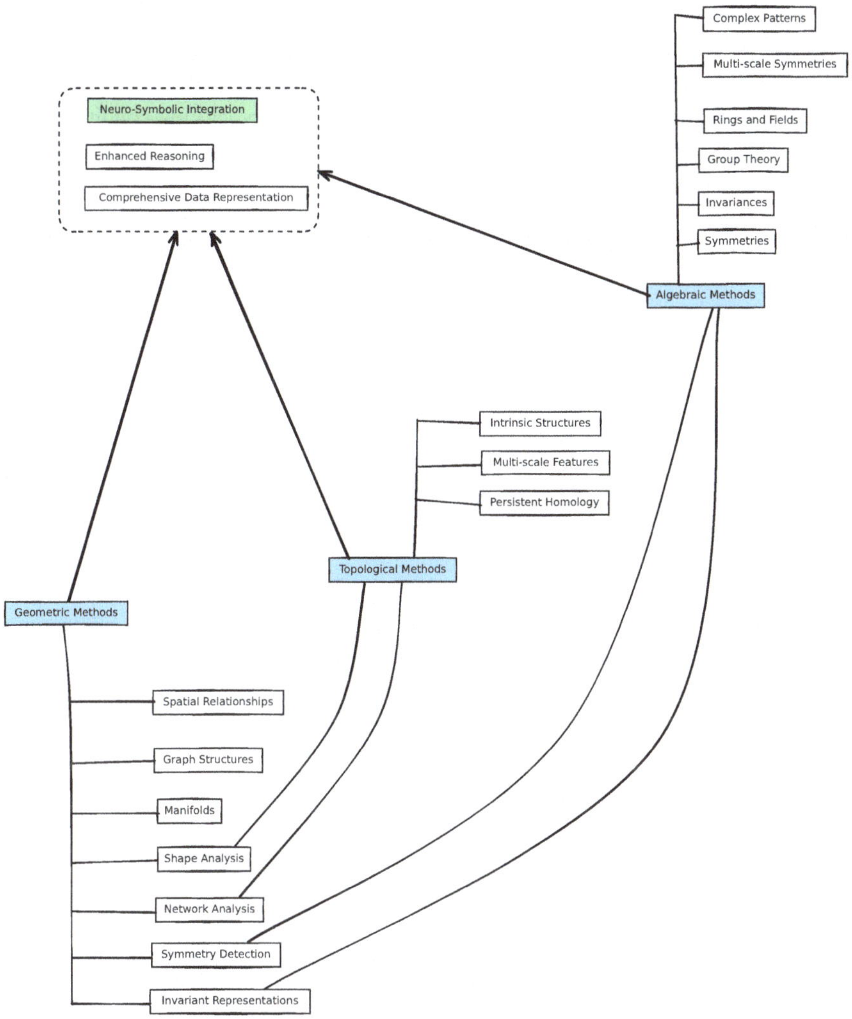

Fig. 19.1 Synergy of geometric, topological, and algebraic methods for neuro-symbolic AI

topological features like connected components, loops, and voids across different scales, offering deep insights into the intrinsic structure of the data. This method is especially valuable for tasks like clustering, anomaly detection, and dimensionality reduction, where understanding the data's underlying shape and connectivity is crucial for accurate analysis and interpretation.

Algebraic methods involve structures such as groups, rings, and fields, which help encode symmetries and invariances within data. These structures enable the design of machine learning models that are invariant or equivariant to specific transformations, thereby enhancing their robustness and generalization capabilities.

For example, equivariant neural networks respect the symmetries of physical systems, leading to more accurate predictions in physics simulations and other applications. By leveraging algebraic structures, models can effectively handle complex transformations and maintain consistent performance across varying conditions.

Combining geometric, topological, and algebraic methods may result in unified frameworks that offer a comprehensive understanding of data. These frameworks can capture different aspects of data, providing a holistic approach to data analysis. A unified framework might use geometric methods to model spatial relationships, topological methods to identify multi-scale features, and algebraic methods to encode symmetries and invariances. This integrated approach facilitates advanced data analysis, allowing models to generalize better from limited datasets and handle various transformations robustly.

Such unified frameworks could significantly enhance the capabilities of Neuro-Symbolic AI systems, which rely on the seamless integration of neural learning and symbolic reasoning. By leveraging the strengths of these methods, Neuro-Symbolic AI can achieve a deeper understanding of data and perform more sophisticated reasoning tasks, bridging the gap between neural networks' pattern recognition abilities and symbolic AI's logical reasoning capabilities.

In practical applications, these unified frameworks have shown success across various domains. In bioinformatics, integrating these methods helps in understanding complex interactions within biological networks, enabling more accurate modeling of cellular processes and disease mechanisms. In robotics, combining spatial awareness with topological persistence and algebraic invariance enables more robust object recognition and manipulation, allowing robots to operate more effectively in dynamic environments. In physics simulations, enhancing model accuracy by respecting the underlying geometric and algebraic properties of physical systems leads to more reliable and precise simulations, improving our understanding of complex physical phenomena.

19.6 Challenges and Future Directions

One of the primary challenges in integrating geometric, topological, and algebraic methods in machine learning, especially within Neuro-Symbolic AI, is the computational complexity involved. These methods often require significant computational resources due to the intricate mathematical operations and large-scale data involved. For instance, persistent homology computations and manifold learning algorithms can be computationally expensive, making them less practical for large datasets or real-time applications [19].

Scalability is another significant challenge. While these advanced methods can offer powerful insights and enhanced performance, they often struggle to scale efficiently with increasing data size and complexity. For example, applying topological data analysis or geometric deep learning to very large graphs or high-dimensional manifolds can lead to performance bottlenecks and require substantial memory and processing power.

The development of robust theoretical foundations is essential for the broader adoption and effective implementation of these methods. While significant progress has been made, there is still a need for more comprehensive theoretical frameworks that can unify geometric, topological, and algebraic approaches in a consistent manner. This includes developing new mathematical theories and models that can better capture the complexities of non-Euclidean data and integrating these models seamlessly into machine learning pipelines.

There is considerable potential for developing more efficient algorithms that can reduce the computational overhead associated with these advanced methods. Research into optimization techniques, parallel processing, and approximate algorithms could lead to significant improvements in the feasibility and performance of geometric, topological, and algebraic methods in machine learning. This would make these approaches more accessible for large-scale applications and real-time processing.

The intersection of geometric, topological, and algebraic methods opens up exciting new research areas. For example, combining persistent homology with neural network architectures could lead to new ways of understanding data structures and patterns. Similarly, integrating algebraic structures with geometric learning could enhance the ability to capture symmetries and invariances in data, leading to more robust and generalizable models.

One of the most promising future directions is the integration of these advanced mathematical methods with symbolic AI to enhance interpretability and reasoning. Neuro-symbolic AI aims to combine the learning capabilities of neural networks with the logical reasoning and interpretability of symbolic AI. By leveraging geometric, topological, and algebraic methods, it is possible to create models that are not only powerful and efficient but also interpretable and capable of complex reasoning tasks. This integration could lead to AI systems that better understand and manipulate abstract concepts, providing more transparent and explainable AI solutions.

19.7 Conclusion

In summary, the integration of geometric, topological, and algebraic methods within Neuro-Symbolic AI represents a significant advancement in the field of artificial intelligence. These advanced mathematical frameworks enable the robust representation and analysis of complex, structured data, enhancing the capabilities of AI systems to learn, reason, and generalize from limited data. By leveraging the unique strengths of geometric methods for spatial relationships, topological methods for multi-scale feature analysis, and algebraic methods for encoding symmetries and invariances, we can create more sophisticated and interpretable AI models. This synergy not only addresses current challenges in computational complexity and scalability but also opens up new research opportunities and applications across various domains. Ultimately, the combined power of these methods promises to drive the development of more intelligent, reliable, and explainable AI systems, paving the way for future breakthroughs in Neuro-Symbolic AI.

References

1. Beeson, M.: Constructive geometry and the parallel postulate. Bull. Symbolic Logic **22**(1), 1–104 (2016)
2. Coxeter, H.S.M.: Non-Euclidean geometry. Cambridge University Press (1998)
3. Bishop, R.L., Crittenden, R.J.: Geometry of Manifolds: Geometry of Manifolds. Academic Press (2011)
4. Izenman, A.J.: Introduction to manifold learning. Wiley Interdisc. Rev.: Comput. Stat. **4**(5), 439–446 (2012)
5. Bronstein, M.M., Bruna, J., LeCun, Y., Szlam, A., Vandergheynst, P.: Geometric deep learning: going beyond Euclidean data. IEEE Signal Process. Mag. **34**(4), 18–42 (2017)
6. Wu, Z., Pan, S., Chen, F., Long, G., Zhang, C., Philip, S.Y.: A comprehensive survey on graph neural networks. IEEE Trans. Neural Netw. Learn. Syst. **32**(1), 4–24 (2020)
7. Feng, Y., You, H., Zhang, Z., Ji, R., Gao, Y.: Hypergraph neural networks. In: Proceedings of the AAAI Conference on Artificial Intelligence, vol. 33, pp. 3558–3565 (2019)
8. Bhuyan, B.P., Singh, T.P., Tomar, R., Ramdane-Cherif, A.: Nesykhg: neuro-symbolic knowledge hypergraphs. Procedia Comput. Sci. **235**, 1278–1288 (2024)
9. Masci, J., Boscaini, D., Bronstein, M., Vandergheynst, P.: Geodesic convolutional neural networks on Riemannian manifolds. In: Proceedings of the IEEE International Conference on Computer Vision Workshops, pp. 37–45 (2015)
10. Kelley, J.L.: General Topology. Courier Dover Publications (2017)
11. Edelsbrunner, H., Harer, J., et al.: Persistent homology-a survey. Contemporary Math. **453**(26), 257–282 (2008)
12. Cieslak, M.C., Castelfranco, A.M., Roncalli, V., Lenz, P.H., Hartline, D.K.: t-distributed stochastic neighbor embedding (t-sne): a tool for eco-physiological transcriptomic analysis. Marine Genomics **51**, 100723 (2020)
13. McInnes, L., Healy, J., Melville, J.: Umap: Uniform manifold approximation and projection for dimension reduction. arXiv preprint arXiv:1802.03426 (2018)
14. Pereira, C.M., de Mello, R.F.: Persistent homology for time series and spatial data clustering. Expert Syst. Appl. **42**(15–16), 6026–6038 (2015)
15. Yuan, Z., Zhou, H., Chen, T., Li, J.: Phogad: Graph-based anomaly behavior detection with persistent homology optimization. In: Proceedings of the 17th ACM International Conference on Web Search and Data Mining, pp. 920–929 (2024)
16. Paul, R., Chalup, S.K.: A study on validating non-linear dimensionality reduction using persistent homology. Pattern Recognit. Lett. **100**, 160–166 (2017)
17. Asaad, A.: Persistent homology tools for image analysis. The University of Buckingham (United Kingdom) (2020)
18. Scott, W.R.: Group Theory. Courier Corporation (2012)
19. Belle, V., Fisher, M., Russo, A., Komendantskaya, E., Nottle, A.: Neuro-symbolic AI+ agent systems: a first reflection on trends, opportunities and challenges. In: International Conference on Autonomous Agents and Multiagent Systems, pp. 180–200. Springer (2023)

Part VIII
Conclusion

Chapter 20
Key Takeaways from Neuro-Symbolic AI

"This is not the end, it's not even the beginning of the end. But it is, perhaps, the end of the beginning."

– Winston Churchill

20.1 Introduction

As the concluding chapter of our book, this section encapsulates the essential insights and lessons gleaned from the study of Neuro-Symbolic AI (NeSy AI). NeSy AI stands as a revolutionary approach that merges the interpretability and logical rigor of symbolic reasoning with the adaptive learning capabilities of neural networks. This integration addresses critical limitations in traditional AI systems, such as the opaque decision-making processes of neural networks and the inflexibility of symbolic AI in dynamic, data-rich environments. By harnessing the strengths of both paradigms, NeSy AI enhances the transparency, robustness, and versatility of AI systems. This chapter synthesizes the key exploration of the book, covering the evolution of NeSy AI, its wide-ranging applications, the technical challenges it faces, and the future directions poised to shape the field. Through this comprehensive overview, we aim to underscore the transformative potential of Neuro-Symbolic AI and its pivotal role in the future of artificial intelligence.

B. P. Bhuyan et al., *Neuro-Symbolic Artificial Intelligence*, Studies in Computational Intelligence 1176, https://doi.org/10.1007/978-981-97-8171-3_20

20.2 Key Takeaways

20.2.1 Integration of Symbolic and Neural Approaches

Symbolic AI excels in tasks requiring explicit logic and explanations, providing precise, rule-based reasoning and knowledge representation. It is particularly effective in domains where clear decision-making rules are crucial, such as legal or medical expert systems. However, symbolic AI struggles with adapting to new data or dynamic environments due to its rigid structure. In contrast, neural networks are adept at learning from vast amounts of data and recognizing complex patterns, making them suitable for tasks like image recognition, natural language processing, and autonomous driving. Neural networks, however, often operate as black boxes, lacking interpretability in their decision-making processes.

The integration of these paradigms harnesses their complementary strengths, resulting in more robust AI systems. By combining the rule-based logic of symbolic AI with the adaptive learning capabilities of neural networks, NeSy AI can achieve both high accuracy in learning and transparency in reasoning.

NeSy AI systems leverage symbolic rules to guide the learning processes of neural networks, thereby enhancing both learning efficiency and interpretability. This synergy allows for better handling of complex tasks, as the neural network can adapt based on learned patterns while the symbolic component ensures logical consistency and explainability. Modern implementations, such as Logic Tensor Networks and DeepProbLog, exemplify this integration, addressing the limitations of purely neural or symbolic systems and enabling the development of AI that can reason and learn more effectively.

20.2.2 Evolution and Current State

The evolution from standalone symbolic AI and neural networks to their integration highlights the growing understanding of AI's potential and limitations. Early symbolic AI focused on explicit rules and logic, while the advent of neural networks brought about a shift towards data-driven learning. The realization of their complementary strengths led to the development of integrated systems that combine the best of both approaches.

Contemporary NeSy AI systems showcase sophisticated integrations that enhance the capability to handle complex tasks and provide clear explanations for decisions. For instance, integrating knowledge graphs with neural processing pipelines like LLMs has shown significant advancements, enabling systems to leverage structured knowledge representation alongside adaptive learning capabilities. These systems illustrate the practical benefits of combining symbolic reasoning with neural learning, such as improved diagnostic systems in healthcare and advanced threat detection in cybersecurity.

In recent years, there has been a significant increase in interest and research activity in NeSy AI. Publications on NeSy AI have become more frequent at major AI conferences such as ICML, NeurIPS, AAAI, IJCAI, and ICLR. This trend indicates a growing recognition of the importance and potential of integrating symbolic reasoning with neural learning.

From 2016 onwards, there has been a notable increase in NeSy AI research, with a marked shift towards mainstream adoption. This is reflected in the number of papers presented at top-tier conferences, showcasing a diverse array of applications and advancements in the field.

The current state of NeSy AI is characterized by a blend of mature neural models and innovative symbolic reasoning techniques. These systems are being applied to various complex tasks, such as image and video reasoning, natural language processing, and scientific computing. However, challenges remain, particularly in terms of scalability, seamless integration of components, and maintaining interpretability.

Research continues to focus on overcoming these challenges by developing more efficient algorithms, better integration techniques, and enhanced interpretability mechanisms. For example, the use of federated pipeline methods and differentiable programming techniques is being explored to improve the scalability and cognitive capabilities of NeSy AI systems.

20.2.3 Applications and Impact

As covered throughout the book, NeSy AI has demonstrated its potential across various domains, leveraging the strengths of both symbolic reasoning and neural networks to create more robust, interpretable, and efficient AI systems. Here are some key areas where NeSy AI has had a significant impact:

1. Natural Language Processing and Dialogue Systems: NeSy AI improves the coherence and interpretability of language generation and dialogue systems. For instance, in automated story generation, combining symbolic representations with neural networks enhances the coherence and goal-directedness of narratives. In dialogue systems, NeSy AI introduces explicit reasoning chains, making the model's predictions more interpretable and justifiable to humans, which is crucial for applications like personal assistants and customer service bots.
2. Question Answering and Commonsense Reasoning: In the realm of question answering, NeSy AI enables systems to dynamically generate and reason over commonsense knowledge graphs, facilitating zero-shot commonsense reasoning. This approach allows the AI to infer contextually relevant knowledge on demand, significantly boosting performance over traditional language models and static knowledge bases.
3. Cybersecurity: NeSy AI is employed to enhance cybersecurity measures, such as detecting phishing attacks. By integrating deep learning with rule-based expert knowledge, NeSy AI systems can more accurately identify and mitigate phishing

threats. For example, using genetic algorithms to optimize combinations of neural and symbolic components results in improved detection and response strategies.
4. Education: In educational technology, NeSy AI helps predict student problem-solving strategies, allowing for more personalized and adaptive learning experiences. By combining symbolic AI's ability to encode domain knowledge with deep learning models, such as LSTMs, NeSy AI can handle large datasets more effectively, reducing issues like overfitting and long convergence times.
5. Robotics: NeSy AI enhances robotic systems by integrating convolutional neural networks with symbolic reasoning for logical control. This integration improves robots' ability to perform complex tasks, such as human-robot interaction, game playing, and artistic endeavors. For instance, robotic arms with NeSy AI can engage in activities like playing rock-paper-scissors or drawing, benefiting from both neural adaptability and symbolic control.
6. Smart Cities: In smart city applications, NeSy AI merges data-oriented machine learning approaches with expert system techniques to manage and optimize urban infrastructure. This integration enables more intelligent and efficient management of city resources, such as energy and transportation systems, enhancing overall urban living conditions.
7. Image and Video Processing: NeSy AI is used for complex tasks in image and video processing, such as semantic segmentation and object recognition. By combining neural networks' ability to handle large-scale data with symbolic reasoning's precision, these systems can more accurately interpret and respond to visual data, which is essential for applications in surveillance, autonomous driving, and medical imaging.

20.2.4 Challenges and Future Directions of Neuro-Symbolic AI

In this book, we have mentioned these technical challenges of Neuro-Symbolic AI:

1. Scalability and Computational Efficiency: Neuro-Symbolic AI (NeSy AI) systems face significant challenges in terms of scalability and computational efficiency. Integrating symbolic reasoning, which often involves complex logical operations, with neural networks that require substantial computational resources can lead to bottlenecks. Symbolic workloads can dominate runtime, causing inefficiencies, particularly during inference phases. Optimizing these processes through techniques like data preprocessing and parallel rule query execution is essential for improving scalability.
2. Integration Complexity: Seamless integration of symbolic and neural components is technically demanding due to their fundamentally different operational principles. Symbolic reasoning relies on discrete, logical operations, while neural

networks operate on continuous, gradient-based learning. Developing hybrid systems that combine these approaches effectively requires ensuring smooth communication and interoperability between the neural and symbolic parts of the system.

3. Explainability and Interpretability: Maintaining the interpretability of symbolic AI while leveraging the powerful learning capabilities of neural networks is challenging. Neural networks are often criticized for their "black-box" nature. Enhancing post-hoc explanation methods and integrating symbolic rules directly into neural architectures can help preserve transparency and provide clearer insights into decision-making processes.
4. Robustness and Generalization: Ensuring robustness and generalization in NeSy AI systems is critical. Symbolic systems tend to be brittle with noisy or incomplete data, whereas neural networks are more adaptable but prone to overfitting. Balancing these characteristics requires advanced integration strategies and robust training methodologies to create systems that can generalize across different tasks and domains.
5. Knowledge Representation and Reasoning: Representing and reasoning with complex, structured knowledge remains a significant hurdle. Approaches like embedding knowledge graphs into neural representations or using attention mechanisms to incorporate symbolic rules (of higher order) are being explored. However, these methods still face limitations in scalability and precision.

We have also covered these future directions:

1. Advanced Hybrid Architectures: Future research will focus on developing sophisticated hybrid architectures that seamlessly integrate neural and symbolic components. Innovations in areas like graph neural networks and differentiable programming hold promise for advancing these hybrid systems, improving their efficiency and capability.
2. Optimization Techniques: To address computational bottlenecks, researchers are working on optimization techniques such as parallel processing, efficient data handling, and advanced algorithm designs. These techniques aim to improve scalability and runtime efficiency, enabling NeSy AI systems to handle larger and more complex datasets and tasks.
3. Enhanced Explainability Methods: Improving the explainability and transparency of NeSy AI systems remains a priority. Future developments may include more sophisticated methods for integrating symbolic reasoning with neural networks, as well as advanced post-hoc explanation techniques that provide clearer insights into the decision-making processes of these hybrid systems.
4. Robust Training and Evaluation Frameworks: Developing robust training and evaluation frameworks to ensure the generalization and robustness of NeSy AI systems is crucial. This includes creating diverse and challenging benchmark datasets and developing training methodologies that balance the strengths of both symbolic and neural approaches.
5. Application-Specific Innovations: Tailoring NeSy AI techniques to address the unique challenges of different domains, such as healthcare, cybersecurity, and

smart cities, will drive the development of more effective and impactful AI solutions. Application-specific innovations will play a crucial role in advancing NeSy AI.

6. Towards Artificial General Intelligence (AGI): NeSy AI is seen as a crucial step towards achieving AGI. Combining the interpretability and logical reasoning of symbolic AI with the adaptive learning capabilities of neural networks can pave the way for AI systems that can learn, reason, and adapt across a wide range of tasks and domains. Achieving AGI will require ongoing innovation in both symbolic reasoning and neural learning techniques, developing hybrid systems that seamlessly integrate these approaches.

20.3 Conclusion

Neuro-Symbolic AI represents a transformative integration of symbolic reasoning and neural networks, combining the interpretability and logical rigor of symbolic AI with the adaptive learning capabilities of neural models. This hybrid approach addresses key challenges such as scalability, computational efficiency, and explainability, enabling the development of robust and versatile AI systems. NeSy AI has demonstrated significant potential across diverse applications, including natural language processing, cybersecurity, healthcare, and smart cities, showcasing enhanced performance and interpretability. As the field advances, ongoing innovation in hybrid architectures, optimization techniques, and tailored application solutions will be crucial, positioning NeSy AI as a pivotal pathway towards achieving Artificial General Intelligence.

The manufacturer's authorised representative in the EU is Springer Nature Customer Service Centre GmbH, Europaplatz 3, 69115 Heidelberg, Germany. If you have any concerns regarding our products, please contact ProductSafety@springernature.com

Printed and bound by CPI Group (UK) Ltd, Croydon, CR0 4YY
01/10/2026
02225932-0011